Introduction to Classical Mathematics I

Mathematics and Its Applications

Volume 70

Introduction to Classical Mathematics I

From the Quadratic Reciprocity Law to the Uniformization Theorem

by

Helmut Koch

Karl-Weierstrass-Institut für Mathematik, Berlin, Germany

Translated by John Stillwell

KLUWER ACADEMIC PUBLISHERS

DORDRECHT / BOSTON / LONDON

Library of Congress Cataloging-in-Publication Data

```
Koch, Helmut, 1932-
   [Vom quadratischen Reziprozitätsgesetz bis zum
  Uniformisierungssatz.  English]
   From the quadratic reciprocity law to the uniformization theorem /
  by Helmut Koch.
      p.    cm. -- (Mathematics and its applications ; v. 70)
   Translation of: Vom quadratischen Reziprozitätsgesetz bis zum
  Uniformisierungssatz.
   At head of title: Introduction to classical mathematics I.
   Includes bibliographical references and indexes.
   ISBN 0-7923-1231-7 (HB : printed on acid free paper)
   1. Mathematics.   I. Title.   II. Title: Introduction to classical
  mathematics I.   III. Title: Introduction to classical mathematics
  1.  IV. Series: Mathematics and its applications (Kluwer Academic
  Publishers) ; v. 70.
  QA37.2.K6313  1991
  510--dc20                                                 91-12580
```

ISBN 0-7923-1231-7 (HB)
ISBN 0-7923-1238-4 (PB)

Published by Kluwer Academic Publishers,
P.O. Box 17, 3300 AA Dordrecht, The Netherlands.

Kluwer Academic Publishers incorporates
the publishing programmes of
D. Reidel, Martinus Nijhoff, Dr W. Junk and MTP Press.

Sold and distributed in the U.S.A. and Canada
by Kluwer Academic Publishers,
101 Philip Drive, Norwell, MA 02061, U.S.A.

In all other countries, sold and distributed
by Kluwer Academic Publishers Group,
P.O. Box 322, 3300 AH Dordrecht, The Netherlands.

Printed on acid-free paper

This edition is a revised edition of
H. Koch, *Einführing in die klassische Mathematik I*
© 1986 Akademie-Verlag Berlin, Germany

Printed in the Netherlands

SERIES EDITOR'S PREFACE

'Et moi, ..., si j'avait su comment en revenir, je n'y serais point allé.'
Jules Verne

The series is divergent; therefore we may be able to do something with it.
O. Heaviside

One service mathematics has rendered the human race. It has put common sense back where it belongs, on the topmost shelf next to the dusty canister labelled 'discarded non-sense'.
Eric T. Bell

Mathematics is a tool for thought. A highly necessary tool in a world where both feedback and non-linearities abound. Similarly, all kinds of parts of mathematics serve as tools for other parts and for other sciences.

Applying a simple rewriting rule to the quote on the right above one finds such statements as: 'One service topology has rendered mathematical physics ...'; 'One service logic has rendered computer science ...'; 'One service category theory has rendered mathematics ...'. All arguably true. And all statements obtainable this way form part of the raison d'être of this series.

This series, *Mathematics and Its Applications*, started in 1977. Now that over one hundred volumes have appeared it seems opportune to reexamine its scope. At the time I wrote

> "Growing specialization and diversification have brought a host of monographs and textbooks on increasingly specialized topics. However, the 'tree' of knowledge of mathematics and related fields does not grow only by putting forth new branches. It also happens, quite often in fact, that branches which were thought to be completely disparate are suddenly seen to be related. Further, the kind and level of sophistication of mathematics applied in various sciences has changed drastically in recent years: measure theory is used (non-trivially) in regional and theoretical economics; algebraic geometry interacts with physics; the Minkowsky lemma, coding theory and the structure of water meet one another in packing and covering theory; quantum fields, crystal defects and mathematical programming profit from homotopy theory; Lie algebras are relevant to filtering; and prediction and electrical engineering can use Stein spaces. And in addition to this there are such new emerging subdisciplines as 'experimental mathematics', 'CFD', 'completely integrable systems', 'chaos, synergetics and large-scale order', which are almost impossible to fit into the existing classification schemes. They draw upon widely different sections of mathematics."

By and large, all this still applies today. It is still true that at first sight mathematics seems rather fragmented and that to find, see, and exploit the deeper underlying interrelations more effort is needed and so are books that can help mathematicians and scientists do so. Accordingly MIA will continue to try to make such books available.

If anything, the description I gave in 1977 is now an understatement. To the examples of interaction areas one should add string theory where Riemann surfaces, algebraic geometry, modular functions, knots, quantum field theory, Kac-Moody algebras, monstrous moonshine (and more) all come together. And to the examples of things which can be usefully applied let me add the topic 'finite geometry'; a combination of words which sounds like it might not even exist, let alone be applicable. And yet it is being applied: to statistics via designs, to radar/sonar detection arrays (via finite projective planes), and to bus connections of VLSI chips (via difference sets). There seems to be no part of (so-called pure) mathematics that is not in immediate danger of being applied. And, accordingly, the applied mathematician needs to be aware of much more. Besides analysis and numerics, the traditional workhorses, he may need all kinds of combinatorics, algebra, probability, and so on.

In addition, the applied scientist needs to cope increasingly with the nonlinear world and the extra mathematical sophistication that this requires. For that is where the rewards are. Linear models are honest and a bit sad and depressing: proportional efforts and results. It is in the non-linear world that infinitesimal inputs may result in macroscopic outputs (or vice versa). To appreciate what I am hinting at: if electronics were linear we would have no fun with transistors and computers; we would have no TV; in fact you would not be reading these lines.

There is also no safety in ignoring such outlandish things as nonstandard analysis, superspace and anticommuting integration, p-adic and ultrametric space. All three have applications in both electrical engineering and physics. Once, complex numbers were equally outlandish, but they frequently proved the shortest path between 'real' results. Similarly, the first two topics named have already provided a number of 'wormhole' paths. There is no telling where all this is leading - fortunately.

Thus the original scope of the series, which for various (sound) reasons now comprises five subseries: white (Japan), yellow (China), red (USSR), blue (Eastern Europe), and green (everything else), still applies. It has been enlarged a bit to include books treating of the tools from one subdiscipline which are used in others. Thus the series still aims at books dealing with:

- a central concept which plays an important role in several different mathematical and/or scientific specialization areas;
- new applications of the results and ideas from one area of scientific endeavour into another;
- influences which the results, problems and concepts of one field of enquiry have, and have had, on the development of another.

There are certain parts of classical mathematics which are so central to the development of the subject and which are so pervasive in their influence, even now, that one really wishes one understood them perfectly. I certainly do, for one. Such are many of the results in Gauss disquisitiones arithmeticae, such is the idea of a Riemann surface, such is Galois theory, and there is much more. These things are part of our cultural heritage and one feels like an ill-educated bohunkus whenever one doesn't have it all at one's fingertips. To do something about that, the remedy is classical: go back and read the Masters: Gauss, Dirichlet, Weyl, Poincaré, Riemann, Hilbert, Dedekind, Galois, ... Easier said than done. That takes time, energy and persistence. Masters they were, but even so there has been progress, and much that was difficult to grasp, understand, prove and even difficult to define at the time, is now crystal clear.

The present volume, the first of two, offers the best of both worlds: the original inspirations and motivations, and modern clarity and technical power.

The shortest path between two truths in the real domain passes through the complex domain.

J. Hadamard

La physique ne nous donne pas seulement l'occasion de résoudre des problèmes ... elle nous fait pressentir la solution.

H. Poincaré

Never lend books, for no one ever returns them; the only books I have in my library are books that other folk have lent me.

Anatole France

The function of an expert is not to be more right than other people, but to be wrong for more sophisticated reasons.

David Butler

Bussum, 5 March 1991

Michiel Hazewinkel

Contents

Series Editor's Preface v
Notation xiii
Foreword xv

Chapter 1. Congruences

1.1 Introduction. Gauss's "*Disquisitiones arithmeticae*" 1
1.2 Basic properties of congruences 2
1.3 Power residues 3
1.4 Quadratic residues 5
1.5 A look at biquadratic residues 6
Exercises 8

Chapter 2. Quadratic forms

2.1 Introduction 11
2.2 Decomposable forms 12
2.3 Equivalence of forms 12
2.4 Primitive representations 13
2.5 Transformations which carry a form into itself 14
2.6 Forms with negative discriminant 16
2.7 Three theorems of Euler 20
2.8 Forms with positive discriminant 21
2.9 Composition of classes of forms 21
Exercises 22

Chapter 3. Division of the circle (cyclotomy)

3.1 Formulation of the problem 23
3.2 Auxiliary theorems about polynomials 24
3.3 Definition of the Gaussian periods and related theorems 26
3.4 Solution of the problem 28
3.5 $p = 17$ 29
3.6 Periods of length $(p-1)/2$ 30
3.7 Proof of the law of quadratic reciprocity 32
Exercises 33

Chapter 4. Theory of surfaces

4.1 Introduction. Gauss's "*Disquisitiones generales circa superficies curvas*" 35
4.2 Curves in the plane 35
4.3 Curves in space 37
4.4 Surfaces 37
4.5 Geometric meaning of the Gaussian curvature 38
4.6 Euler's contribution to the theory of surfaces 40
4.7 The intrinsic geometry of surfaces 41
Exercises 42

Chapter 5. Harmonic analysis

5.1 The equation of the vibrating string 44
5.2 Formulation of Dirichlet's theorem and preliminaries to the proof 46
5.3 Proof of Lemma 2 48

5.4 The Fourier formula 51
5.5 Harmonic analysis for complex-valued functions 53
5.6 Application: computation of the ζ-function for even positive arguments 53
Exercises 56

Chapter 6. Prime numbers in arithmetic progressions

6.1 The distribution of primes 59
6.2 Characters of finite abelian groups 60
6.3 Dirichlet L-functions 62
6.4 Proof of Theorem 1 64
Exercises 65

Chapter 7. Theory of algebraic equations

7.1 Third and fourth degree equations 67
7.2 Solution of equations by radicals 69
7.3 The general equation of nth degree and the theory of symmetric functions 70
7.4 The Galois memoir on the theory of equations 74
7.5 The theorem of the primitive element 75
7.6 The Galois group of a polynomial 76
7.7 An irreducibility criterion 78
7.8 Irreducible equations of prime degree with cyclic group 79
7.9 The cyclotomic equation 80
7.10 The fundamental theorem on the solution of equations by radicals 81
7.11 Permutation groups 82
7.12 On irreducible equations of prime degree 84
7.13 Equations of fifth degree with symmetric group 87
Exercises 88

Chapter 8. The beginnings of complex function theory

8.1 Introduction: from a letter of Gauss to Bessel 90
8.2 Basic concepts 91
8.3 The line integral 93
8.4 The Cauchy integral formula 95
8.5 Power series expansion 96
8.6 Coefficient estimates 97
8.7 The maximum-modulus principle 98
8.8 The Laurent expansion 99
8.9 Residues 102
8.10 The double series theorem 104
8.11 Analytic continuation 105
Exercises 106

Chapter 9. Entire functions

9.1 Functions with finitely many singular points 108
9.2 The Weierstrass product theorem 108
9.3 The Γ-function 111
9.4 Stirling's formula 113
Exercises 117

Chapter 10. Riemann surfaces

10.1 The 19th century view of the problems of function theory 118
10.2 The extended complex plane 120
10.3 The nth root 120

10.4 Definition of a Riemann surface 123
10.5 The Riemann surface of an algebraic function 126
10.6 The topology of closed Riemann surfaces 128
10.7 Polygon complexes 129
10.8 Classification of closed orientable surfaces 132
Exercises 135

Chapter 11. Meromorphic differentials on closed Riemann surfaces

11.1 Differentials and integrals on Riemann surfaces 137
11.2 The Riemann existence theorem for differentials 138
11.3 The Riemann period relations 141
11.4 Meromorphic functions 143
11.5 Riemann surfaces of genus 0 145
11.6 The Riemann-Roch theorem 146
11.7 The field of meromorphic functions on a closed Riemann surface 150
Exercises 152

Chapter 12. The theorems of Abel and Jacobi

12.1 Abel's theorem 154
12.2 Non-special divisors 158
12.3 The analytic nature of ΨA 159
12.4 Jacobi's theorem and the Jacobi inversion problem 161
Exercises 163

Chapter 13. Elliptic functions

13.1 Elliptic functions in the framework of Riemann's function theory 165
13.2 Construction of elliptic functions 168
13.3 Classification of Riemann surfaces of genus 1 173
13.4 The elliptic modular function 175
13.5 Picard's theorem in the theory of entire functions 179
Exercises 180

Chapter 14. Riemannian geometry

14.1 Riemann's inaugural lecture 182
14.2 n-dimensional Riemannian manifolds 183
14.3 Tangent vectors 185
14.4 Geodesics 186
14.5 Riemannian normal coordinates 189
14.6 The Riemann curvature tensor 193
14.7 Tensors as multilinear forms 197
14.8 The relation between the curvature tensor and the form Q(a,b) 198
14.9 Orthonormal bases 198
14.10 The Gaussian curvature 200
14.11 Spaces of constant curvature 202
14.12 Conformal mapping 206
14.13 Non-euclidean geometry 206
Exercises 208

Chapter 15. On the number of primes less than a given magnitude

15.1 Formulation of the problem 210
15.2 The functional equation of the ζ-function 210
15.3 On the zeros of the ζ-function 213
15.4 Riemann's exact formula 214
Exercises 217

Chapter 16. The origins of algebraic number theory

16.1 The Gaussian integers 219
16.2 Introduction to Chapters 17 to 21 220
Exercises 222

Chapter 17. Field theory

17.1 Field isomorphisms 223
17.2 Normal extensions and Galois groups 224
17.3 The fundamental theorem of Galois theory 225
17.4 The group of an equation 226
17.5 The composite of two fields. 227
17.6 Trace, norm, different and discriminant 228
Exercises 231

Chapter 18. Dedekind's theory of ideals

18.1 Integral elements 232
18.2 Lattices in finite extensions of P 235
18.3 The integers of quadratic fields 236
18.4 The ideals of O_K 237
18.5 The fundamental theorem of ideal theory 240
18.6 Consequences of the fundamental theorem 242
18.7 The norm of an ideal 244
18.8 Congruences 246
18.9 Prime ideal decomposition in quadratic fields 247
Exercises 248

Chapter 19. The ideal class group and the group of units

19.1 The finiteness of the class number 251
19.2 Units in quadratic fields 252
19.3 The structure of the group of units in the orders of algebraic number fields 253
19.4 The logarithmic components 254
19.5 The kernel of l 254
19.6 The image of l 255
19.7 The rank of $l(E)$ 256
19.8 Dirichlet's unit theorem as an assertion about diophantine equations 260
Exercises 261

Chapter 20. The Dedekind ζ-function

20.1 Definition of the Dedekind ζ-function 262
20.2 Preliminaries to the proof of Theorem 1 263
20.3 Reduction to a volume calculation 264
20.4 Volume calculation 266
20.5 Proof of Theorem 2 268
20.6 Application 270
Exercises 270

Chapter 21. Quadratic forms and quadratic fields

21.1 Modules in quadratic fields 272
21.2 Comparison with the ideal group 276
21.3 Forms and modules 278
Exercises 279

Chapter 22. The different and the discriminant

22.1 Relative extensions 281
22.2 Complementary modules 283
22.3 The second fundamental theorem of Dedekind 286
Exercises 289

Chapter 23. Theory of algebraic functions of one variable

23.1 Algebraic function fields 291
23.2 The Riemann surface 292
23.3 The order of a function at a point 293
23.4 Normal bases 295
23.5 The function space associated with a divisor 299
23.6 Differentials 299
23.7 The Riemann-Roch theorem 304
Exercises 305

Chapter 24. The geometry of numbers

24.1 The lattice point theorem 307
24.2 Application to the ideals of an algebraic number field 308
Exercises 312

Chapter 25. Normal extensions of algebraic number- and function fields

25.1 Normal extensions 313
25.2 Proof of the Dedekind different theorem 317
25.3 Cyclotomic fields 318
25.4 The ζ-function of a cyclotomic field 322
25.5 The theorem of Kronecker and Weber 324
Exercises 329

Chapter 26. Entire functions with growth of finite order

26.1 Formulation of the problem 331
26.2 Entire functions of finite order 331
26.3 Application to the Riemann ζ-function 338
Exercises 340

Chapter 27. Proof of the prime number theorem

27.1 Hadamard and de la Vallée Poussin 342
27.2 The Chebyshev function 343
27.3 The method of complex integration 345
27.4 Commencement of the proof of Theorem 3 349
27.5 On the zeros of the ζ-function 350
Exercises 355

Chapter 28. Combinatorial topology

28.1 Polyhedra in $\mathbb{R}^n$ 359
28.2 Topological polyhedra 361
28.3 The homology groups of a polyhedron 362
28.4 Computation of homology groups in simple cases 364
28.5 Betti numbers and Euler characteristic 367
28.6 The fundamental group 368
28.7 The edge path group of a polygon complex 369
28.8 Presentation of the edge path group by generators and relations 371
28.9 Covering spaces 373
28.10 Covering transformations 375

28.11 Quotient spaces 376
Exercises 376

Chapter 29. The idea of a Riemann surface

29.1 The concept of a real n-dimensional manifold 378
29.2 Definition of a Riemann surface 379
29.3 Orientability of Riemann surfaces 380
29.4 Meromorphic differentials 381
29.5 The Dirichlet integral on a Riemann surface 383
29.6 Dirichlet's principle 385
29.7 The Poisson integral 386
29.8 Dirichlet's principle for the circle 388
29.9 The smoothing process 390
29.10 Idea of the proof of the existence theorem for differentials 394
29.11 Proof of Dirichlet's principle 396
29.12 Proof of the Riemann existence theorem for differentials on closed Riemann surfaces 403
Exercises 406

Chapter 30. Uniformisation

30.1 The concept of uniformisation 407
30.2 The Riemann mapping theorem 407
30.3 The automorphisms of simply connected Riemann surfaces 412
30.4 Normal form of a Riemann surface 413
Exercises 415

Appendix 1. Rings

A1.1 Basic ring concepts 417
A1.2 Euclidean rings 419
A1.3 The characteristic of a ring 420
A1.4 Modules over euclidean rings 421
A1.5 Construction of fields 426
A1.6 Polynomials over fields 427

Appendix 2. Set theoretic topology

A2.1 Definition of a topological space 429
A2.2 Compact spaces 430

Appendix 3. Green's theorem 432

Appendix 4. Euclidean vector and point spaces 434

Appendix 5. Projective spaces 439

Bibliography 441

Name index 446

General index 448

Notation

$\mathbb{Z}$ Ring of integers.

$\mathbb{Q}$ Field of rational numbers.

$\mathbb{R}$ Field of real numbers.

$\mathbb{C}$ Field of complex numbers.

If $z = x + iy$ is a complex number then $\operatorname{Re} z = x$ resp. $\operatorname{Im} z = y$ denotes the real resp. imaginary part of z.

If Λ is an arbitrary ring then Λ^+ denotes the additive group and $\Lambda^\times$ the multiplicative group of Λ (cf. Section A.1.1).

$A := B$ and $B =: A$ mean that A is defined by B.

The end of a proof is signalled by the sign □. If □ stands immediately after a theorem, this means the proof has been carried out before the formulation of the theorem, or that the proof is left as an exercise to the reader. The sign ▣ means that the proof of a theorem is deferred until later.

If a is a real number, $[a]$ denotes the greatest integer $\leq a$.

For two real numbers a, b with $a < b$, $[a,b] := \{x \in \mathbb{R} \mid a \leq x \leq b\}$ resp. $(a,b) := \{x \in \mathbb{R} \mid a < x < b\}$ are the closed resp. open interval between a and b.

log always denotes the natural logarithm. For a complex number $z = re^{i\phi}$, $-\pi < \phi \leq \pi$, $\log z$ generally denotes the principal value $\log r + i\phi$ of the logarithm.

The exponential function is often denoted by exp.

If f, g are complex-valued functions and $h > 0$ is a real-valued function then

$$f = g + O(h),$$

denotes that there is a constant $c > 0$ with $|f-g| \leq ch$.

Foreword

Was du ererbt von deinen Vätern hast,
Erwirb es, um es zu besitzen.

Goethe, Faust I

This book is directed towards all those who have mastered two years of university mathematics. It aims to convey an overview of classical mathematics, particularly that of the 19th century and the first half of the 20th century.

Size limitations and the author's own mathematical inclinations prescribed the choice of material in the following way: the results which come to the fore are those of essential influence on the development of present-day mathematical structures in the sense of Bourbaki. The structural view of mathematics began in the 19th century with Gauss and Galois. However, Riemann and Dedekind are responsible for the further development of structural mathematics in the 20th century sense. The work of these two mathematicians is therefore central to the present volume, which is devoted mainly to the mathematics of the 19th century. In present-day mathematics the theory of Lie groups undoubtedly occupies a prominent place. However, the ideas of Lie first found their proper place in the edifice of mathematics only after further development, especially in the hands of E. Cartan. This theory is therefore reserved for a second volume.

The basic idea of the book is to present mathematical results in the spirit of their time of origin. The advantages of such a presentation are, apart from the historical dimension, the direct penetration to essentials without the ballast of many preparatory chapters typical of modern mathematical texts, and the direct motivation of readers by presentation of the main problems of the historical moment under consideration, often at the beginning of chapters or sections. The disadvantages of historical presentation are also obvious. The original form of presentation often deviates so much from present modes of thought that the effort required to understand it is justified only in historical research, which this book does not attempt. But progress in mathematics also consists of the simplification of results, which originally appeared complicated, by constructing an adequate framework for them (though this often requires the above-mentioned ballast of preparations). The way out of this situation adopted in this book is to rely on modern mathematical language and in a few places to defer proofs until later in the book, when the ideas of a succeeding historical period give them their present simplicity. This is true, e.g., of the Galois theory of equations, whose main theorems take their present form within the framework of Dedekind's theory of field extensions. The attraction of the theory lies, however, in its application to the theory of equations, following the original approach of Galois. We therefore present Galois theory first in Chapter 7 as a theory of equations, and then in Chapter 17 as field theory.

The order of material generally follows the historical development, beginning with Gauss's *Disquisitiones arithmeticae* and ending the first volume with the *Idee der Riemannschen Fläche* of Weyl. A number of chapters open with flashbacks to the mathematics of the 16th-18th centuries. One thing that becomes clear is that all the fundamental questions of 19th century mathematics (as far as they are treated here) were already touched on by Euler.

When mathematical ideas originate they often do not have the clarity which is required for a textbook. In such cases we begin at the point where clarity was reached. Thus we take up the theory of abelian integrals (Chapter 12) following the introduction of the concept of Riemann surface (Chapter 10). Later simplifications within the framework of the original idea are introduced without comment, since in any case the works of the great mathematicians serve only as a guideline for this book. The main theorems and proofs of certain theories seem to reach a final form, being reproduced in many textbooks with little conceptual change since the time of their origin. This is true, e.g., of the theory of functions with growth of finite order (Chapter 26). We present these in their usual form. The most important exception is the prime number theorem (Chapter 27), whose formulation and proof stems from knowledge of the zeros of the Riemann ζ-function, though nowadays this seems quite provisional. However, this theorem represents a high point in the development of mathematics which could not be omitted from this book.

The book is not intended to be a complete historical appreciation of 19th century mathematics. Things which the reader is assumed to know are passed over in silence. These include Cantor's set theory, the development of "Weierstrassian rigour" and others of great importance from the present-day viewpoint. Such an appreciation, from various angles, may be found in the following works:

Klein, F., *Die Entwicklung der Mathematik im 19. Jahrhundert, Teil I, Springer-Verlag* 1926.

Various authors, edited by A.N. Kolmogorov and A.P. Yushkevich, *Mathematics in the 19th Century* (Russian), vol. I, *Nauka* 1978, vol. II, *Nauka* 1981, vol. III in preparation.

Various authors, directed by J. Dieudonné, *Abrégé d'histoire des mathématiques 1700-1900*, vols. I, II, *Hermann* 1978.

Some of the mathematical prerequisites assumed of the reader have been collected in appendices at the end of the book, as an aid to memory. They are followed by a list of cited literature and references for further reading. The references to the original literature given in the text, if they do not prompt the reader to peruse them, at least provide additional scientific historical information.

My teachers H. Reichardt and I.R. Shafarevich, in their lectures, books and conversations have strongly influenced the picture of mathematics I have attempted to present here. For this I offer them my hearty thanks. Further thanks go, above all, to D. Schwarz, who worked through almost the whole manuscript and made numerous improvements and corrections. Among other colleagues who read parts of the manuscript and made valuable suggestions I mention my wife, H. Bothe, E. Krauss, W. Narkiewicz, O. Neumann, H. Pieper and I. Schiemann.

Helmut Koch

1. Congruences

1.1 Introduction. Gauss's "Disquisitiones arithmeticae"

When Gauss entered Göttingen University in 1795, he was still not sure whether to become a mathematician or a philologist. The scales were tipped in favour of mathematics by the discovery that the regular 17-gon is constructible with ruler and compass, which he made on 29 March 1796, while still lying in bed, after having been occupied with the question for a long time. This discovery rested on an algebraic theory, and he used it to illustrate that theory in his book, written in Latin in 1799, the *Disquisitiones Arithmeticae*. The subject of the other sections of this work is the "*higher arithmetic*", by which Gauss understood all mathematical investigations "*which have to do with the integers*". Among his predecessors Gauss mentioned, in his foreword to the *Disquisitiones Arithmeticae*, Euclid, Diophantus, Fermat, Euler, Lagrange and Legendre.

In the first three sections Gauss summarised the knowledge of his time. In the fourth section he proved the *quadratic reciprocity law* as the first fundamental achievement of the book. A second is the development of the theory of quadratic forms in the fifth, and by far the most voluminous, section. The short sixth section brings "*various applications of the preceding investigations*". Finally, the seventh and last chapter deals with division of the circle.

The *Disquisitiones arithmeticae* were translated into German by Maser and published under the title *Arithmetische Untersuchungen* in 1889. We have based our account on this edition. The contents of the first three sections is nowadays largely contained in the secondary school algebra course. We can therefore confine ourselves to a few remarks about it. The proofs of the theorems in sections 4 and 5 are long and opaque. This remained the case until the last third of the 19th century, when Dedekind and Kronecker arrived at clearer proofs on the basis of their theory of algebraic numbers (Section 25.3, Chapter 21). The circle division of section 7 of the *Disquisitiones arithmeticae* will be presented in detail in Chapter 3 of this book. In the present chapter we deal with sections 1 to 4. Our chapter 2 contains some details of the theory of quadratic forms. The results derived there are due mainly to Lagrange (*Recherches d'arithmétique, Nouv. Mém. Acad. roy. Sci. Belles-Lettres Berlin* 1773, 1775).

1.2 Basic properties of congruences

Two integers b, c are called *congruent modulo a natural number* a when a is a divisor of the difference b-c. We write this as
$b \equiv c \pmod{a}$.

With $b \equiv c \pmod{a}$ and $d \equiv e \pmod{a}$ one also has $b + d \equiv c + e \pmod{a}$ and $bd \equiv ce \pmod{a}$, i.e. one computes with congruences analogously as with equations. The following holds concerning division:

Theorem 1. *For a natural number* a *and integers* b,c *the congruence*

$$bx \equiv c \pmod{a} \tag{1}$$

is solvable if and only if the greatest common divisor (a,b) *of* a *and* b *divides* c : $(a,b)\,|\,c$.

Proof. The solvability of (1) means that there are integers x_1,y_1 with $bx_1 = c + ay_1$. It follows from this that $(a,b)\,|\,c$. Conversely, suppose that $(a,b)\,|\,c$. Then one finds, from the representation $(a,b) = bx_2 + ay_2$ by integers x_2,y_2 (euclidean algorithm, Section A 1.2), the solution $x = x_2c/(a,b)$ of (1). □

Two solutions x_1,x_2 of the congruence (1) are considered the same if $x_1 \equiv x_2 \pmod{a}$. As one easily sees, the solution of (1) in this sense is unique if $(a,b) = 1$.

Pairwise relatively prime natural numbers $a_1,\dots,a_s$ satisfy what is known today as the *Chinese remainder theorem:*

Theorem 2. *If* $b_1,\dots,b_s$ *are arbitrary integers and* $a = a_1\dots a_s$, *then there is an integer* b *with*

$$b \equiv b_i \pmod{a_i} \text{ for } i = 1,\dots,s. \tag{2}$$

b *is uniquely determined modulo* a.

In the language of present-day algebra, Theorem 2 is restated in the following context.

Computation with congruences modulo a is equivalent to computation in the congruence class ring $\mathbb{Z}/a\mathbb{Z}$ (one often writes $a\mathbb{Z} := \{ax \mid x \in \mathbb{Z}\}$ as (a), the principal ideal generated by a).

Theorem 2′. *The correspondence* $\psi : b + a\mathbb{Z} \mapsto (b + a_1\mathbb{Z},\dots,b + a_s\mathbb{Z})$ *defines a ring isomorphism from* $\mathbb{Z}/a\mathbb{Z}$ *onto the direct sum* $\mathbb{Z}/a_1\mathbb{Z} + \dots + \mathbb{Z}/a_s\mathbb{Z}$.

Proof. ψ is obviously a homomorphism of rings, and the kernel of ψ consists only of the congruence class $a\mathbb{Z}$. Therefore ψ is injective. Since ψ is a mapping of finite rings with the same number a of elements, ψ is also surjective. □

The following theorem goes back to Lagrange (*Mém. Acad. Sci. Berlin* 1768).

Theorem 3. *If* $a_0,a_1,\dots,a_s$ *are arbitrary integers, and if* p *is a prime number which does not divide* a_0, *then the congruence*

$$a_0x^s + a_1x^{s-1} + \dots + a_s \equiv 0 \pmod{p} \tag{3}$$

has at most s *solutions.* □

Today we understand this congruence to be an equation with coefficients in the field $\mathbb{Z}/p\mathbb{Z}$ with p elements. The corresponding result for real numbers was already known to Descartes and Newton.

1.3 Power residues

The third section of the *Disquisitiones arithmeticae* deals with power residues, i.e the powers $a, a^2,\dots$ are investigated in terms of their congruence properties for a prime modulus p, where $a \not\equiv 0 \pmod{p}$. The first considerations of Gauss concern the fact that the $p-1$ nonzero congruence classes mod p form a group under multiplication (Theorem 1). This yields in particular

Theorem 4. *If* $a \not\equiv 0$ $(mod$ $p)$ *then* $a^{p-1} \equiv 1$ $(mod$ $p)$. □

Gauss gives the following commentary: "*This theorem, which is highly noteworthy for its elegance as well as its usefulness, is named Fermat's theorem after its discoverer*" (*Fermatii Opera Math., Tolosae* 1679). The first published proof was given by Euler (*Comm. Acad. Sci. Imp. Petropolitanae* VIII, 1736).

A later work of Euler (*Novi Comm. Acad. Sci. Petropolitanae* 1760/61) contains the generalisation $a^{\phi(m)} \equiv 1 \pmod{m}$ for an arbitrary natural number m prime to a, where $\phi(m)$ is the number of integers prime to m between 0 and m. Following Gauss's proposal, $\phi(m)$ is called the *Euler function*.

Closely connected with Fermat's theorem is the following theorem, which says that the multiplicative group of congruence classes is cyclic for a prime modulus.

Theorem 5. *There is a number* a *for which* $a^i \not\equiv 1 \pmod p$ for $i = 1,\dots,p-2$.

Euler called such a number a a *primitive root* mod p.

Gauss gave two proofs of Theorem 5. They depend on Theorem 3, according to which the congruence $x^d \equiv 1 \pmod p$ has at most d solutions. If there were no primitive roots, then the number of solutions of $x^d \equiv 1 \pmod p$ for a certain divisor d of p-1 would be too large.

The first proof of Gauss, which one can regard as the canonical proof of Theorem 5, runs as follows.

For any natural number n we have

$$\sum_{d|n} \phi(d) = n, \tag{4}$$

where the sum is taken over all divisors d of n.

We consider the multiplicative group G of congruence classes mod p. There are at most $\phi(d)$ elements $g \in G$ of order d. For if g is such an element, one obtains the d solutions of $x^d = 1$ as $g, g^2, \dots, g^d$, and exactly $\phi(d)$ of these d elements have order d, by definition of ϕ. We now apply (4) to the order p-1 of G. If G had no element of order p-1 then the number of elements of G would be strictly smaller than

$$\sum_{d|p-1} \phi(d) = p-1.$$

This is a contradiction. □

Let h be a primitive root mod p. Then for each $a \not\equiv 0 \pmod p$ there is a unique ν, $0 \le \nu < p-1$ such that $a \equiv h^\nu \pmod p$. ν is called the *index of* a : ν = ind a. This construction is analogous to the logarithm.

EXAMPLE. p = 17, h = 3.

Power table	ν	0	1	2	3	4	5	6	7	8	9	10	11	12	13	14	15
	a	1	3	9	10	13	5	15	11	16	14	8	7	4	12	2	6

Index table	a	1	2	3	4	5	6	7	8	9	10	11	12	13	14	15	16
	ν	0	14	1	12	5	15	11	10	2	3	7	13	4	9	6	8

1.4 Quadratic residues

Section 4 of the *Disquisitiones arithmeticae* deals with quadratic residues and contains the first complete proof of the law of quadratic reciprocity, which Gauss calls the "*fundamental theorem of the theory of quadratic residues*". Altogether he gave seven proofs of this theorem, however they should all be regarded as verifications, which give no insight into the background of the law. Here we shall content ourselves with the formulation of the theorem, and in Section 3.7 we shall provide a short proof, going back to Gauss. Later we shall see how one can obtain the law of quadratic reciprocity as a corollary of a theorem concerning algebraic numbers related to division of the circle (Section 25.3).

The number a is called a *quadratic residue mod* p when there is a number x with $x^2 \equiv a \pmod p$. Since the case $p = 2$ is uninteresting, we assume $p \neq 2$ in what follows. For $a \not\equiv 0 \pmod p$ Legendre in 1808 introduced the symbol $(\frac{a}{p})$, which is equal to 1 when a is a quadratic residue mod p, and otherwise equals -1. One easily sees that

$$(\frac{ab}{p}) = (\frac{a}{p})(\frac{b}{p}),$$

$$(\frac{a}{p}) \equiv a^{\frac{p-1}{2}} \pmod p \quad (\textit{Euler's criterion}) \tag{5}$$

One sets $a \equiv h^{\nu}$, $b \equiv h^{\mu} \pmod p$ with a primitive root h (Theorem 5). Then obviously $(\frac{a}{p}) = (-1)^{\nu}$, $(\frac{b}{p}) = (-1)^{\mu}$. This immediately gives the following theorem, which Gauss proved in a different way.

Theorem 6. *Half the numbers* 1,...,p-1 *are quadratic residues, the others nonresidues.* □

(In the cyclic group of congruence classes mod p the squares form a subgroup of index 2.)

In what follows, we are interested in the dependence of $(\frac{a}{p})$ on the prime p. The following is called the *first supplement to the quadratic reciprocity law.*

Theorem 7. $(\frac{-1}{p}) = \begin{cases} +1 & \text{for } p \equiv 1 \pmod 4, \\ -1 & \text{for } p \equiv 3 \pmod 4. \end{cases}$

Proof. This is a special case of (5). □

The *second supplement* is

Theorem 8. $\left(\frac{2}{p}\right) = \begin{cases} +1 & \text{for } p \equiv 1,7 \pmod 8, \\ -1 & \text{for } p \equiv 3,5 \pmod 8. \end{cases}$ ◼ (Sect. 3.7)

Because of (5), combination of Theorem 7 and Theorem 8 yields

Theorem 9. $\left(\frac{-2}{p}\right) = \begin{cases} +1 & \text{for } p \equiv 1,3 \pmod 8, \\ -1 & \text{for } p \equiv 5,7 \pmod 8. \end{cases}$ □

The group of congruence classes prime to 8 mod 8 has three subgroups of order 2. The prime numbers p for which $-1,2,-2$ are quadratic residues are distributed among these subgroups.

Finally, the *quadratic reciprocity law* says:

Theorem 10. *If* p,q *are prime numbers* $\neq 2$, *then*

$\left(\frac{p}{q}\right) = \left(\frac{q}{p}\right)$ *when* p *or* $q \equiv 1$ (*mod* 4),

$\left(\frac{p}{q}\right) = -\left(\frac{q}{p}\right)$ *when* p *and* $q \equiv 3$ (*mod* 4). ◼ (Sect. 3.7)

It is left to the reader to generalise the assertion in Theorem 9 to a theorem on the dependence of $\left(\frac{a}{p}\right)$ on the denominator p, when a is an integer and p runs through the prime numbers relatively prime to a.

1.5 A look at biquadratic residues

After studying quadratic residues it is natural to wonder whether a similar theory is possible for other power residues. In 1828 and 1832 Gauss published two works on *biquadratic residues* (i.e. fourth power residues), which went back to research he had done since 1805. He had quickly come to the realisation that a satisfactory theory of biquadratic residues could only be carried through when one passed from the domain of integers to a larger one, known today as the *ring of Gaussian integers*, which consists of all numbers of the form $a + b\sqrt{-1}$ where a and b are arbitrary integers (Chap. 16). Gauss wrote that "*the establishment of a general theory ... necessarily requires the domain of higher arithmetic to be extended, in some sense infinitely many times*".

The reason it becomes necessary *to extend the domain of arithmetic* can easily be seen from the present-day standpoint, when one reflects that the theory of quadratic residues is based on the homomorphism of the congruence class group $(\mathbb{Z}/p\mathbb{Z})^{\times}$ onto the group of square roots of unity given by the Legendre symbol $(\frac{a}{p})$. The theory of biquadratic residues is based on a corresponding homomorphism onto the group of fourth roots of unity. The latter lie in the ring of Gaussian integers.

In Article 67 of the second work, Gauss formulated the "*fundamental theorem of biquadratic residues*" as an analogue of the law of quadratic reciprocity. Gauss then wrote: "*Despite the great simplicity of this theorem, its proof is one of the most hidden mysteries of higher arithmetic*". He left the proof for a third work which, however, was never written. A sketch of the proof was found in Gauss's *Nachlass*. The first proof is due to Jacobi (*Vorlesungen über Zahlentheorie, Universität Königsberg* 1835/36).

In the second half of the 19th century the theory of Gaussian integers became a part of the general theory of algebraic numbers (Chap. 18). In the 1920's the law of biquadratic reciprocity was absorbed by the *general Artin reciprocity law*, which will be treated in the second volume of this book.

Exercises

1.1 Let n be a natural number with prime decomposition $n = p_1^{a_1} \dots p_s^{a_s}$. Show that the group $(\mathbb{Z}/n\mathbb{Z})^\times$ of congruence classes prime to n mod n equals the direct product of the groups $(\mathbb{Z}/p_i^{a_i}\mathbb{Z})^\times$ for $i = 1,\dots,s$. Show also that the group $(\mathbb{Z}/p^a\mathbb{Z})^\times$ is cyclic for a prime $p \neq 2$. What is the structure of $(\mathbb{Z}/2^a\mathbb{Z})^\times$?

1.2 Solve the congruence $7x \equiv 12 \pmod{17}$ with the help of the index and power table in Sect. 1.4. Also investigate the solvability of the congruences $5^x \equiv 7 \pmod{17}$, $8^x \equiv 9 \pmod{17}$, $4^x \equiv 6 \pmod{17}$.

1.3 Determine a primitive root for 41 and construct the power and index table.

1.4 (*Wilson's theorem*). Let p be a prime number. Show $(p-1)! \equiv -1 \pmod{p}$.

1.5 Suppose p is prime and $n \le p$. Show that the congruence $f(x) \equiv 0 \pmod{p}$, whence $f(x) = x^n + a_1x^{n-1} + \dots + a_n$ with $a_1,\dots,a_n \in \mathbb{Z}$ has n different solutions if and only if there is a $g(x) \in \mathbb{Z}[x]$ with $f(x)g(x) \equiv x^p - x \pmod{p}$.

1.6 A function $h(a)$ which is defined for all natural numbers a and takes values in a ring is called *multiplicative* when $h(a_1a_2) = h(a_1)h(a_2)$ for all relatively prime a_1, a_2. Show that the Euler function $\phi(a)$ is multiplicative.

1.7 Let $\mu(a)$ be the multiplicative function which for prime powers p^α takes the value $\mu(p^\alpha) = -1$ when $\alpha = 1$ and $\mu(p^a) = 0$ for $\alpha > 1$. $\mu(a)$ is called the *Möbius function.* Show that

$$\sum_{d \mid a} \mu(d) = \begin{cases} 0 & \text{for } a > 1 \\ 1 & \text{for } a = 1, \end{cases} \tag{6}$$

where the sum is taken over all divisors d of a.

1.8 Let $f(a)$ and $g(a)$ be functions, defined for all natural numbers a, with values in an additive abelian group. Then

$$\sum_{d|a} f(d) = g(a), \text{ for all } a, \text{ if and only if}$$

$$\sum_{d|a} \mu(a/d)g(d) = f(a) \text{ for all } a.$$

(*Möbius inversion formula*)

1.9 Which formulae correspond to (4) and (6) for an arbitrary multiplicative function?

1.10 Let a_1,a_2 be integers and let m_1,m_2 be relatively prime natural numbers. Show that the system of congruences $x \equiv a_1 \pmod{m_1}$, $x \equiv a_2 \pmod{m_2}$ may be solved with the help of the euclidean algorithm (Sect. A 1.2).

1.11 Let α be a real number and let $q_1 = [\alpha]$ be the greatest integer which is less than or equal to α. In the case $\alpha_1 \neq [\alpha_1]$ we continue the process and put α_1 in the form $\alpha_1 = q_2 + 1/\alpha_2$, etc.. Show that the *continued fraction algorithm* so defined terminates in a finite number of steps for rational numbers α, and establish its connection with the euclidean algorithm (Sect. A 1.2).

1.12 The continued fraction $q_1 + \cfrac{1}{q_2 + \cfrac{\ddots}{+\cfrac{1}{q_s}}}$

is abbreviated $[q_1;q_2,...,q_s]$. When $\alpha = [q_1;q_2,...,q_n,\alpha_n]$, $[q_1;q_2,...,q_s]$ for $s \leq n$ is called the sth *convergent* of α. Let P_s/Q_s with $Q_s \geq 1$ be the representation of $[q_1;q_2,...,q_s]$ as a reduced fraction.

Additionally, we set $P_0 = 1$, $Q_0 = 0$. Show:

a) $P_s = q_sP_{s-1} + P_{s-2}$, $Q_s = q_sQ_{s-1} + Q_{s-2}$ for $s \geq 2$.

b) $P_sQ_{s-1} - Q_sP_{s-1} = (-1)^s$ for $s \geq 1$.

c) $P_1/Q_1 < P_3/Q_3 < ...\alpha... < P_4/Q_4 < P_2/Q_2$.

d) For α irrational, $\lim_{s\to\infty} P_s/Q_s = \alpha$.

1.13 Let a,b be integers and let m be a natural number prime to a. Also, let $m/a = [q_1;q_2,...,q_n]$. Show that the congruence $ax \equiv b \pmod{m}$ has the solution $x \equiv (-1)^{n-1}P_{n-1}b \pmod{m}$, where P_{n-1}/Q_{n-1} is the $(n-1)^{th}$ convergent of m/a.

1.14 Let $p \neq 2$ be a prime number such that $p = a^2 + b^2$ for natural numbers a,b with $a \equiv 1 \pmod 2$ (cf. Chap. Theorem 10). Show that $(\frac{a}{p}) = 1$.

(Hint: Use the quadratic reciprocity law.)

2. Quadratic forms

2.1 Introduction

Let a,b,c be integers and let x,y be indeterminates. A *binary quadratic form* is a polynomial

$$ax^2 + 2bxy + cy^2. \tag{1}$$

The theory of integral binary quadratic forms is mainly concerned with the question of which integer values (1) can take when integers are substituted for x and y. This is therefore a subfield of the theory of diophantine equations, which deals with the possible values of an arbitrary polynomial with integer coefficients when integers are substituted for the indeterminates.

Before Gauss, quadratic forms were treated above all by Lagrange, "*and much of the nature of forms was either discovered by this great geometer* [i.e. mathematician], *or else proved by him after being previously discovered by Euler or Fermat.*"[1)] However, Gauss gives an independent presentation of the material, in order to provide a basis for his own, more far-reaching, results. Gauss abbreviates the form (1) by (a,b,c). We shall use vector notation, and hence set

$$ax^2 + 2bxy + cy^2 = zAz^T, \tag{2}$$

where $z := (x,y)$, $A := \binom{a\ b}{b\ c}$ and $z^T = \binom{x}{y}$ is the transposed vector. We shall further abbreviate the form (1) by (A).

An important rôle is played by the number - det A, the *discriminant of the quadratic form.*

A simplified presentation of Gauss's theory of quadratic forms was given by Dirichlet in his *Vorlesungen über Zahlentheorie*, first published in 1871 by Dedekind. We shall partly follow these *Vorlesungen* below.

---

1) A careful analysis of the number theoretic results of the mathematicians Fermat, Euler, Lagrange and Legendre, illustrating the above remark of Gauss, is the subject of the book Weil [1].

2.2 Decomposable forms

Theorem 1. *The quadratic form* (1) *decomposes into the product of a rational number and two linear forms in* x,y *with integer coefficients if and only if the discriminant* D *of the form is a square.*

The proof of Theorem 1 comes from the identity

$$ax^2 + 2bxy + cy^2 = \frac{1}{a}(ax + (b + \sqrt{D})y)(ax + (b - \sqrt{D})y). \quad \square \tag{3}$$

When the form (1) decomposes as in Theorem 1, one can reduce its number-theoretic properties to those of linear forms. In what follows we therefore assume that the discriminant of the form in question is not a square.

2.3 Equivalence of forms

If one makes an integral linear transformation of the indeterminates,

$$\mathbf{z}^T = \binom{x}{y} = \mathbf{B}\binom{x'}{y'} = \mathbf{B}\mathbf{z}'^T \tag{4}$$

then the form (**A**) goes over to the form ($\mathbf{B}^T\mathbf{A}\mathbf{B}$), whose discriminant equals - det **A** (det **B**)^2 (cf. (2)). In particular, when **B** is invertible, i.e. det **B** = ± 1, then the two forms are called *equivalent*. Two forms are called *properly equivalent* when one can be converted to the other by a transformation with det **B** = 1.[1)]

Equivalent forms represent the same numbers, and hence are not essentially different for the purposes of our question. They have the same discriminant. The following problems therefore present themselves as immediately important:

I. *Given two forms with the same discriminant, decide whether they are equivalent or not.*

[1)] The concept of proper equivalence did not appear in the work of Lagrange. It is of decisive importance for the composition of classes of forms, as this can only be carried out for proper equivalence classes (Sect. 2.9, Chap. 21).

II. *Given any form, find whether a given number can be represented by it, and find all its representations.*

To solve the first problem, Gauss introduced the concept of adjacent forms: the forms (a,b,c) = (A) and (a′,b′,c′) = (A′) are called *adjacent*, when they have the same discriminant, $c = a'$ and $b + b' \equiv 0 \pmod c$.

Theorem 2. *Two adjacent forms are properly equivalent.*

Proof. Let $B = \begin{pmatrix} 0 & -1 \\ 1 & d \end{pmatrix}$ with $d = \frac{(b+b')}{c}$. Then $\det B = 1$ and $A' = B^T A B$. □

2.4 Primitive representations

A representation

$$m = au^2 + 2buv + cv^2 \tag{5}$$

of the number m by the form (a,b,c) is called *primitive* when u and v are relatively prime.

The following theorem establishes the connection between quadratic forms and quadratic residues.

Theorem 3. *If the number* m *admits a primitive representation by the form* (a,b,c), *then the discriminant* D *of the form is a quadratic residue* mod m.

Proof. Let (5) be the primitive representation of m. Then there are integers u', v' with $uv' - vu' = 1$. We set $B = \begin{pmatrix} u & u' \\ v & v' \end{pmatrix}$. The quadratic form $(B^T AB)$ has first coefficient m. The assertion therefore follows from $D = -\det A = -\det(B^T AB)$. □

Let $n = u'(ua + vb) + v'(ub + vc)$ be the second coefficient of $(B^T AB)$. The numbers u', v' are uniquely determined by the condition $uv' - vu' = 1$ up to summands tu, tv with an arbitrary integer t. In passing from u', v' to $u' + tu$, $v' + tv$, n goes over to $n + tm$. Thus the congruence class of n mod m is uniquely determined by the representation (5) of m. We say that *the representation* (5) *belongs to the root* n.

These considerations lead to

Theorem 4. *Suppose the representation* (5) *belongs to root* n. *Then the forms* (a,b,c) *and* $(m,n,(n^2-D)/m)$ *are properly equivalent.* □

Theorem 4 shows the connection between the transformations of the quadratic form (a,b,c) and the representability of a number m by this form. In order to gain an overview of the various possible representations of m by (a,b,c) which belong to root n, one has to determine all transformations $\mathbf{B}$ which carry (a,b,c) into $(m,n,(n^2-D)/m)$. This leads to the problem of finding the transformations which carry (a,b,c) into itself, to which we turn next.

2.5 Transformations which carry a form into itself

Let the greatest common divisor of a, $2b$ and c be s.

Theorem 5. *A transformation* $\mathbf{B} = \begin{pmatrix} u & u' \\ v & v' \end{pmatrix}$ *with* $\det \mathbf{B} = 1$ *carries* (a,b,c) *into itself if and only if*

$$u = \frac{t-bw}{s},\quad u' = \frac{-cw}{s},\quad v = \frac{aw}{s},\quad v' = \frac{t+bw}{s} \tag{6}$$

where t *and* w *are integers satisfying the conditions*

$$t \equiv bw \pmod{s} \tag{7}$$

and

$$t^2 - Dw^2 = s^2 \tag{8}$$

(6) defines a one-to-one mapping of all pairs s, w which satisfy the conditions (7), (8) onto the transformations $\mathbf{B} = \begin{pmatrix} u & u' \\ v & v' \end{pmatrix}$ with $\det \mathbf{B} = 1$ which carry (a,b,c) into itself.

Remark. Equation (8), which already appears in a famous problem of Archimedes, is (incorrectly) called *Pell's equation* in the case $s = 1$.

Proof of Theorem 5. If $\mathbf{B}$ has the form (6), then $\det \mathbf{B} = (t^2 - Dw^2)/s^2 = 1$ and (a,b,c) is carried to itself. Conversely, suppose $\mathbf{B} = \binom{u\ u'}{v\ v'}$, with $\det \mathbf{B} = 1$, carries the form (a,b,c) into itself. Then

$$au^2 + 2buv + cv^2 = a, \tag{9}$$

$$auu' + b(uv' + u'v) + cvv' = b. \tag{10}$$

We can also write (10) as

$$auu' + 2bu'v + cvv' = 0. \tag{11}$$

Elimination of b and c respectively from (9), (11) yields

$$au' = -cv \quad \text{and} \quad a(u-v') = -2bv.$$

This shows that a/s is a divisor of the greatest common divisor $\gcd(cv/s, 2bv/s)$ of cv/s and $2bv/s$. Since a/s is relatively prime to $\gcd(c/s, 2b/s)$, a/s must be a divisor of v. We set $w = sv/a$, which is permissible, since $a \neq 0$ otherwise D would be a square. Then $u - v' = -2bw/s$, which implies that $s(u+v')$ is divisible by 2. We set $t = s(u+v')/2$. Then the equations (6) are satisfied. Therefore $t \equiv bw \pmod s$ and $s^2 \det\mathbf{B} = t^2 - Du^2 = s^2$. □

Since consideration of (a,b,c) and (da,db,dc) is essentially the same for an integer d, we can confine ourselves to forms (a,b,c) with $\gcd(a,b,c) = 1$. The interesting cases are therefore $s = 1$ and $s = 2$. Admission of $s = 2$ also permits the handling of forms whose middle coefficient is odd, by multiplying all coefficients by 2. The restriction to even middle coefficients, which the reader may have wondered at, is therefore inessential.

In what follows we confine ourselves to quadratic forms with $\gcd(a,b,c) = 1$, and hence $s = 1$ or $s = 2$, which will be called *primitive forms.* In the case $s = 2$, $D \equiv 1 \pmod 4$. In the case $s = 1$ obviously all integers can appear as discriminant.

2.6 Forms with negative discriminant

The solution of problems I and II proceeds differently according as the discriminant is positive or negative. We first consider the second case. In this section the discriminant $D = -\det A$ of the form (A) in question is always negative.

Theorem 6. *Let* (a,b,c) *and* (a',b',c') *be equivalent forms with negative discriminant. Then the numbers* a,c,a',c' *have the same sign.*

Proof. Since $ac = b^2 - D > 0$, a and c have the same sign. The corresponding fact holds for a' and c'. Also, because of the equivalence of the forms (a,b,c) and (a',b',c'),

$$a' = au^2 + 2buv + cv^2 \tag{12}$$

for certain integers u, v. From this it follows that

$$a'a = (au+bv)^2 - Dv^2 > 0. \quad \square \tag{13}$$

Because of Theorem 6, two forms (A) and (-A) can never be equivalent. It therefore suffices to consider forms (a,b,c) with positive a for the solution of problems I and II. We call such a form *positive*.

Theorem 7. *For each positive form* (a_1,b_1,c_1) *with negative discriminant* D *there is a properly equivalent form* (a,b,c) *with*

$$2|b| \le a \le \min(\sqrt{(4/3)|D|},c). \tag{14}$$

Proof. A form (a,b,c) satisfying (14) is called *reduced*. Suppose (a_1,b_1,c_1) is not reduced. We have to show that (a_1,b_1,c_1) is properly equivalent to a reduced form. Let b' be the absolutely smallest residue of $-b_1$ with respect to the modulus $a' := c_1$ and let $a'' := (b'^2-D)/a'$. Then a'' is an integer because

$$b'^2 - D \equiv b_1^2 - D = a_1a' \equiv 0 \pmod{a'}.$$

The form (a',b',a'') is adjacent to the form (a_1,b_1,a'). Also, $|b'| \le a'/2$. Thus we have found for (a_1,b_1,a') a properly equivalent form (a',b',a'') for which the left inequality in (14) is satisfied (Theorem 2). When $a' > a''$, we repeat the process and obtain a properly equivalent form (a'',b'',a'''), etc.. The procedure must terminate in finitely many steps, otherwise there would be an infinite sequence $a' > \ldots > a^{(i)} > \ldots$ of positive integers, suppose therefore that $a^{(n)} \le a^{(n+1)}$. Then

$$a^{(n)2} \le a^{(n)}a^{(n+1)} = b^{(n)2} - D \le a^{(n)2}/4 - D.$$

It follows from this that $a^{(n)} \le \sqrt{(4/3)|D|}$. Thus the form $(a,b,c) = (a^{(n)},b^{(n)},a^{(n+1)})$ achieves what was desired in Theorem 7. □

It remains to investigate the equivalence of reduced forms.

Theorem 8. *A reduced form* (a,b,c) *which is properly equivalent to another reduced form satisfies one of the conditions*

$$a = 2|b| \quad \text{or} \quad a = c.$$

If one of these conditions is satisfied, then (a,b,c) *is properly equivalent to* $(a,\pm b,c)$ *but to no other reduced form.*

Proof. Let $a = 2b$ resp. $a = -2b$. Then

$$\begin{pmatrix} a & -b \\ -b & c \end{pmatrix} = \begin{pmatrix} 1 & 0 \\ -1 & 1 \end{pmatrix}\begin{pmatrix} a & b \\ b & c \end{pmatrix}\begin{pmatrix} 1 & -1 \\ 0 & 1 \end{pmatrix} \quad \text{resp} \quad \begin{pmatrix} a & -b \\ -b & c \end{pmatrix} = \begin{pmatrix} 1 & 0 \\ 1 & 1 \end{pmatrix}\begin{pmatrix} a & b \\ b & c \end{pmatrix}\begin{pmatrix} 1 & 1 \\ 0 & 1 \end{pmatrix}.$$

Let $a = c$. Then

$$\begin{pmatrix} a & -b \\ -b & c \end{pmatrix} = \begin{pmatrix} 0 & 1 \\ -1 & 0 \end{pmatrix}\begin{pmatrix} a & b \\ b & c \end{pmatrix}\begin{pmatrix} 0 & -1 \\ 1 & 0 \end{pmatrix}.$$

Now we suppose, conversely, that the reduced form (a,b,c) is carried to the reduced form (a',b',c') by the transformation $\mathbf{B} = \begin{pmatrix} u & u' \\ v & v' \end{pmatrix}$ with $\det \mathbf{B} = 1$. Then

$$a' = au^2 + 2buv + cv^2 \tag{15}$$

and

$$b' = auu' + b(uv'+u'v) + cvv' \tag{16}$$

Multiplication of (15) by a gives

$$aa' = (au+bv)^2 - Dv^2.$$

By (14), $aa' \le (4/3)|D|$ and hence $v = 0$ or $|v| = 1$.

We first consider the case $v = 0$. Then $uv' = 1$, $a' = au^2$ and $b' = auu' + b$. This implies $u = v' = \pm 1$ and hence $a' = a$ and $b' - b = \pm au'$. Thus, by (14) $a|u'| \le |b| + |b'| \le a$.

This means either $u' = 0$, i.e. $b' = b$, $c' = c$ or else $|u'| = 1$. In the latter case $|b| + |b'| = a$, and because of (14) we have $|b| + |b' = a/2$, $c' = c$, i.e. the first case of Theorem 8 holds.

Now suppose $|v| = 1$. Then (15) has the form

$$a' = au^2 \pm 2bu + c.$$

Without loss of generality we can assume that $a' \le a$. Then by (14) we also have $a' \le c$,

$$au^2 \le au^2 + c - a' = 2|bu| \le a|u| \le au^2.$$

This implies $c = a' = a$ and $au^2 = 2|bu|$. One can now write (16) in the form $b + b' = auu' + 2buv + cvv' = a(uu' \pm u^2v \pm v')$.

This implies that $b + b' = 0$ or $|b+b'| = a$, similarly to the case $v = 0$. In the second case $b = b'$, $c = c'$, and in the first $b = -b'$, $c = c'$, i.e. the second case of Theorem 8 holds. □

In order to obtain an overview of the classes of properly equivalent positive forms with given negative discriminant, one has to find the corresponding reduced forms (a,b,c). This can be done as follows: take all b with $|b| \le \frac{1}{2}\sqrt{\frac{4}{3}|D|}$ and decompose $b^2 - D$ into two factors $a \le c$ which are both $\ge 2|b|$. One obviously finds all the desired forms in this way. In particular

Theorem 9. *The number of reduced forms with given negative discriminant is finite.* □

For the smallest value of $|D|$ one obtains the following table of reduced forms, each of which represents one class of properly equivalent forms.

$-D = 1\,|\,(1,0,1)$	$-D = 7\,|\,(1,0,7),(2,1,4)$
$2\,|\,(1,0,2)$	$8\,|\,(1,0,8),(2,0,4),(3,1,3)$
$3\,|\,(1,0,3),(2,1,2)$	$9\,|\,(1,0,9),(2,1,5),(3,0,3)$
$4\,|\,(1,0,4),(2,0,2)$	$10\,|\,(1,0,10),(2,0,5)$
$5\,|\,(1,0,5),(2,1,3)$	$11\,|\,(1,0,11),(2,1,6),(3,1,4),(3,-1,4)$
$6\,|\,(1,0,6),(2,0,3)$	$12\,|\,(1,0,12),(2,0,6),(3,0,4),(4,2,4)$

In order to gain an overview of the different primitive representations of a number m by the quadratic form (a,b,c) one can now proceed as follows.

If m is to be representable by (a,b,c), then by Theorem 3, D must be a quadratic residue mod m. Suppose this holds. Then one determines all n with $-m/2 < n \le m/2$ and $n^2 \equiv D(\text{mod } m)$ and establishes whether (a,b,c) and $(m,n,(n^2-D)/m)$ are properly equivalent, which is possible by the reduction to reduced forms described above and Theorem 8. This is the case precisely when there is a representation of m by (a,b,c) which belongs to root n (Theorem 4). In order to find the different representations of m by (a,b,c) which belong to root n, one has to solve equation (8) under the side condition (7). The equations (6) then allow us to go from one solution x_0,y_0 of the equation $ax^2 + 2bxy + cy^2 = m$, belonging to root n, to all others, by means of the relation $(x_1,y_1)^T = B(x_0,y_0)^T$.

Solution of equation (8) poses no difficulty for the negative discriminant. Instead of (8) we consider

$$4(t^2-Dw^2) = 4s^2. \tag{17}$$

Since $4D$ is divisible by s^2, the same is true of $4t^2$. Thus $2t$ is divisible by s. Let $v = 2t/s$. Then (17) becomes

$$v^2 - (4D/s^2)w^2 = 4. \tag{18}$$

We distinguish three cases.

1. $|4D/s^2| > 4$. Then (18) has two solutions $v = \pm 2$, $w = 0$.

2. $|4D/s^2| = 4$. Then (18) has four solutions $v = \pm 2$, $w = 0$ and $v = 0$, $w = \pm 1$.

3. $|4D/s^2| < 4$. We have $-4D/s^2 \equiv -(2b/s)^2 \pmod 4$. Therefore $4D/s^2 = -3$. There are six solutions, $v = \pm 2$, $w = 0$ and $v = \pm 1$, $w = \pm 1$.

2.7 Three theorems of Euler

As an application we prove the following theorem, conjectured by Fermat and first proved by Euler.

Theorem 10. *A prime number* $p \neq 2$ *is the sum of two squares if and only if* $p \equiv 1$ (mod 4), *and the decomposition is unique.*

Proof. By Theorem 7 of Chap. 1 and Theorem 3, $p \equiv 1 \pmod 4$ when p is representable by the form $(1,0,1)$, as one also easily sees directly. The forms $(p,n,(n^2+1)/p)$ and $(p,-n,(n^2+1)/p)$ are properly equivalent to $(1,0,1)$, since there is only one reduced positive form with discriminant -1. Thus p may be represented by $(1,0,1)$.

A solution x_0,y_0 immediately yields eight solutions $\pm x_0, \pm y_0$ and $\pm y_0, \pm x_0$. On the other hand, equation (8) has four solutions. Since we have two roots n and $-n$, there are altogether eight solutions of the equation $x^2 + y^2 = p$, which we have already found above, and which all yield the same representation of p as a sum of two squares. □

Euler also proved the following two theorems, which can be proved in the same way as Theorem 10 when it is recalled that by Theorem 9 of Chap. 1 the number -2 is a quadratic residue modulo a prime $p \neq 2$ if and only if $p \equiv 1 \pmod 8$ or $p \equiv 3 \pmod 8$, and by Theorem 10 of Chap. 1 the number -3 is a quadratic residue modulo a prime $p \neq 3$ if and only if $p \equiv 1 \pmod 3$.

Theorem 11. *A prime number* $p \neq 2$ *is the sum of a square and twice a square if and only if* $p \equiv 1 \pmod 8$ *or* $p \equiv 3 \pmod 8$, *and the decomposition is unique.* □

Theorem 12. *A prime number* $p \neq 3$ *is the sum of a square and three times a square if and only if* $p \equiv 1 \pmod 3$, *and the decomposition is unique.* □

In the *Disquisitiones arithmeticae* Gauss speaks of the "*outstanding elegance*" of these three theorems, and says that "*they have acquired a certain classical aspect, since Euler investigated them in detail*".

2.8 Forms with positive discriminant

The theory of forms with positive discriminant is more difficult. Above all, equation (8) always has infinitely many solutions. This was first proved rigorously by Lagrange. We do not go into these questions here, returning to it instead in Chap. 19 in a more general context. As in the case of negative discriminant, there are also only finitely many classes of forms with a fixed positive discriminant. We shall prove this in Chap. 21. There we shall also solve the problem of *composition of classes of forms.* The following section merely explains what this problem, raised by Gauss, is about.

2.9 Composition of classes of forms

Composition of classes of forms arises from the following problem: given forms (A_1) and (A_2), find a third (A_3) such that, when (A_1) represents the number m_1 and (A_2) represents the number m_2, then (A_3) represents the number m_1m_2. Gauss solved this problem, but his presentation is very complicated. A definitive simplification was first brought about as a consequence of the absorption of the theory of quadratic forms into the theory of quadratic number fields by Dedekind (Chap. 21).

A simplified presentation of the Gaussian composition of forms was given by Dirichlet. We follow the *X. Supplement* to his *Vorlesungen über Zahlentheorie* in the 1894 edition.

We confine ourselves to classes of forms with fixed discriminant and $s = 1$. If two classes K_1 and K_2 are given, then one can always find in them forms

$$(a,b,a'c),\ (a',b,ac) \tag{19}$$

where $a, a', 2b$ have no common divisor. If $X = xx' - cyy'$ and $Y = (ax+by)y' + (a'x'+by')y$ then

$$(ax^2+2bxy+a'cy^2)(a'x'^2+2bx'y'+acy'^2) = aa'X^2 + 2bXY + CY^2. \qquad (20)$$

The class K_3 of (aa',b,c) does not depend on the choice of forms (19), and it is defined to be the product of K_1 and K_2.

In this way one obtains an abelian group, whose identity element is the class of $(1,0,-D)$. $(1,0,-D)$ is properly equivalent to $(1,b,b^2-D)$ via $B = \begin{pmatrix} 1 & b \\ 0 & 1 \end{pmatrix}$. Therefore the forms (a,b,c) and $(1,b,b^2-D)$ are suitable representatives of their classes for the multiplication (20). The inverse of the class of (a,b,c) is the class of (c,b,a).

Exercises

2.1 Show that a prime number p different from 2 and 5 is representable by the form $x^2 + 5y^2$ resp. $2x^2 + 2xy + 3y^2$ if and only if $p \equiv 1$ or $9 \pmod{20}$ resp. $p \equiv 3$ or $7 \pmod{20}$.

2.2 Show that a prime number p different from 2 and 3 is representable by the form $x^2 + 6y^2$ resp. $2x^2 + 3y^2$ if and only if $p \equiv 1$ or $7 \pmod{24}$ resp. $p \equiv 5$ or $11 \pmod{24}$.

2.3 Show that a prime number p different from 2 and 7 is representable by the form $x^2 + 7y^2$ if and only if $p \equiv 1, 2$ or $4 \pmod 7$.

2.4 Which prime numbers are representable by the form $x^2 + xy + 2y^2$?

2.5 Show that three relatively prime natural numbers x,y,z satisfy the equation $x^2 + y^2 = z^2$ if and only if x,y are of the form $2uv, u^2 - v^2$ and z is of the form $u^2 + v^2$, where u and v are relatively prime natural numbers, one of which is even.

2.6 Use exercise 2.5 to show that the equation $x^4 + y^4 = z^2$ has no solution in natural numbers.

3. Division of the circle (Cyclotomy)

3.1 Formulation of the problem

Section 7 of the *Disquisitiones arithmeticae* deals with division of the circle. We shall give a detailed presentation of the main result, which played an important rôle in Gauss's personal development (Sect. 1.1) as well as in the development of algebra.

The ancient Greeks were interested in the problem of dividing the circle into n equal parts with ruler and compasses. They succeeded in doing this for n of the form $n = 2^a.3.5$, where a is an arbitrary nonnegative integer.[1] Gauss was able to show that it is also possible for n of the form $n = 2^a.p_1...p_s$, where the p_i are distinct primes of the form $p_i = 1 + 2^{h_i}$. (One easily sees that $1 + 2^h$ can only be a prime when h is a power of 2.) The smallest new prime of this kind is 17.

If we consider the unit circle in the plane of complex numbers, then for n-fold division we have to represent the numbers

$$\cos\frac{2\pi k}{n} + i\sin\frac{2\pi k}{n} = \exp\left[\frac{2\pi ki}{n}\right] \quad \text{for } k = 1,...,n-1. \tag{1}$$

Construction by ruler and compasses means that these numbers can be obtained from integers by the four arithmetic operations and extraction of square roots of real expressions. Since the numbers of the form (1) are the zeros of the polynomial $x^n - 1$, we are led to consider this polynomial. Moreover, one sees easily that when division of the circle is possible for relatively prime n_1, n_2, then it is also possible for n_1n_2. Division of the circle into 2^a parts comes from the well-known angle bisection process. In what follows we therefore confine ourselves to n a prime number $p \neq 2$.

---

[1] For more details see the article by R. Böker, "*Winkel und Kreisteilung*" in: *Pauly-Wissowa Realenzyklopädie der Klassischen Altertumswissenschaften, 2. Reihe, Halbband* 17 (1961), 127-150.

We first convince ourselves that the square root of a complex number can be managed via square roots of real numbers. Suppose

$$\sqrt{a+bi} = c + di.$$

Then $a = c^2 - d^2$, $b = 2cd$ and hence c^2 and $-d^2$ satisfy the quadratic equation $x^2 - ax - b^2/4 = 0$. The assertion follows from this.

Thus a definitive formulation of the problem is: for a prime number p, find a sequence of complex numbers $\alpha_1,\ldots,\alpha_t$ such that $\alpha_t = \exp\frac{2\pi i}{p}$ and, for $k = 1,\ldots,t$, α_k is the solution of a quadratic equation whose coefficients are polynomials in $\alpha_1,\ldots,\alpha_{k-1}$ with rational coefficients. In the language of field extensions (Sect. A 1.5) one formulates this condition more simply as $[\mathbb{Q}(\alpha_1,\ldots,\alpha_k) : \mathbb{Q}[\alpha_1,\ldots,\alpha_{k-1})] | 2$ for $k = 1,\ldots,t$.

3.2 Auxiliary theorems about polynomials

We need a few theorems about polynomials with integer coefficients, which are also of interest in another connection. A polynomial $f(x)$ of degree n is called *monic* when the coefficient of x^n is 1. The *content* $I(f)$ of a polynomial f is the greatest common divisor of its coefficients.

Theorem 1. *(Gauss's lemma). Let* g *and* h *be polynomials with integer coefficients. Then*

$$I(gh) = I(g)I(h).$$

Proof. Without loss of generality we can assume that g and h both have content 1. Let

$$g(x) = b_0 + \ldots + b_r x^r, \quad h(x) = c_0 + \ldots + c_s x^s.$$

Suppose that there is a prime number p which divides $I(gh)$. Let i resp. j be the smallest subscript such that b_i resp. c_j is not divisible by p. Then

$$b_ic_j + b_{i+1}c_{j-1} + ... + b_{i-1}c_{j+1} + ...,$$

the coefficient of x^{i+j} in $g(x)h(x)$, is not divisible by p, contrary to the assumption that p divides the content of gh. □

Theorem 2. *Let* g *and* h *be monic polynomials with rational coefficients. If* gh *has integer coefficients, then so too have* g *and* h.

Proof. Multiply g and h by a number a so that ag and ah have integer coefficients. If gh has integer coefficients, then $I(a^2gh) = a^2$. On the other hand, $I(A^2gh) = I(ag)I(ah)$ by Theorem 1. Since $I(ag)$ and $I(ah)$ are divisors of a, it follows that $I(ag) = I(ah) = a$. □

Theorem 3. *(Eisenstein's irreducibility criterion). Let* $f(x) = x^n + a_1x^{n-1} + ... + a_n$ *be a polynomial with integer coefficients which are all divisible by a prime number* p. *Suppose* p *occurs in* a_n *to the first power. Then* $f(x)$ *is irreducible over* $\mathbb{Q}$, *i.e. there are no nonconstant polynomials* $g(x)$, $h(x)$ *with rational coefficients such that* $f(x) = g(x)h(x)$.

Proof. Suppose there were such a decomposition. Then, by Theorem 2, g and h would have integer coefficients. We can therefore pass to the corresponding polynomials $\bar{f}$, $\bar{g}$, $\bar{h}$ with coefficients in the field $\mathbb{Z}/p\mathbb{Z}$. By hypothesis, $\bar{f}(x) = x^h$. Since the theorem on unique decomposition of a polynomial into irreducibles holds in the polynomial ring $\mathbb{Z}/p\mathbb{Z}[x]$ (Sect. A 1.2), $\bar{g}(x) = x^r$, $\bar{h}(x) = x^s$. It follows from this that a_n is divisible by p^2, contrary to hypothesis. □

Theorem 4. *Let* p *be a prime number. Then the polynomial* $f(x) := x^{p-1} + x^{p-2} + ... + x + 1$ *is irreducible over* $\mathbb{Q}$.

Proof. When $y = x - 1$ the binomial theorem gives

$$f(x) = (x^p-1)/(x-1) = \sum_{i=1}^{p} \binom{p}{i} y^{i-1}$$

The polynomial on the right-hand side satisfies the hypotheses of Theorem 3. It is therefore irreducible, and the same holds for $f(x)$. □

Gauss gave a different proof for Theorem 4.

3.3 Definition of the Gaussian periods and related theorems

Let ζ be a pth root of unity, $\zeta \neq 1$. Then $\zeta, \zeta^2, \ldots, \zeta^p = 1$ are all the zeros of $x^p - 1$. Hence by the Viète root theorem

$$\sum_{i=0}^{p-1} \zeta^{il} = \begin{cases} p & \text{for } p \mid l \\ 0 & \text{otherwise} \end{cases}$$

Let g be a primitive root mod p, i.e. $g, g^2, \ldots, g^{p-1} \equiv 1 \pmod p$ run through the congruence classes mod p (Sect. 1.3). It follows that the numbers $\zeta, \zeta^g, \ldots, \zeta^{g^{p-2}}$ are all the pth roots of unity apart from 1. More generally, the same holds for $\zeta^l, \zeta^{lg^{p-2}}$ when $l \not\equiv 0 \pmod p$.

Let e,f be natural numbers with $ef = p - 1$, and let $h = g^e$. By the *Gaussian period* (f,l) one means the number

$$(f,l) = \zeta^l + \zeta^{lh} + \ldots + \zeta^{lh^{f-1}}.$$

Example. $p = 17,\ g = 3,\ f = 8,\ l = 1,3$

$$(8,1) = \zeta + \zeta^{-1} + \zeta^2 + \zeta^{-2} + \zeta^4 + \zeta^{-4} + \zeta^8 + \zeta^{-8}$$

$$(8,3) = \zeta^3 + \zeta^{-3} + \zeta^5 + \zeta^{-5} + \zeta^6 + \zeta^{-6} + \zeta^7 + \zeta^{-7}.$$

Obviously $(f,l) = (f,lh)$, and (f,g^i) runs through all periods (f,l) apart from $(f,0) = f$ when $i = 0,1,\ldots,e-1$.

Theorem 5. $(f,l)(f,m) = \sum_{i=0}^{f-1} (f, lh^i + m)$.

Proof. $$\left(\sum_{i=0}^{f-1} \zeta^{lh^i}\right)\left(\sum_{j=0}^{f-1} \zeta^{mh^j}\right) = \sum_{i,j=0}^{f-1} \zeta^{lh^i+mh^j} = \sum_{i,j=0}^{f-1} \zeta^{lh^{i+j}+mh^j}$$

$$= \sum_{i=0}^{f-1} (f,lh^i+m). \qquad \square$$

Theorem 6. *Suppose* $l \not\equiv 0 \pmod p$. *Then* (f,m) *is equal to a polynomial in* (f,l) *with rational coefficients of degree at most* $e-1$.

Proof. By Theorem 5

$$(f,l)^s = \sum_{i=0}^{e-1} a_{is}(f,lg^i) \quad \text{with integer } a_{is},\ s = 2,\dots,e-1.$$

Together with

$$1 + (f,l) = -\sum_{i=1}^{e-1} (f,lg^i),$$

this gives a system of $e-1$ linear equations in the $e-1$ unknowns (f,lg^i), $i = 1,\dots,e-1$.

Suppose that the determinant of the system is equal to 0. Then there is a linear dependence between the powers $(f,l)^s$, $s = 0,\dots,e-1$:

$$\sum_{s=0}^{e-1} b_s(f,l)^s = 0.$$

If we carry through these considerations with ζ^a in place of ζ, for $a = 2,\dots,p-1$, then we obtain the same equation. Thus the latter is also satisfied by (f,la). It follows that two of the e periods of length f are equal. Therefore there is a linear dependence between $\zeta, \zeta^2,\dots,\zeta^{p-1}$. However, this contradicts the irreducibility of the cyclotomic polynomial $(x^p-1)/(x-1) = x^{p-1} + x^{p-2} + \dots + 1$ (Theorem 4). $\square$

Theorem 7. *Let* $\phi(x_1,\dots,x_f)$ *be a symmetric polynomial. Then* $\phi(\zeta^l,\zeta^{lh},\dots,\zeta^{lh^{f-1}})$ *is a linear combination of the* (f,g^i), $i = 0,\dots,e-1$, *with rational coefficients.*

Proof. By the theory of symmetric polynomials (Sect. 7.3) and by Theorem 5 it suffices to prove Theorem 7 for power sums: $\phi(x_1,\ldots,x_f) = \sum_{i=1}^{f} x_i^j$, $\phi(\zeta^l,\ldots,\zeta^{lh^{f-1}}) = \sum_{i=0}^{f-1} \zeta^{ljh^i} = (f,lj)$. □

Theorem 8. *Let* $p-1 = abc$ *and let* $\phi(x_1,\ldots,x_b)$ *be a symmetric function. Then* $\phi((c,l),(c,lg^a),\ldots,(c,lg^{a(b-1)}))$ *is a linear combination of the periods* (bc,g^i), $i = 0,\ldots,a-1$, *with rational coefficients.*

Proof. Again suppose $\phi(x_1,\ldots,x_b) = \sum_{i=1}^{b} x_i^j$. By Theorem 5

$$(c,l)^j = \sum_{h=0}^{ab-1} a_k (c,g^k) \text{ with integers } a_k, \tag{2}$$

hence

$$\sum_{i=0}^{b-1} (c,lg^{ai})^j = \sum_{i=0}^{b-1} \sum_{k=0}^{ab-1} a_k (c,g^{k+ai})$$

$$= \sum_{k=0}^{ab-1} a_k \sum_{i=0}^{b-1} (c,g^{k+ai}) = \sum_{k=0}^{ab-1} a_k (bc,g^k).$$

Here we have used the fact that (2) continues to hold when ζ is replaced by $\zeta^{g^{ai}}$. □

3.4 Solution of the problem

Let $p_1 \ldots p_s$ be the prime factorisation of $p-1 = p_1 f_1$. The symmetric functions of the periods (f_1,g^i), $i = 0,\ldots,p_1-1$, are rational numbers. In order to see this, consider Theorem 8 in the case $a = 1$, $b = p_1$, $c = f_1$. The periods $(p_1 f_1,m)$ are rational by (2), so the assertion follows.

The periods (f_1,g^i), $i = 0,\ldots,p_1-1$, are therefore solutions of an equation of degree p_1 with rational coefficients. In passing from ζ to ζ^g the periods are cyclically permuted. If we set $(f_1,1)$ equal to an arbitrary one of these solutions, then ζ is determined up to exponentiation by a power of g^{p_1}. By Theorem 6, the remaining periods may be expressed in terms of $(f_1,1)$.

Now let $f_1 = p_2 f_2$. By Theorem 8 again, the symmetric functions of the periods $(f_2, g^{p_1 i})$, $i = 0,\dots,p_2-1$, may be expressed as linear combinations of periods (f_1, g^i), $i = 0,\dots,p_1-1$. One now has to set $a = p_1$, $b = p_2$, $c = f_2$. In particular, the periods $(f_2, g^{p_1 i})$, $i = 0,\dots,p_2-1$, are solutions of an equation of degree p_2 with coefficients in $\mathbb{Q}((f_1,1))$.

One continues in this way and finally arrives at the periods $(1,l) = \zeta^l$, thereby reducing the solution of the cyclotomic equation to successive solutions of equations of prime degree, corresponding to all prime divisors of $p-1$.

If in particular $p = 1 + 2^h$, then one has h quadratic equations to solve. Thus the original problem of constructing a regular p-gon by ruler and compasses is solved for $p = 1 + 2^h$.

3.5 $p = 17$

We now work completely through the example $p = 17$, $g = 3$. In Sect. 3.3 we have exhibited the periods of length 8. By (2) of Theorem 5 we have

$$(8,1) + (8,3) = -1, \quad (8,1)(8,3) = \sum_{i=0}^{7} (8,(-8)^i+3).$$

Since $(-8)^i + 3 \not\equiv 0 \pmod{17}$ for all i, the summands of $\sum_{i=0}^{7} (8,(-8)^i+3)$ are equal to $(8,1)$ or $(8,3)$. Since $(8,1)(8,3)$ is rational, $(8,1)$ and $(8,3)$ must appear equally often, so $(8,1)(8,3) = -4$. $(8,1)$ is therefore a zero of the polynomial

$$x^2 + x - 4 \qquad (3)$$

We have four periods of length 4:

$$(4,1) = (4,4),\ (4,2) = (4,8),\ (4,3) = (4,5),\ (4,6) = (4,7).$$

By Theorem 6, these are polynomials in $(4,1)$ with rational coefficients. However, we prefer to express $(4,2)$, $(4,3)$ and $(4,6)$ in terms of $(4,1)$ and $(8,1)$. We have

$$(4,1) + (4,2) = (8,1),$$
$$(4,3) + (4,6) = (8,3) = -1 - (8,1),$$
$$(4,1)^2 = (4,2) + 2(4,3) + 4.$$

From this one obtains the expressions desired. We also have

$$(4,1)(4,2) = \sum_{i=0}^{3} (4,(-4)^i+2) = -1.$$

Thus (4,1) is a zero of the polynomial

$$x^2 - (8,1)x - 1. \tag{4}$$

Of the eight periods of length 2 we need just (2,1) and (2,4). Since

$$(2,1) + (2,4) = (4,1),$$
$$(2,1)(2,4) = (2,3) + (2,5) = (4,3) = \frac{1}{2}((4,1)^2 + (4,1) - (8,1) - 4),$$

(2,1) is a zero of

$$x^2 - (4,1)x + \frac{1}{2}((4,1)^2 + (4,1) - (8,1) - 4). \tag{5}$$

Finally, $(2,1) = \zeta + \zeta^{-1}$, and hence ζ is a zero of

$$x^2 - (2,1)x + 1. \tag{6}$$

Since one has a choice between the two roots of (3), (4), (5), (6) for (8,1), (4,1), (2,1), ζ, this determines one of the 16 primitive 17th roots of unity.

3.6 Periods of length (p-1)/2

We now consider the special case of periods of length $f = (p-1)/2$.
We have

$$(f,1) + (f,g) = -1$$

and

$$(f,1)(f,g) = \sum_{i=0}^{f-1} (f,g^{2i}+g) = a(f,1) + b(f,g) + c(f,0). \tag{7}$$

Obviously c is equal to 1 or 0 according as the congruence $g^{2i} + g \equiv 0 \pmod p$ has a solution i or not. Since the left side of (7) is rational, $a = b$. On the other hand, $a + b + c = f$ and therefore $c \equiv f \pmod 2$. Thus $c = 0$ for $f \equiv 0 \pmod 2$ and $c = 1$ for $f \equiv 1 \pmod 2$. Putting all this together we get

$$(f,1)(f,g) = \begin{cases} -f/2 \text{ for } f \equiv 0 \pmod 2, \\ (f+1)/2 \text{ for } f \equiv 1 \pmod 2. \end{cases}$$

It follows that

$$(f,1) - (f,g) = \sum_{i=1}^{p-1} \left(\frac{i}{p}\right)\zeta^i = \pm\sqrt{(-1)^f p}. \tag{8}$$

The sign depends on the choice of the root ζ of unity. If one sets $\zeta := \exp(2\pi i/p)$ then the sign is determined. Actually finding it proves to be extremely difficult. Gauss was very interested in this question, which is of significance for various number-theoretic problems, and he was able to answer it after a long struggle. In his work on this question in the year 1811 he wrote:

"That is why it is the goal of this work to give a rigorous proof of this highly elegant theorem, which we have sought in vain for many years by different methods, finally succeeding through curious and seemingly subtle considerations, without damage to its elegance but rather by raising it to a far higher level of generality".

Gauss gives the exact value of the sum $\tau_n := \sum_{u=0}^{n-1} \exp(2\pi i u^2/n)$ for an arbitrary natural number n. For odd n he finds

$$\tau_n = \begin{cases} \sqrt{n} \text{ for } n \equiv 1 \pmod 4, \\ i\sqrt{n} \text{ for } n \equiv 3 \pmod 4. \end{cases} \tag{9}$$

Obviously when n is a prime p and $\zeta := \exp(2\pi i/p)$

$$\tau_p = 1 + 2(f,1) = (f,1) - (f,g).$$

3.7 Proof of the law of quadratic reciprocity

We now want to prove the law of quadratic reciprocity (Chap. 1, Theorem 10), proceeding from (8). Let q be an odd prime $\neq p$. We have

$$\tau_p^2 = (-1)^{(p-1)/2}p$$

and hence

$$\tau_p^q = (-1)^{(p-1)(q-1)/4}p^{(q-1)/2}\tau_p. \tag{10}$$

By Chap. 1 (5),

$$p^{(q-1)/2} \equiv \left(\frac{p}{q}\right) \pmod q. \tag{11}$$

Also

$$\tau_p^q = \left[\sum_{i=1}^{p-1}\left(\frac{i}{p}\right)\zeta^i\right]^q = \sum_{i=1}^{p-1}\left(\frac{i}{p}\right)\zeta^{iq} + q\alpha = \left(\frac{q}{p}\right)\tau_p + q\alpha \tag{12}$$

for some α in $\mathbb{Z}[\zeta]$.

It follows from (10) to (12) that

$$c\tau_p = q\beta \tag{13}$$

for some β in $\mathbb{Z}[\zeta]$ and $c = (-1)^{(p-1)(q-1)/4}\left(\frac{p}{q}\right) - \left(\frac{q}{p}\right)$.

The law of quadratic reciprocity is equivalent to $c = 0$, and the latter follows immediately from (13) when one recalls that $|c| \leq 2$ and that elements γ of $\mathbb{Z}[\zeta]$ can be written uniquely in the form

$$\gamma = \sum_{i=1}^{p-1} a_i\zeta^i \quad \text{with } a_i \in \mathbb{Z}$$

because of Theorem 4 and the fact that $1 + \zeta + \ldots + \zeta^{p-1} = 0$.

The second supplement (Chap. 1, Theorem 8) is proved analogously by proceeding with a primitive eighth root of unity ζ from the relation

$$\tau_2'^2 = 2 \quad \text{with} \quad \tau_2' := \zeta + \zeta^{-1}.$$

This implies

$$\left(\frac{2}{q}\right)\tau_2' = 2^{(q-1)/2}\tau_2' + q\alpha = \tau_2'^q + q\alpha = \zeta^q + \zeta^{-q} + q\beta$$

for some $\alpha, \beta \in \mathbb{Z}[\zeta]$. Also

$$\zeta^q + \zeta^{-q} = (-1)^{(q^2-1)/8}\tau_2'.$$

Now the assertion follows as above.

Exercises

3.1 Let n be a natural number. A *primitive* n^{th} *root of unity* is a complex number ζ with $\zeta^n = 1$ and $\zeta^i \neq 1$ for $i = 1,2,\ldots,n-1$. Show that one obtains all primitive n^{th} roots of unity in the form ζ^i, where ζ is a fixed primitive n^{th} root of unity and i runs through the set M_n of natural numbers $< n$ and prime to n.

3.2 $\Phi_n(x) = \prod_{i \in M_n} (x-\zeta^i)$ is called the nth *cyclotomic polynomial*. Show that $\Phi_n(x)$ has integer coefficients.

3.3 Suppose $f(x) \in \mathbb{Z}[x]$ and p is prime. Show that $f(x)^p \equiv f(x^p) \pmod p$ where the congruence is understood coefficient-wise.

3.4 Let ξ_0 be an arbitrary n^{th} root of unity. Show that $n\xi_0^{n-1} = \prod_{\xi}(\xi_0-\xi)$, where the product is taken over all n^{th} roots of unity different from ξ_0.

3.5 Let ζ be a primitive n^{th} root of unity, and let $f_\zeta(x)$ be the associated polynomial, irreducible over $\mathbb{Q}$. Show that $f_\zeta(x)$ has integer coefficients.

3.6 Let p be a prime number with $f_\zeta(\zeta^p) \neq 0$. Show that the divisibility relation $p \,|\, f_\zeta(\zeta^p) \,|\, n$ holds in the ring $\mathbb{Z}[\zeta]$. Deduce that $f_\zeta(\zeta^q) = 0$ for all numbers q prime to n and show that $\Phi_n(x)$ is irreducible over $\mathbb{Q}$.

3.7 Compute $\Phi_n(x)$ in the case where n is a prime power.

3.8 Let ζ be a primitive 7th resp. 9th root of unity. Compute the irreducible polynomial $f(x) \in \mathbb{Z}[x]$ associated with $\zeta + \zeta^{-1}$, and investigate the prime numbers which appear as divisors of $f(a)$ for $a \in \mathbb{Z}$.

4. Theory of surfaces

4.1 Introduction. Gauss's "Disquisitiones generales circa superficies curvas"

In 1828 Gauss published his investigations on surfaces (*Disquisitiones generales circa superficies curvas. Soc. Reg. Sci. Gottingensis Rec.* 6 (1828)). This great work, which as late as 1900 was regarded by Darboux as the most self-contained and useful introduction to the study of differential geometry, was Gauss's most important published contribution to geometry.

Gauss's motivation for the geometry of curved surfaces came from astronomy and the Hannover land survey, with which he was occupied for several years.

Riemann developed Gauss's thoughts on geometry further. In this book we present them in the framework given by Riemann. The present chapter gives only an introduction.

4.2 Curves in the plane

We begin with a few remarks about curves in the plane, which constitute a starting point of Gauss's reflections.

All functions f appearing in this chapter will be assumed twice continuously differentiable. The derivative of f with respect to the variable x will frequently be denoted by f_x.

We represent a curve in the euclidean plane (A.4) with the help of coordinates x, y. The curve goes through the point $(0,0)$. We are interested only in the neighbourhood of the point $(0,0)$. In the latter, let the curve be given by the equation $y = f(x)$ with a twice continuously differentiable function $f(x)$. Without loss of generality we assume that the first derivative vanishes at the point $(0,0)$, i.e. that the tangent to the curve at the origin is the x-axis. We want to determine the circle which osculates the curve at the origin (Fig. 1).

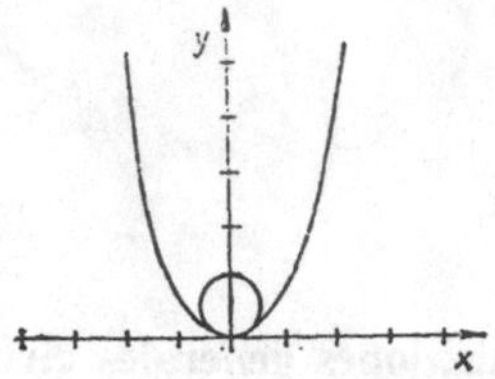

Fig. 1

First suppose that $f_{xx}(0,0) \neq 0$. Then the curve has a maximum or a minimum at $(0,0)$ according as $f_{xx}(0,0)$ is negative or positive. The osculating circle lies below the x-axis in the first case and above it in the second, and it touches the x-axis at $(0,0)$. If r is the radius of the osculating circle, provided with the sign of $f_{xx}(0,0)$, then the equation of the circle is

$$x^2 + (y-r)^2 = r^2.$$

For points (x,y) in the neighbourhood of $(0,0)$ one therefore has

$$y = r - \sqrt{r^2-x^2} = x^2/2r + \dots,$$

where the dots indicate terms with higher powers of x. Since the circle osculates the curve we must have, since $f(x) = f_{xx}(0,0)x^2/2 + \dots$, that

$$1/r = f_{xx}(0,0). \tag{1}$$

This number, even in the case $f_{xx}(0,0) = 0$, is called the *curvature of the curve at the point* $(0,0)$. The sign of the curve has an invariant meaning when one considers the plane to be oriented and the curve to have a direction. In our description of the curve as a function of x the positive direction corresponds to increasing x. Let $\mathbf{t}$ be the unit tangent vector at the origin in the positive x direction and let $\mathbf{n}$ be the normal vector in the direction of the centre of the osculating circle. The curvature is positive if and only if $\mathbf{t}, \mathbf{n}$ have the same direction as the coordinate axes, i.e. when $\mathbf{n}$ lies in the upper half plane in our set-up.

4.3 Curves in space

We now consider curves in three-dimensional euclidean space E^3 with coordinates x,y,z (A.4) (relative to the origin O and the orthonormal basis e_1,e_2,e_3 of the associated vector space V^3). A curve C is given parametrically by a mapping of the interval [a,b] of the real axis into E^3 which is twice continuously differentiable, i.e. when C is given by the coordinate functions x(t), y(t), z(t) for $a \le t \le b$ then these functions have to be twice continuously differentiable. We also assume that the derivatives $\dot{x}(t)$, $\dot{y}(t)$, $\dot{z}(t)$ do not vanish simultaneously. We think of the curve C as being oriented according to increasing t. Two curves C and C′ are considered to be the *same* when there is a function t′ of t with $dt'/dt > 0$ and $C(t') = C(t)$, i.e. when C and C′ differ only in their parametrisation and have the same orientation. By definition of the derivative the vector

$$\dot{C}(t) = \dot{x}(t)e_1 + \dot{y}(t)e_2 + \dot{z}(t)e_3 \tag{2}$$

hugs the curve C at the point C(t). It is called the *tangent vector at the point* C(t).

4.4 Surfaces

We now come to the real subject of Gauss's work, the investigation of surfaces in E^3. We are interested only in the behaviour of a surface $\mathcal{F}$ in the neighbourhood of a point $O \in E^3$, which we choose to be the origin of the coordinate system. Let $\mathcal{F}$ be given by the equation $z = f(x,y)$ in the coordinates x,y,z.

Let C be a curve on the surface which goes through the point $P = C(t_0)$. The tangent vector $\dot{C}(t_0)$ to the curve at the point P then has the coordinates

$$\dot{x}(t_0),\ \dot{y}(t_0),\ \dot{z}(t_0) = f_x(P)\dot{x}(t_0) + f_y(P)\dot{y}(t_0) \tag{3}$$

The coordinates of the tangent vectors to curves at the point P on the surface $\mathcal{F}$ therefore satisfy the same linear equation and therefore lie in a plane, the *tangent plane to the surface at the point* P. Let the coordinate system be chosen so that the tangent plane to $\mathcal{F}$ at O is the xy-plane, i.e. suppose $f_x(0,0) = f_y(0,0) = 0$.

We set

$$a := f_{xx}(0,0).\ b := f_{xy}(0,0),\ c := f_{yy}(0,0) \quad \text{and} \quad A = \begin{pmatrix} a & b \\ b & c \end{pmatrix}.$$

Then the Taylor expansion of $f(x,y)$ begins with the quadratic form

$$\frac{1}{2}(x,y)A(x,y)^T. \tag{4}$$

We define the *Gaussian curvature* R *of* $\mathcal{F}$ *at the origin* by

$$R = \det A. \tag{5}$$

This definition is independent of the choice of admissible coordinate system. Since we have fixed the origin and the xy-plane, the only transformations which come under consideration are rotations about the z-axis. Suppose one such is given by $(x,y)^T = B(x',y')^T$. In the coordinates x',y',z the surface is given by the equation

$$z = f(B(x',y')^T) = f'(x',y').$$

When $\frac{1}{2}(x',y')A'(x',y')^T$ is the quadratic form (4) corresponding to $f'(x',y')$, we get

$$(x',y')A'(x',y')^T = (x,y)A(x,y)^T = (x',y')B^TAB(x',y')^T,$$

hence $A' = B^TAB$ and $\det A' = (\det B)^2 \det A = \det A$ (Sect. 2.3). Obviously, (5) is also invariant under reflection in the xy-plane. The Gaussian curvature is therefore an invariant of metric geometry.

4.5 Geometric meaning of the Gaussian curvature

Gauss gave a beautiful geometric interpretation of R, which at the same time shows its invariance. The critical reader should regard the following as a heuristic discussion.

Of the two normal directions to the tangent plane at the origin we distinguish the one in the positive z direction. The corresponding unit normal vector **n**(P) to the surface at a point P in the neighbourhood of the origin has coordinates

$$x_1 = -\frac{f_x}{\sqrt{1+f_x^2+f_y^2}}, \; y_1 = -\frac{f_y}{\sqrt{1+f_x^2+f_y^2}}, \; z_1 = \frac{1}{\sqrt{1+f_x^2+f_y^2}}, \tag{6}$$

because this unit vector is perpendicular to the tangent vectors given by (3), and it becomes (0,0,1) as P → O.

We now consider the unit sphere, on which we let the point **n**(P) represent the point P. Then an element $d\mathcal{F}$ of our surface in E^3 becomes represented by a surface element $d\mathcal{G}$ on the unit sphere. The element $d\mathcal{G}$ is called the *spherical image* of $d\mathcal{F}$. We consider in particular the test triangle with the vertices O,dP = (dx,dy,dz), ∂P = (∂x,∂y,∂z) in a small neighbourhood of the origin. Since the tangent plane to the surface at O is the xy-plane, the oriented area $d\mathcal{F}$ of this test triangle is approximated by

$$\tfrac{1}{2}(dx\,\partial y - dy\,\partial x).$$

The area is positive or negative according as the vectors OdP, O∂P, e_3 have the same orientation as e_1,e_2,e_3 or not. Suppose that the points dP and ∂P correspond to the points $(dx_1,dy_1,1+dz_1)$ and $(\partial x_1,\partial y_1,1+\partial z_1)$ on the unit sphere. Then, by (6)

$$dx_1 = -f_{xx}(0,0)dx - f_{xy}(0,0)dy + \dots,$$

$$dy_1 = -f_{xy}(0,0)dx - f_{yy}(0,0)dy + \dots,$$

$$dz_1 = 0 + \dots,$$

where the dots indicate terms of higher order in dx, dy, and there are corresponding formulae for $\partial x_1,\partial y_1,\partial z_1$.

The oriented area $d\mathcal{G}$ of the spherical image of our test triangle is therefore approximated by

$$\frac{1}{2}\det\begin{pmatrix} dx_1 & \partial x_1 \\ dy_1 & \partial y_1 \end{pmatrix} = \frac{1}{2}\det\left(-A\begin{pmatrix} dx_1 & \partial x_1 \\ dy_1 & \partial y_1 \end{pmatrix}\right).$$

If we let the test triangle tend to 0, then we obtain

$$\frac{d\mathcal{F}}{d\mathcal{G}} = R. \tag{7}$$

4.6 Euler's contribution to the theory of surfaces

Euler had already studied the curvature of a surface at a point by considering the curvature of the curves obtained by cutting the surface by planes through the normal vector at the point (*Recherches sur la courbure des surfaces, Mém. Acad. Sci. Berlin* 16 (1760))

A normal section of our surface at the point O is given by its intersection with the xy-plane (the tangent plane to the surface at the point O). Let this be given parametrically by $x = t\cos\alpha$, $y = t\sin\alpha$ for $t \in \mathbb{R}$, so $z = f(t\cos\alpha, t\sin\alpha) =: g(t,\alpha)$. The associated curvature is found, using (1), to be

$$K(\alpha) = g_{tt}(0,0) = f_{xx}(0,0)\cos^2\alpha + f_{xy}(0,0)2\cos\alpha\sin\alpha + f_{yy}(0,0)\sin^2\alpha.$$

By rotating the coordinate system about the z-axis one can always arrange to have $f_{xy}(0,0) = 0$, which we shall assume done in what follows. Two cases are possible. Either we have $f_{xx}(0,0) = f_{yy}(0,0)$, in which case $K(\alpha) = f_{xx}(0,0)$ is independent of the angle. Or else $f_{xx}(0,0) \neq f_{yy}(0,0)$, in which case

$$K(\alpha) = f_{xx}(0,0) + (f_{yy}(0,0) - f_{xx}(0,0))\sin^2\alpha$$

has extreme values $f_{xx}(0,0)$ and $f_{yy}(0,0)$, for $\alpha = 0$ and $\alpha = \pi/2$ respectively. These values are called the *principal curvatures*. Thus we have

Theorem 1. (Euler). *The normal sections of a surface at a point* O *either all have the same curvature at* O, *or else the tangent plane at* O *contains two orthogonal lines, which meet at* O, *for which the corresponding normal sections realise the maximal and minimal curvatures.* □

These results and (5) immediately yield

Theorem 2. *The Gaussian curvature is equal to the product of the principal curvatures.* □

Rodrigues, in his work "*Recherches sur la theorie analytique de lignes et rayons de courbure des surfaces*" (*Bull. Soc. Philomatique, Paris*) of 1815, already had defined the "*Gaussian curvature*" corresponding to (7) and proved Theorem 2.

4.7 The intrinsic geometry of surfaces

For the later results of Gauss it was of great importance to pass to a more general description of the surface: let x, y and z be given as functions of two parameters p and q. Let $C = C(t)$ be a curve on the surface. The length l of the path along C between two points P and Q is given by the integral of the line element $ds := \sqrt{dx^2+dy^2+dz^2}$ along C:

$$l = \int_P^Q \sqrt{\dot{x}^2+\dot{y}^2+\dot{z}^2}\,dt.$$

We can express $dx^2 + dy^2 + dz^2$ in terms of p and q in the form

$$dx^2 + dy^2 + dz^2 = Edp^2 + 2Fdpdq + Gdq^2 \tag{8}$$

where E, F and G are functions of p and q.

If one varies the surface so that all path lengths remain invariant, then the form (8) must also remain invariant. Conversely, all path lengths remain invariant when the form (8) does, under a variation of the surface. We call such a variation of the surface a *bending*. The intrinsic geometry of the surface concerns those geometric quantities which remain invariant under bending.

As one of the most important results of his work, Gauss obtained the theorem that the curvature R defined above is invariant under bending. He called this theorem the "*Theorema egregium*". More precisely, he proved

Theorem 3. R *satisfies*

$$4(EG-F^2)R = E(E_q G_q - 2F_p G_q + G_p^2)$$
$$+ F(E_p G_q - E_q G_p - 2E_q F_q + 4E_p F_q - 2F_p G_p)$$
$$+ G(E_p G_p - 2E_p F_q + E_q^2)$$
$$- 2(EG-F^2)(E_{qq} - 2F_{pq} + G_{pp}). \qquad \blacksquare \tag{9}$$

Since (8) is positive definite, $EG-F^2 > 0$, i.e. R is determined by (9).

Riemann placed the formula (9) in a general context and gave it a structural significance. We shall come back to this in Chap. 14. The proof of Theorem 3 will be given in Sect. 14.10.

As a special case one obtains the theorem stated by Monge, but not completely proved by him, that a developable surface satisfies $f_{xx}f_{yy} = f_{xy}^2$ (in the notation of Sect. 4).

The further considerations of Gauss were concerned with geodesics, i.e. shortest paths between points on the surface, and he developed a triangle theory as a generalisation of ordinary and spherical trigonometry. One of his most interesting results is the generalisation of the theorem that in a plane triangle the sum of the interior angles is π : for a triangle on a surface, whose sides are geodesics, the sum of the interior angles equals π plus the total curvature of the triangle, where the total curvature is the surface integral of the Gaussian curvature over the triangle.

Geodesics will be considered in Chap. 14 in the framework of Riemannian geometry.

Exercises

4.1 Let $x(t)$ and $y(t)$ be continuously differentiable functions of t in the interval $[a,b]$ and let C be the oriented plane curve with the points $C(t) = (x(t),y(t))$, $a \le t \le b$, (cf. Sect. 4.3). The arc length of the curve C is defined by

$$b(c) := \int_a^b \sqrt{\dot{x}(t)^2+\dot{y}(t)^2}dt.$$

a) Show that $b(c)$ is independent of the choice of parameter t.

b) Let $a = t_0 < t_1 < \ldots < t_n = b$ and let Γ_n be the corresponding polygon inscribed in C, which results from connecting the points $C(t_0), C(t_1), \ldots, C(t_n)$ by straight lines. Let $b(R_n)$ be the length of this polygon. Show that $\lim_{n\to\infty} b(\Gamma_n) = b(C)$ when the parameters t_ν are chosen so that the distances $t_\nu - t_{\nu-1}$ for $\nu = 1,\ldots,n$ become arbitrarily small with increasing n.

c) C is called *regular* (for the parameter t) at the point $C(t)$ when $\dot{C}(t) \neq (0,0)$, and *singular* when $\dot{C}(t) = (0,0)$. Let C be regular in the interval $[a,b]$. Show that C admits parametrisation by arc length, i.e. the function $s(h) = \int_a^h \sqrt{\dot{x}(t)^2+\dot{y}(t)^2}dt$ has a single-valued inverse for $a \leq h \leq b$, and the inverse function is continuously differentiable.

4.2 Let the curve C be parametrised by arc length s, and let $C(s)$ be twice continuously differentiable.

a) Show that the tangent vector $\mathbf{t} = \dot{C}(s)$ has length 1.

b) Let $\mathbf{n}$ be the unit normal vector of C at the point $C(s)$, i.e. $\mathbf{n}$ is perpendicular to $\mathbf{t}$ and the orthonormal basis $\mathbf{t},\mathbf{n}$ is positively oriented. Show that the vector $\dot{\mathbf{t}} = \frac{d\mathbf{t}}{ds}$ is proportional to $\mathbf{n}$ and that the factor of proportionality is equal to the curvature r at the point $C(s)$.

c) Show $\dot{\mathbf{n}} = -r\mathbf{t}$. (The formulae $\dot{\mathbf{t}} = r\mathbf{n}$ and $\dot{\mathbf{n}} = -r\mathbf{t}$ are known as the Frenet formulae in the theory of plane curves.)

4.3 Let r, ϕ be polar coordinates in the xy-plane, and let a surface of rotation be given by $z = f(r)$, $0 \leq r \leq r_1$. Determine the Gaussian curvature R of the surface as a function of R. When is R constant?

5. Harmonic analysis

5.1 The equation of the vibrating string

In the middle of the 18th century the most important mathematicians were concerned with the solution of the equation of the vibrating string (in the xy-plane)

$$\frac{\partial^2 y}{\partial t^2} = c^2 \frac{\partial^2 y}{dx^2}, \tag{1}$$

fixed at the points (0,0) and $(0,l)$. Here t denotes time and c is a parameter depending on the unit of mass.

D'Alembert (*Mém. Acad. Sci. Berlin* 1747) solved this equation by passing to the new variables $v^+ = x + ct$, $v^- = x - ct$, which transforms (1) into

$$\frac{\partial^2 y}{\partial v^+ \partial v^-} = 0. \tag{2}$$

The general solution of the latter equation has the form

$$y = f(v^+) + g(v^-)$$

for two functions f, g which satisfy the boundary conditions $y = 0$ for $x = 0, l$ and for all t. This means

$$f(w) + g(-w) = 0, \; f(l+w) + g(l-w) = 0 \tag{3}$$

for all $w \in \mathbb{R}$. These conditions are satisfied if and only if f is a periodic, but otherwise arbitrary, function with period $2l$ and $g(w) = -f(-w)$. Thus the d'Alembert solution reads

$$y = f(ct+x) - f(ct-x). \tag{4}$$

The nature of the function f gave rise to a dispute between d'Alembert and Euler. The former was of the opinion that y had to be expressible "analytically" as a function of x and t, while the latter thought that f(w) could be prescribed "arbitrarily" between $-l$ and $+l$.

Even earlier, in 1713, Taylor (*Philos. Trans.*) had given the special solutions

$$y = \sin \frac{n\pi x}{l} \cos \frac{n\pi ct}{l} \quad \text{for } n = 1,2,\ldots$$

of (1), which have a certain connection with Pythagoras, who discovered the relation between consonance of tones and the ratios of string lengths which produce them on a monochord. Following Taylor, D. Bernoulli (*Mém. Acad. Sci. Berlin* 1753) gave the solution consisting of superimposed harmonics

$$y = \sum_{n=1}^{\infty} a_n \sin \frac{n\pi x}{l} \cos \frac{n\pi c}{l} (t-b_n),$$

where the a_n, b_n are real parameters, and held this solution to be the most general. Euler replied immediately (*Mém. Acad. Sci. Berlin* 1753) that this could only be the case if every periodic function f with period 2π were representable in the form

$$f(x) = b_0/2 + \sum_{n=1}^{\infty} (a_n \sin nx + b_n \cos nx). \tag{5}$$

Only 50 years later was this question essentially advanced, with the discovery of Fourier (report to the Paris Academy on 21.12.1807) that when a function is representable in the form (5) the coefficients a_n, b_n must be of the form

$$a_n = \frac{1}{\pi} \int_{-\pi}^{\pi} f(x) \sin nx \, dx, \quad b_n = \frac{1}{\pi} \int_{-\pi}^{\pi} f(x) \cos nx \, dx. \tag{6}$$

The a_n, b_n are called the *Fourier coefficients* of f(x). For an even function only the coefficients $b_0, b_1, b_2, \ldots$ appear. In the latter case (6) was already known to Euler (*Novi Comm. Acad. Sci. Imp. Petropolitanae* 11 (1793)). However, this posthumous publication went unnoticed.

In 1829 there appeared in *J. reine angew. Math.* 4 a work of Dirichlet, "*Sur la convergence des séries trigonométriques qui servent à représentir une fonction arbitraire entre des limites donnés*", in which it was proved that, under very mild conditions on the function f, the latter was representable in the form (5), where the coefficients a_n, b_n were given by (6). Riemann described this work in his *Habilitationsschrift* (Göttingen 1854) as "*the first fundamental work on this subject*". We give an exposition of Dirichlet's work below. For further details on the early history of harmonic analysis we refer to the first section of Riemann's *Habilitationsschrift.*

5.2 Formulation of Dirichlet's theorem and preliminaries to the proof

Theorem 1. (Dirichlet's theorem). *Let* f *be a function (i.e. a mapping of* $\mathbb{R}$ *into* $\mathbb{R}$*) with period* 2π, *which satisfies the following conditions:*

1. *The interval* $[-\pi,\pi]$ *may be divided into finitely many subintervals* $[a,b]$ *such that* f *is continuous in* (a,b) *and either monotonic increasing or monotonic decreasing there.*

2. *At the initial point* a *of this interval the limits*

$$f^-(a) = \lim_{\substack{x<a \\ x\to a}} f(x) \quad \text{and} \quad f^+(a) = \lim_{\substack{x>a \\ x\to a}} f(x)$$

are finite and

$$f(a) = \frac{1}{2}(f^-(a) + f^+(a)). \tag{7}$$

Then f *has a representation* (5), *where the coefficients* a,b *are given by* (6).

The starting point of the proof of Theorem 1 is the following: we consider the functions

$$f_n(x) = \frac{1}{2}b_0 + \sum_{j=1}^{n} (a_j \sin jx + b_j \cos jx),\ n = 1,2,\dots . \tag{8}$$

Substituting the integrals (6), which exist because of the hypotheses on f, one obtains

$$f_n(x) = \frac{1}{2\pi}\int_{-\pi}^{\pi} f(y)dy + \frac{1}{\pi}\sum_{j=1}^{n}\int_{-\pi}^{\pi} f(y)(\sin jy \sin jx + \cos jy \cos jx)dy$$

$$= \frac{1}{\pi}\int_{-\pi}^{\pi} f(y)\left(\frac{1}{2} + \sum_{j=1}^{n} \cos j(y-x)\right)dy. \tag{9}$$

We have to show $\lim_{n\to\infty} f_n(x) = f(x)$.

Lemma 1. *For real* z

$$\left(\frac{1}{2} + \sum_{j=1}^{n} \cos jz\right)2 \sin \frac{1}{2}z = \sin(n + \frac{1}{2})z. \tag{10}$$

Proof. It follows from the addition theorem for the sine function that

$$2 \cos jz \sin \frac{1}{2}z = \sin(j + \frac{1}{2})z - \sin(j - \frac{1}{2})z$$

and hence

$$\sin \frac{1}{2}z + 2 \sum_{j=1}^{n} \cos jz \sin \frac{1}{2}z$$

$$= \sin \frac{1}{2}z + \sum_{j=1}^{n} (\sin(j+\frac{1}{2})z - \sin(j-\frac{1}{2})z)) = \sin(n+\frac{1}{2})z. \qquad \square$$

We substitute (10) in (9) and obtain

$$f_n(x) = \frac{1}{\pi}\int_{-\pi}^{\pi} f(y)\frac{\sin(n+\frac{1}{2})(y-x)}{2\sin(y-x)/2}dy,$$

where the integrand has to be set equal to $f(y)(n+\frac{1}{2})$ at the point $y = x$. We change the variable of integration to $u = y - x$ and set $2n + 1 = k$. Then

$$f_n(x) = \frac{1}{\pi}\int_{-(\pi+x)}^{0} f(x+u)\frac{\sin ku/2}{2\sin u/2}du + \frac{1}{\pi}\int_{0}^{\pi-x} f(x+u)\frac{\sin ku/2}{2\sin u/2}du.$$

Finally we replace u by -2u in the first integral and u by 2u in the second:

$$f_n(x) = \frac{1}{\pi}\int_0^{\frac{\pi+x}{2}} f(x-2u)\frac{\sin ku}{\sin u}du + \frac{1}{\pi}\int_0^{\frac{\pi-x}{2}} f(x+2u)\frac{\sin ku}{\sin u}du$$

Lemma 2. Let $0 \le a < b < \pi$ and suppose $g(u)$ is a continuous monotonic function for $a \le u \le b$. Then

$$\lim_{n\to\infty}\int_a^b g(u)\frac{\sin ku}{\sin u}du = 0 \quad \text{for } a > 0, \tag{11}$$

$$\lim_{n\to\infty}\int_0^b g(u)\frac{\sin ku}{\sin u}du = \frac{\pi}{2}g(0). \tag{12}$$

Since we are able to divide the interval of integration into subintervals in which $f(x \pm 2u)$ is monotonic, Lemma 2 immediately implies $\lim_{n\to\infty} f_n(x) = f(x)$. Hence to prove Theorem 1 it suffices to prove Lemma 2.

5.3 Proof of Lemma 2

We first reduce Lemma 2 to a special case. By decomposition into subintervals we can arrange that $a \ge \pi/2$ or $b \le \pi/2$. In the first case we set $u = \pi-u'$ and obtain

$$\int_a^b g(u)\frac{\sin ku}{\sin u}du = \int_{\pi-b}^{\pi-a} g(\pi-u')\frac{\sin ku'}{\sin u'}du'.$$

The function $g(\pi-u')$ is monotonic along with $g(u)$, and continuous in the corresponding interval, where $0 \le \pi-b \le \pi-a \le \pi/2$. We can therefore assume $b \le \pi/2$ in Lemma 2.

It suffices to prove (12). Namely, if $g(u)$ is defined for $a \le u \le b$ and continuous and monotonic, then we define $g(u)$ for $0 \le u < a$ by $g(u) = g(a)$. The extended function is again continuous and monotonic, and

$$\int_a^b g(u)\frac{\sin ku}{\sin u}du = \int_0^b g(u)\frac{\sin ku}{\sin u}du - \int_0^a g(u)\frac{\sin ku}{\sin u}du = 0.$$

One also sees easily that it suffices to consider monotonic decreasing positive functions.

We note that

$$\int_0^{\pi/2} \frac{\sin ku}{\sin u} du = \int_0^{\pi/2} (1 + \sum_{j=1}^{n} 2 \cos 2ju) du = \frac{\pi}{2} \tag{13}$$

by Lemma 1. We now divide $\frac{\sin ku}{\sin u}$ into positive and negative parts; the function is ≥ 0 for u between $2j\pi/k$ and $(2j+1)\pi/k$, and ≤ 0 for u between $(2j+1)\pi/k$ and $(2j+2)\pi/k$, $j = 0,1,\ldots$. Let

$$r_i := \int_{(i-1)\pi/k}^{i\pi/k} \left|\frac{\sin ku}{\sin u}\right| du \quad \text{for } i = 1,\ldots,n,$$

$$r_{n+1} := \int_{n\pi/k}^{\pi/k} \left|\frac{\sin ku}{\sin u}\right| du.$$

When u is increased by $i\pi/k$ the numerator of the integrand remains unchanged, while the denominator increases. Therefore

$$r_1 > r_2 > \ldots > r_{n+1}.$$

In addition, (13) gives

$$r_1 - r_2 + \ldots + (-1)^n r_{n+1} = \frac{\pi}{2}$$

and hence

$$r_1 - r_2 + \ldots - r_{2m} < \frac{\pi}{2},\ r_1 - r_2 + \ldots + r_{2m-1} > \frac{\pi}{2}$$

for $2m < n$.

Also let

$$S_k := \int_0^b g(u) \frac{\sin ku}{\sin u} du, \quad R_i := \int_{(i-1)\pi/k}^{i\pi/k} g(u) \left|\frac{\sin ku}{\sin u}\right| du.$$

Then $R_1 > R_2 > \ldots$, $S_k \geq R_1 - R_2 + \ldots - R_{2m}$, $S_k \leq R_1 - R_2 + \ldots + R_{2m-1}$ for $2m < bk/\pi$ and

$$r_i g\left[\frac{(i-1)\pi}{k}\right] \geq R_i \geq r_i g\left[\frac{i\pi}{k}\right].$$

It follows that

$$\begin{aligned} S_k &\geq (r_1-r_2)g(\tfrac{\pi}{k}) + (r_3-r_4)g(\tfrac{3\pi}{k}) + \ldots + (r_{2m-1}-r_{2m})g\left[\frac{(2m-1)\pi}{k}\right] \\ &\geq (r_1-r_2+\ldots-r_{2m})g\left[\frac{2m\pi}{k}\right] \\ &\geq \frac{\pi}{2}g\left[\frac{2m\pi}{k}\right] - r_{2m}g\left[\frac{2m\pi}{k}\right], \end{aligned}$$

$$\begin{aligned} S_k &\leq r_1g(0) - (r_2-r_3)g(\tfrac{2\pi}{k}) - \ldots - (r_{2m-2}-r_{2m-1})g\left[\frac{(2m-2)\pi}{k}\right] \\ &\leq r_1g(0) - (r_2-r_3+\ldots-r_{2m-1})g\left[\frac{2m\pi}{k}\right] \\ &\leq r_1(g(0)-g\left[\frac{2m\pi}{k}\right]) + \frac{\pi}{2}g\left[\frac{2m\pi}{k}\right] + r_{2m}g\left[\frac{2m\pi}{k}\right]. \end{aligned}$$

Now we let m go to ∞ independently of n, so that m/l goes to 0 and at the same time $2m < bk/\pi$ (e.g. set $m = [b \log k/\pi]$).

For r_m we have the following estimate:

$$r_m = \int_{(m-1)\pi/k}^{m\pi/k} \left|\frac{\sin ku}{\sin u}\right| du \leq \frac{\pi}{k \sin(m-1)\pi/k}$$

$$= \frac{1}{(m-1)} \frac{(m-1)\pi/k}{\sin(m-1)\pi/k}.$$

It follows that $r_m \to 0$ as $m/k \to 0$ and $m \to \infty$.

It also follows easily from Lemma 1 that r_1 remains bounded as $n \to \infty$.

The estimates for S_k now imply that $\lim_{n\to\infty} S_k = \frac{\pi}{2}g(0)$. □

So much for Dirichlet's result. In the sections which follow we give some extensions and examples.

5.4 The Fourier formula

Up to now we have considered functions with period 2π. Instead of this we can also investigate functions with period $2l$. In the present section we wish to find out what one gets when l is allowed to tend to ∞.

Theorem 2. (Fourier formula) *Let* $f : \mathbb{R} \to \mathbb{R}$ *be a function which satisfies the following conditions:*

1. *Each finite interval may be divided into a finite number of subintervals* $[a,b]$ *such that* f *is continuous in* (a,b) *and monotonic increasing or decreasing.*

2. *At the initial point* a *of the latter interval the limit values* $f^-(a)$ *and* $f^+(a)$ *are finite, and*

$$f(a) = \frac{1}{2}(f^-(a) + f^+(a)).$$

3. *The improper integral*

$$\int_{-\infty}^{\infty} |f(t)|\,dt$$

exists.

Then the integral

$$\frac{1}{\pi}\int_0^{\infty}\int_{-\infty}^{\infty} f(t)\cos u(t-x)\,dt\,du \tag{14}$$

also exists for all $x \in \mathbb{R}$ *and equals* $f(x)$.

Proof. We first consider functions f with compact support, i.e. we assume that there is an $h \geq 0$ with $f(x) = 0$ for $|x| > k$. In this case it follows from Theorem 1 for $l > h$ and $|x| \leq l$ that

$$f(x) = \frac{1}{2}b_0 + \sum_{j=1}^{\infty} \left(a_j \sin\frac{j\pi x}{l} + b_j \cos\frac{j\pi x}{l}\right)$$

with

$$a_j = \frac{1}{l}\int_{-h}^{h} f(t)\sin\frac{j\pi t}{l}dt,\ b_j = \frac{1}{l}\int_{-h}^{h} f(t)\cos\frac{j\pi t}{l}dt.$$

It follows from this that

$$f(x) = \frac{1}{2l}\int_{-h}^{h} f(t)dt + \frac{1}{l}\sum_{j=1}^{\infty}\int_{-h}^{h} f(t)\cos\frac{j\pi(t-x)}{l}dt.$$

As $l \to \infty$ the right-hand side becomes

$$\frac{1}{\pi}\int_{0}^{\infty} f(t)\cos u(t-x)dt\ du,$$

whence Theorem 2 is proved for functions with compact support.

Now let f be an arbitrary function which satisfies conditions 1-3. Condition 3 allows us to change the order of integration in

$$I(v,f,x) := \frac{1}{\pi}\int_{0}^{v}\int_{-\infty}^{\infty} f(t)\cos u(t-x)dt\ du,$$

and thus obtain

$$I(v,f,x) = \frac{1}{\pi}\int_{-\infty}^{\infty} f(t)\frac{\sin v(t-x)}{t-x}dt.$$

We define

$$f_l(t) := \begin{cases} f(t) & \text{for } |t| \le l \\ 0 & \text{for } |t| > l. \end{cases}$$

Let $\varepsilon > 0$ be given arbitrarily. Then for fixed x condition 3 implies that for sufficiently large l and arbitrary v

$$|v(v,f-f_l,x)| < \frac{\varepsilon}{2}. \tag{15}$$

On the other hand, we have already seen that, for sufficiently large v

$$|I(v,f_l,x) - f(x)| < \frac{\varepsilon}{2}. \tag{16}$$

The assertion of Theorem 2 follows easily from (15) and (16). □

5.5 Harmonic analysis for complex-valued functions

We now consider functions f with complex values, whose real and imaginary parts satisfy the conditions of Theorem 1 resp. Theorem 2. For these one obtains the representation

$$f(x) = \sum_{j=-\infty}^{\infty} c_j \exp ijx \text{ with } c_j = \frac{1}{2\pi}\int_{-\pi}^{\pi} f(t)\exp(-ijt)dt \tag{17}$$

resp.

$$f(x) = \frac{1}{2\pi}\int_{-\infty}^{\infty}\int_{-\infty}^{\infty} f(t)\exp iu(t-x)dt\, du. \tag{18}$$

(Here $\sum_{j=-\infty}^{\infty}$ resp. $\int_{-\infty}^{\infty}$ are to be understood as $\lim_{n\to\infty}\sum_{j=-n}^{n}$ and $\lim_{n\to\infty}\int_{-n}^{n}$.)

We define the *Fourier transform* $\Phi(f)$ by

$$\Phi(f)(u) := \frac{1}{\sqrt{2\pi}}\int_{-\infty}^{\infty} f(t)\exp itu\, dt. \tag{19}$$

Then we can write (18) in the form

$$\Phi(\Phi(f))(x) = f(-x), \tag{20}$$

which is known as the *Fourier inversion formula.*

The formulae (17) and (20) form a starting point for the further development and generalisation of harmonic analysis. The exponential functions appearing in (17) and (19) may be characterised as all the continuous group homomorphisms of $\mathbb{R}/2\pi\mathbb{Z}$ resp. $\mathbb{R}$ into the multiplicative group of complex numbers of absolute value 1.

5.6 Application : computation of the ζ-function for even positive arguments

We consider the example of the functions x^h for $-\pi \le x \le \pi$, $h = 1,2,\ldots$, and thereby compute the values of the function

$$\zeta(s) := \sum_{n=1}^{\infty} \frac{1}{n^s}$$

for even positive integers s. This function was studied very deeply by Riemann in connection with the distribution of primes (Chap. 15), and it is therefore known as the *Riemann ζ-function.*

We have

$$x^h = \frac{1}{2}b_0(h) + \sum_{j=1}^{\infty} (a_j(h)\sin jx + b_j(h)\cos jx) \text{ for } -\pi < x < \pi.$$

Since x^h is an even or odd function according as h is even or odd, $a_j(h) = b_j(h+1) = 0$ for even h.

Also, for even h, integration by parts yields

$$b_j(h) = \int_{-\pi}^{\pi} x^h \cos jx\, dx = \left[\frac{1}{\pi j} x^h \sin jx\right]_{-\pi}^{\pi} - \frac{h}{\pi j}\int_{-\pi}^{\pi} x^{h-1}\sin jx\, dx,$$

i.e.

$$b_j(h) = -\frac{h}{j}\, a_j(h-1),$$

and

$$a_j(h-1) = \frac{1}{\pi}\int_{-\pi}^{\pi} x^{h-1}\sin jx\, dx$$

$$= \left[-\frac{1}{\pi j} x^{h-1}\cos jx\right]_{-\pi}^{\pi} + \frac{h-1}{\pi j}\int_{-\pi}^{\pi} x^{h-2}\cos jx\, dx,$$

i.e.

$$a_j(h-1) = (-1)^{j+1} 2\pi^{h-2}/j + \frac{(h-1)}{j} b_j(h-2).$$

Since $b_j(0) = 0$, one obtains the sawtooth function as

$$x = \sum_{j=1}^{\infty} (-1)^{j+1} \frac{2}{j} \sin jx \quad \text{for } -\pi < x < \pi.$$

For $x = \frac{\pi}{2}$ one finds the series

$$\frac{\pi}{4} = 1 - \frac{1}{3} + \frac{1}{5} - \ldots,$$

which was already known to Leibniz and inspired him to declare "*Numero deus impare gaudet*" (God is pleased by odd numbers).[1]

For $b_j(h)$ one obtains the recursion formula

$$b_j(h) = (-1)^j 2h\pi^{h-2}\frac{1}{j^2} - \frac{h(h-1)}{j^2}b_j(h-2). \tag{21}$$

Whence

$$x^2 = \frac{x^3}{3} + \sum_{j=1}^{\infty}(-1)^j\frac{4}{j^2}\cos jx \quad \text{for } -\pi \le x \le \pi.$$

For $x = \pi$ one finds

$$\pi^2 = \frac{\pi^3}{3} + \sum_{j=1}^{\infty}\frac{4}{j^2}$$

and hence $\zeta(2) = \pi^2/6$.[2]

In general we proceed as follows: it follows from (21) that

$$b_j(h) = \sum_{l=1}^{h/2}(-1)^{j+k+1}\cdot\frac{2\pi^{h-2k}h!}{j^{2k}(h-2k+1)!}$$

and therefore

$$x^h = \frac{\pi^h}{h+1} + \sum_{j=1}^{\infty}\sum_{k=1}^{h/2}(-1)^{j+k+1}\cdot\frac{2\pi^{h-2k}k!}{j^{2k}(h-2k+1)!}\cos jx$$

for $-\pi \le x \le \pi$, $2|h$.

---

[1] According to Kummer, *Vortrag über Leibniz* on 4.7.1867 in the Prussian Academy of Sciences. The saying is on a piece of paper and comes from Virgil, *Ecloga VIII*, 75.

[2] $\zeta(2) = \pi^2/6$ was first proved by Euler. This was one of his most sensational discoveries, since Leibniz and the Bernoullis, among others, had tried without success to compute the value of $\zeta(2)$. See Weil [1], Chap. III, §§5, 17, 18, 19, 20.

For $x = \pi$ we get

$$\pi^h = \frac{\pi^h}{h+1} + \sum_{j=1}^{\infty} \sum_{k=1}^{h/2} (-1)^{j+k+1} \cdot \frac{2\pi^{h-2k} k!}{j^{2k}(h-2k+1)!}. \tag{22}$$

We set

$$B_{2k} := (-1)^{k+1} \frac{2(2k)!}{(2\pi)^{2k}} \zeta(2k) \quad \text{for} \quad k = 1,2,... \tag{23}$$

The numbers B_{2k} are called the *Bernoulli numbers.*

Since the right-hand side of (22) converges absolutely, we can change the order of summation, and after multiplication by $(h+1)/\pi^h$ and substitution of B_{2k} we obtain

$$h = \sum_{k=1}^{h/2} B_{2k} 2^{2k} \binom{h+1}{2k}, \quad h = 2,4,... . \tag{24}$$

It follows from this that all the B_{2k} are rational numbers. One finds from (24) that $B_2 = \frac{1}{6}$, $B_4 = -\frac{1}{30}$, $B_6 = \frac{1}{42}$, $B_8 = -\frac{1}{30}$, $B_{10} = \frac{5}{66}$, $B_{12} = -\frac{691}{2730}$, $B_{14} = \frac{7}{6}$, $B_{16} = -\frac{3617}{510}$.

Exercises

5.1 Let y be a real number $0 < y < 1$. Show that for $-\pi < x < \pi$ we have the series expansion

$$\cos yx = \frac{2y \sin \pi y}{\pi} \left[\frac{1}{2y^2} - \sum_{n=1}^{\infty} (-1)^n \frac{\cos nx}{n^2 - y^2} \right].$$

Derive from this the partial fraction decomposition of the function $\cot \pi y$:

$$\pi \cot \pi y = \frac{1}{y} - \sum_{n=1}^{\infty} \frac{2y}{n^2 - y^2}.$$

5.2 Let $f(x)$ be a function which is defined for $-\pi \le x \le \pi$ and which satisfies the conditions 1 and 2 of Theorem 1. Also, let $c_1,...,c_n$, $d_0,...,d_n$ be arbitrary real numbers and let

$$\Delta_n(x) = f(x) - \left(\frac{d_0}{2} + \sum_{k=1}^{n} (c_k \sin kx + d_k \cos kx)\right).$$

As a measure of the goodness of the approximation to $f(x)$ by the trigonometric polynomial $\frac{d_0}{2} + \sum_{k=1}^{n} (c_k \sin kx + d_k \cos kx)$ one uses the *mean square error* $\frac{1}{2\pi}\int_{-\pi}^{\pi} \Delta_n^2(x)dx$.

a) Show that this error is smallest, for fixed n, when the coefficients c_k, d_k equal to Fourier coefficients a_k, b_k, and prove the inequality

$$\frac{b_0^2}{2} + \sum_{k=1}^{n} (a_k^2 + b_k^2) \le \frac{1}{\pi}\int_{-\pi}^{\pi} f(x)^2 dx.$$

b) (Parseval's equation) Show that

$$\frac{b_0^2}{2} + \sum_{k=1}^{\infty} (a_k^2 + b_k^2) = \frac{1}{\pi}\int_{-\pi}^{\pi} f(x)^2 dx.$$

5.3 Let f be a function which satisfies the conditions of Theorem 1.

a) Suppose that $f(x)$ is continuous in the interval $a < x < b$. Show that in each interval $a' \le x \le b'$ with $a < a' < b' < b$ the partial sums $f_n(x)$ converge uniformly to f.

b) Let

$$\tilde{f}_n(x) = \frac{1}{\pi}\int_{-\pi}^{\pi} f(x+2u)\frac{\sin ku}{u}du.$$

Show that $f_n(x) - \tilde{f}_n(x)$ converges to 0 uniformly in x.

5.4 Let $G(v) := \int_0^v \frac{\sin u}{u}du.$

a) Show that $G(\infty) = \frac{\pi}{2}$.

b) Show that $G(v)$ has its maximum at $v = \pi$.

c) Show that $\frac{2}{\pi}G(\pi) > 1179.$

5.5 Let $f(x) := (-\pi-x)/2$ for $-\pi \le x < 0$,

$f(0) := 0$,

$f(x) := (\pi-x)/2$ for $0 < x \le \pi$.

a) Show that $f(x) = \sum_{n=1}^{\infty} \frac{\sin nx}{n}$.

b) Suppose $0 \le x \le \pi$. Show that

$$\lim_{n\to\infty} f_n\left(\frac{2\pi}{k}\right) = \lim_{n\to\infty} f_n\left(\frac{\pi}{k}\right) = G(\pi).$$

c) Show that $\limsup_{\substack{n\to\infty \\ x\to 0}} f_n(x) = G(\pi)$.

Remark. $G(\pi)$ is greater than $\lim_{x\to 0} f(x) = \frac{\pi}{2}$. This phenomenon, which by Exercise 5.3 occurs at each jump discontinuity of a function f satisfying the conditions of Theorem 1, is called the Gibbs phenomenon. It was first discovered in 1848 by H. Wilbraham, *Cambridge and Dublin Math. J.*, 3, and rediscovered by J.W. Gibbs, *Nature* 59, 1898.

5.6 Let $B_0 := 1$, $B_1 := -\frac{1}{2}$, $B_h = 0$ for odd $h > 1$.

a) Show that

$$\sum_{k=0}^{h-1} \binom{h}{k} B_k = 0 \quad \text{for } h = 2,3,\ldots .$$

b) Show that the function $\frac{x}{e^x - 1}$ may be expanded in the Taylor series $\sum_{k=0}^{\infty} \frac{B_k}{k!} x^k$.

c) Show that $\sum_{k=1}^{n-1} k^m = \frac{1}{m+1} \sum_{k=0}^{m} \binom{m+1}{k} B_k n^{m+1-k}$.

6. Prime numbers in arithmetic progressions

6.1 The distribution of primes

Already in the *Elements*, Euclid found the theorem that there are infinitely many prime numbers. Suppose that there are only finitely many and that these are the numbers $p_1,\dots,p_s$. Then we consider the number $p_1p_2\dots p_s+1$. It contains none of the numbers $p_1,\dots,p_s$ as a prime factor. Therefore, there must be still more primes.

A sharpening of Euclid's theorem is due to Euler. He proved in 1737 (*Varia observationes circa series infinites, Comm. Acad. Sci. Imp. Petropolitanae* 9 (1737)) that the sum $\sum_p 1/p$ over all prime numbers diverges.

However, he was pessimistic about ever knowing the distribution of prime numbers within the series of natural numbers: "*Mathematicians down to the present day have tried without success to discover an order in the series of prime numbers, and one has reason to believe that this is a mystery which human intelligence will never penetrate.*"

Legendre made some conjectures about the distribution of prime numbers. One of them, which dates from 1785 (*Recherches d'analyse indéterminée, Histoire de l'Academie Royale des Sciences de Paris* 1788) says that the number $\pi(x)$ of prime numbers which are less than or equal to the positive real number x is approximated by the function

$$\frac{x}{\log x - 1.08366} \tag{1}$$

In Chap. 27, after preparations in Chapters 15 and 26, we shall show that the function $\int_2^x \frac{dt}{\log t}$ approximates the *prime number function* $\pi(x)$ even better than (1). Such a conjecture was made by Gauss in 1792, but never published.

Another conjecture of Legendre concerns the distribution of prime numbers in arithmetic progressions. Let k be a natural number. Then the prime numbers are distributed in a certain way among the $\phi(k)$ congruence classes prime to k mod k (Sect. 13). In the work cited above Legendre gave a "proof" of the law of quadratic reciprocity (Chap. 1, Theorem 10) under the hypothesis that for each prime $p \equiv 1 \pmod 4$ there is a prime number $q \equiv 3 \pmod 4$ with $\left(\frac{p}{q}\right) = -1$. This follows, as one easily sees, from the assumption that each prime congruence class contains prime numbers, which was conjectured by Legendre, but first proved in 1837 by Dirichlet (Abh. Preuss. Akad. Wiss.). More precisely, Dirichlet showed the following

Theorem 1. *Let* k *and* l *be relatively prime natural numbers and let* $\bar{l}$ *be the congruence class of* l*, mod* k*. Then the series*

$$\sum_{p \in \bar{l}} \frac{1}{p}$$

diverges, where the sum is taken over all prime numbers p in $\bar{l}$.

The theorem of Euler cited above is a special case of Theorem 1 (k = 1). Dirichlet's proof of Theorem 1, which we essentially follow here, starts from Euler's proof, but contributes several fundamental new ideas. One of these is to bring into play the characters of the group $(\mathbb{Z}/k\mathbb{Z})^\times$ of prime congruence classes mod k. In the next section we consider characters of finite abelian groups quite generally.

6.2 Characters of finite abelian groups

Let G be a finite abelian group. We write G multiplicatively and denote the identity element of G by 1. By a *character* of G one means a homomorphism χ of G into $\mathbb{C}^\times$. We denote the set of all characters of G by $\hat{G}$. Two characters χ, χ' of G are multiplied by the rule

$$(\chi \cdot \chi')(g) = \chi(g)\chi'(g) \quad \text{for} \quad g \in G.$$

Under this multiplication $\hat{G}$ itself becomes an abelian group, whose identity element is the character χ_0 which sends each $g \in G$ to the number 1. We call χ_0 the *identity character*.

Let $m = |G|$ be the order of G. Then

$$(\chi(g))^m = \chi(g^m) = \chi(1) = 1 \quad \text{for all} \quad g \in G.$$

It follows that the absolute value $|\chi(g)|$ of $\chi(g)$ is equal to 1, whence it follows that $\chi^{-1}(g) = \overline{\chi(g)}$ where the bar denotes the complex conjugate.

Theorem 2. $\hat{G}$ *is isomorphic to* G. *In particular, the number of characters of* G *equals the order of* G.

Proof. By the fundamental theorem on abelian groups (A.1, Theorem 7), G is isomorphic to the direct product of cyclic groups $C_1,\dots,C_s$ of orders $m_1,\dots,m_s$. Let $c_1,\dots,c_s$ be the generators of these groups and let χ_i be the character of C_i which associates c_i with the primitive m_i^{th} root of unity $\exp\left(\frac{2\pi\sqrt{-1}}{m_i}\right)$, $i = 1,\dots,s$. Then χ_i is a generator of the cyclic character group $\hat{C}_i$, and the isomorphism of G onto $\hat{G}$ is given by

$$c_1^{a_1}\dots c_s^{a_s} \mapsto \chi_1^{a_1}\dots\chi_s^{a_s} \quad \text{with} \quad 0 \le a_i < m_i \quad \text{for} \quad i = 1,\dots,s.$$

This isomorphism depends on the choice of a basis of G. But there is the canonical isomorphism of G onto the character group $\hat{\hat{G}}$ of $\hat{G}$. This maps $g \in G$ onto the character $\chi \mapsto \chi(g)$ of $\hat{G}$. □

As a corollary of the proof of Theorem 1 we have the following

Theorem 3. *Let* $\chi \in \hat{G}$. *Then*

$$\sum_{g\in G} \chi(g) = \begin{cases} |G| & \text{for } \chi = \chi_0 \\ 0 & \text{for } \chi \ne \chi_0 \end{cases} \tag{2}$$

Let $g \in G$. *Then*

$$\sum_{\chi\in \hat{G}} \chi(g) = \begin{cases} |G| & \text{for } g = 1 \\ 0 & \text{for } g \ne 1. \end{cases} \tag{3}$$

Proof. (3) results from (2) by passing from G to $\hat{G}$. It therefore suffices to prove (2). For $\chi = \chi_0$, (2) is trivial. Therefore suppose $\chi \ne \chi_0$, i.e. there is an $a \in G$ with $\chi(a) \ne 1$. Then

$$\sum_{g\in G} \chi(g) = \sum_{g\in G} \chi(ag) = \chi(a) \sum_{g\in G} \chi(g) = 0. \qquad \square$$

It follows easily from Theorem 3 that each function $f : G \to \mathbb{C}$ is uniquely expressible in the form

$$f(x) = \sum_{\chi \in \hat{G}} a_\chi \chi(x) \quad \text{with} \quad a_\chi = \frac{1}{|G|} \sum_{g \in G} f(g)\chi(g)^{-1}. \tag{4}$$

This is the analogue to harmonic analysis on $\mathbb{R}/2\pi\mathbb{Z}$ (Chap. 5, (17)).

6.3 Dirichlet L-functions

Let σ be a real number, k a natural number, and χ a character of $(\mathbb{Z}/k\mathbb{Z})^\times$. We define $\chi(x)$ for all natural numbers x by setting

$$\chi(x) = \chi(\bar{x}) \quad \text{for} \quad (x,k) = 1,$$
$$\chi(x) = 0 \quad \text{for} \quad (x,k) \neq 1.$$

Obviously, $\chi(x_1 x_2) = \chi(x_1)\chi(x_2)$ then holds for all natural numbers x_1, x_2. One calls χ a *character modulo* k.

By the Dirichlet *L-function for the character* χ one means the series

$$L(\sigma,\chi) := \sum_{n=1}^{\infty} \frac{\chi(n)}{n^\sigma}, \tag{5}$$

where $n^\sigma = \exp(\sigma \log n)$.

The series converges absolutely and uniformly for $\sigma > \delta > 1$:

$$\sum_{n=1}^{\infty} \frac{|\chi(n)|}{n^\sigma} \le \sum_{n=1}^{\infty} \frac{1}{n^\sigma} \le 1 + \int_1^\infty \frac{dx}{x^\sigma} = 1 + \left[\frac{x^{1-\sigma}}{\sigma-1}\right]_1^\infty \le 1 + \frac{1}{\delta-1}. \tag{6}$$

$L(\sigma,\chi)$ is therefore a continuous function for $\sigma > 1$.

For $\chi \neq \chi_0$ the series (5) even converges uniformly for $\sigma > \delta > 0$. In order to show this, we use the Cauchy convergence criterion: a sequence $b_1, b_2, \ldots$ converges if for each $\varepsilon > 0$ there is an N with $|b_\nu - b_\mu| < \varepsilon$ for $\nu, \mu > N$. Here we can confine ourselves to ν, μ with $\nu \le \mu$.

We have to consider the partial sums $\sum_{n=\nu}^{\mu} \chi(n)n^{-\sigma}$, and we apply a trick of Abel. We set

$$A_\mu = \sum_{n=\nu}^{\mu} \chi(n), \quad A_{\nu-1} = 0.$$

Then

$$\sum_{n=\nu}^{\mu} \chi(n)n^{-\sigma} = \sum_{n=\nu}^{\mu} (A_n - A_{n-1})n^{-\sigma}$$

$$= \sum_{n=\nu}^{\mu} A_n n^{-\sigma} - \sum_{n=\nu}^{\mu-1} A_n (n+1)^{-\sigma}$$

$$= \sum_{n=\nu}^{\mu-1} A_n (n^{-\sigma} - (n+1)^{-\sigma}) + A_\mu \mu^{-\sigma}.$$

By Theorem 3, $|A_n| \le |G|$, hence

$$\left|\sum_{n=\nu}^{\mu} \chi(n)n^{-\sigma}\right| \le \sum_{n=\nu}^{\mu} |G|(n^{-\sigma} - (n+1)^{-\sigma}) + |G|\mu^{-\sigma}$$

$$= |G|\nu^{-\sigma}$$

$$\le |G|\nu^{-\delta}.$$

The uniform convergence of (5) is now obvious.

For the proof of Theorem 1 one needs

Theorem 4. *For* $\chi \ne \chi_0$, $L(1,\chi) \ne 0$. ■

We prove Theorem 4 in Sect. 25.4.

Theorem 5. (Euler product formula) *For* $\sigma > 1$

$$L(\sigma,\chi) = \prod_p \frac{1}{1-\chi(p)p^{-\sigma}}$$

where the product is taken over all prime numbers p.

Proof. Let S be a finite set of prime numbers and let N(S) be the set of natural numbers which are products of primes in S. Because of the absolute convergence of (5) for $\sigma > 1$ we have

$$\sum_{n\in N(S)} \chi(n)n^{-\sigma} = \prod_{p\in S} \left(\sum_{i=1}^{\infty} \xi(p)^i p^{-\sigma i}\right).$$

As S increases the left-hand side tends to $L(\sigma,\chi)$. □

6.4 Proof of Theorem 1

Suppose $\sigma > 1$. Then $1 - \chi(p)p^{-\sigma}$ has a positive real part, and the principal value of the logarithm is determined by

$$\log(1-\chi(p)p^{-\sigma}) = - \sum_{m=1}^{\infty} \frac{\chi(p^m)}{mp^{m\sigma}}.$$

Theorem 5 gives

$$\log L(\sigma,\chi) = -\sum_p \log(1-\chi(p)p^{-\sigma}) = \sum_p \sum_{m=1}^{\infty} \frac{\chi(p^m)}{mp^{m\sigma}}$$

with absolutely convergent series, where $\log L(\sigma,\chi)$ denotes a certain logarithm of $L(\sigma,\chi)$.

$$g(\sigma,\chi) := \sum_p \sum_{m=2}^{\infty} \frac{\chi(p^m)}{mp^{m\sigma}}$$

is absolutely and uniformly convergent for $\sigma > \delta > \frac{1}{2}$:

$$\sum_p \sum_{m=2}^{\infty} \frac{1}{mp^{m\sigma}} \leq \sum_{n=2}^{\infty} \sum_{m=2}^{\infty} \frac{1}{n^{m\sigma}} = \sum_{n=2}^{\infty} \frac{1}{n^{2\sigma}(1-n^{-\sigma})}$$

$$< \frac{1}{1-2^{-\delta}} \sum_{n=2}^{\infty} \frac{1}{n^{2\delta}}$$

$$< \frac{1}{1-2^{-\delta}} \int_1^{\infty} \frac{dx}{x^{2\delta}}$$

$$= \frac{1}{(1-2^{-\delta})(2\delta-1)}.$$

This shows that $g(\sigma,\chi)$ is convergent as $\sigma \to 1$ and therefore harmless.

Now suppose $a \in \mathbb{Z}$ with $la \equiv 1 \pmod k$. Then by Theorem 3

$$\sum_{\chi} \chi(a)\log(\sigma,\chi) = \sum_{\chi} \chi(a) \sum_{p} \frac{\chi(p)}{p^{\sigma}} + \sum_{\chi} \chi(a)g(\sigma,\chi)$$

$$= \phi(k) \sum_{p\in l} \frac{1}{p^{\sigma}} + \sum_{\chi} \chi(a)g(\sigma,\chi), \tag{7}$$

where $\sum_{\chi}$ denotes the sum over all characters χ of $(\mathbb{Z}/k\mathbb{Z})^{\times}$. By Theorem 4, $\log L(1,\chi)$ is finite for $\chi \neq \chi_0$. Also, $\log L(\sigma,\chi)$ is continuous as $\sigma \to 1$. On the other hand, $L(\sigma,\chi_0)$ diverges as $\sigma \to 1$.

For $\sigma > 1$ we have

$$L(\sigma,\chi_0) = \prod_{p \nmid k} \left(\frac{1}{1-p^{-\sigma}}\right) = \zeta(\sigma) \prod_{p \mid k} (1-p^{-\sigma})$$

with $\zeta(\sigma) = \sum_{n=1}^{\infty} \frac{1}{n^{\sigma}} > \int_{1}^{\infty} \frac{dx}{x^{\sigma}} = \frac{1}{\sigma-1}$ and therefore

$$\lim_{\sigma\to 1} L(\sigma,\chi_0) = \infty.$$

Thus it follows from (7) that

$$\lim_{\sigma\to 1} \sum_{p\in l} \frac{1}{p^{\sigma}} = \infty. \qquad \square$$

Exercises

6.1 As a generalisation of Euclid's proof of the existence of infinitely many primes, show that there are infinitely many prime numbers $p \equiv 3 \pmod 4$ and infinitely many prime numbers $p \equiv 1 \pmod 4$. (Hint: In the case $p \equiv 1 \pmod 4$ use Theorem 7 of Chap. 1).

6.2 Show that there are infinitely many prime numbers $p \equiv 1 \pmod 3$ and infinitely many prime numbers $p \equiv 2 \pmod 3$.

6.3 Let k be a natural number, $\zeta = \exp(2\pi i/k)$, and let χ be a character modulo k. The sum

$$\tau_a(\chi) := \sum_{x=1}^{k} \chi(x)\zeta^{ax}$$

is called the *Gauss sum belonging to character* χ *and integer* a (cf. Sect. 3.6). Show that for $\sigma > 1$

$$L(\sigma,\chi) = \frac{1}{k}\sum_{a=1}^{k-1} \tau_a(\chi) \sum_{n=1}^{\infty} \frac{\zeta^{-na}}{n^\sigma}.$$

6.4 Let $a \in \mathbb{Z}$ be a nonmultiple of k. Show that for each $\delta > 0$ the series $\sum_{n=1}^{\infty} \frac{\zeta^{-na}}{n^\sigma}$ converges uniformly in the interval $\delta \le \sigma < \infty$.

6.5 Show that $L(1,\chi) = \frac{1}{k}\sum_{a=1}^{k-1} \tau_a(\chi) \sum_{n=1}^{\infty} \frac{\zeta^{-na}}{n}$.

6.6 Show that $\sum_{n=1}^{\infty} \frac{\zeta^{-na}}{n} = -\log(1-\zeta^{-a})$, where $\log$ denotes the principal value of the logarithm.

6.7 Let χ be a character modulo k. Then χ is called an *imprimitive character* when there is a divisor d of k with $d \neq k$ and a character χ_d modulo d such that $\chi(x) = \chi_d(x)$ for $(x,k) = 1$.

Show that for a primitive character

$$L(1,\chi) = -\frac{\tau_1(\chi)}{k}\sum_{a=1}^{k-1} \chi(a)^{-1}\log(1-\zeta^{-a})$$

and

$$L(1,\chi) = \begin{cases} -\dfrac{\tau_1(\chi)}{k}\displaystyle\sum_{a=1}^{k-1} \chi(a)^{-1}\log 2 \sin\frac{\pi a}{k} & \text{if } \chi(-1) = 1, \\ \dfrac{\pi i \tau_1(\chi)}{k^2}\displaystyle\sum_{a=1}^{k-1} \chi(a)^{-1} a & \text{if } \chi(-1) = -1. \end{cases}$$

7. Theory of algebraic equations

7.1 Third and fourth degree equations

Up to the middle of the 19th century, algebra was almost identical with the question of solving algebraic equations, particularly the nth degree equation in one indeterminate with real or complex coefficients. Hence the name "*Fundamental theorem of algebra*" for the theorem that each such equation has a complex number as solution. To solve an equation meant to present the solutions in terms of radicals. We make this concept precise in Sect. 7.1.

The solution of quadratic equations was known in a few ancient civilisations, but it was not until around 1500 with del Ferro, and 1535 with Tartaglia, that the third degree equation was solved. Not much later, Ferrari found the solution of the fourth degree equation. The formula for the solution of the third degree equation was published by Cardano. As a result, it became incorrectly known as the Cardano formula. Cardano named Tartaglia as the discoverer of the method in his *Artis magnae sive de regulis algebraicus liber unus* of 1545, and described it as "*a very beautiful and wonderful thing, which surpasses all the subtlety and splendour of the human mind, a truly heavenly gift, a demonstration of intelligence so excellent that he who masters it will feel that nothing is beyond his powers*".

Suppose the equation of third degree is given in the form

$$x^3 + a_1x^2 + a_2x + a_3 = 0, \tag{1}$$

where a_1,a_2,a_3 are complex numbers. When we pass to $y = x + a_1/3$, (1) takes the form

$$y^3 + py + q = 0, \tag{2}$$

where p and q are polynomials in a_1,a_2,a_3. One now seeks u and v with $y = u + v$, whereby (2) takes the form

$$u^3 + 3u^2v + 3uv^2 + v^3 + p(u+v) + q = 0.$$

This equation is satisfied when one can determine u and v so that $u^3 + v^3 = -q$ and $3uv = -p$. To do this we consider first the equations

$$u^3 + v^3 = -q, \; u^3v^3 = -(p/3)^3. \tag{3}$$

By Viète's root theorem, u^3 and v^3 are solutions of the equation $z^2 + qz - (p/3)^3 = 0$. Thus if we set

$$u_1 = \sqrt[3]{-\frac{q}{2}+\sqrt{\left(\frac{q}{2}\right)^2+\left(\frac{p}{3}\right)^3}}, \; v_1 = \sqrt[3]{-\frac{q}{2}-\sqrt{\left(\frac{q}{2}\right)^2+\left(\frac{p}{3}\right)^3}},$$

where an arbitrary one of the three values of the cube root is taken for u_1 and v_1 is determined by $u_1v_1 = -p/3$. then we obtain a solution of (2) in the form $u_1 + v_1$. When ρ is a primitive cube root of unity, then one obtains the other two solutions of (2) in the form $\rho u_1 + \rho^2 v_1$, $\rho^2 u_1 + \rho v_1$.

In particular, if $(q/2)^2 + (p/3)^3 \geq 0$ we can choose the real value of the cube root. We see that in this case (2) has one real and two complex solutions, or a multiple solution. On the other hand, if $(q/2)^2 + (p/3)^3 < 0$, then one is forced to extract the cube root from a complex number; u_1 and v_1 are complex conjugates, whence it follows that all three solutions of (2) are real. This case is called the *casus irreducibilis* (cf. Exercise 7.9).

With the equation of fourth degree one proceeds similarly. We shall be more brief about it: in the equation

$$x^4 + px^2 + qx + r = 0 \tag{4}$$

we set $x = u + v + w$ and obtain

$$(u^2+v^2+w^2+2(uv+uw+vw))^2$$

$$+ \; p(u^2+v^2+w^2+2(uv+uw+vw)) + q(u+v+w) + r = 0. \tag{5}$$

We set $u^2 + v^2 + w^2 = -b_1$, $u^2v^2 + u^2w^2 + v^2w^2 = b_2$, $u^2v^2w^2 = -b_3$, so (5) becomes

$$b_1^2 - 4b_1(uv+uw+vw) + 4b_2 + 8(u+v+w)uvw - pb_1$$

$$+ \; 2p(uv+uw+vw) + q(u+v+w) + r = 0.$$

It suffices to determine u, v, w so that

$$b_1^2 + 4b_2 - pb_1 + r = 0, \ -4b_1 + 2p = 0, \ 8uvw + q = 0.$$

This is equivalent to

$$b_1 = p/2, \ b_2 = -r/4 + p^2/16, \ b_3 = -q^2/64, \ uvw = -q/8.$$

We obtain a cubic equation for u^2, v^2, w^2. The signs of u, v, w are determined so that $uvw = -q/8$. This gives exactly four solutions of (4). They are of the form

$$u + v + w = \pm \sqrt{\sqrt{h_1} + \sqrt[3]{\sqrt{h_2} + \sqrt{h_3}} + \sqrt[3]{\sqrt{h_2} - \sqrt{h_3}}}$$

$$\pm \sqrt{\sqrt{h_1} + \rho\sqrt[3]{\sqrt{h_2} + \sqrt{h_3}} + \rho^2\sqrt[3]{\sqrt{h_2} - \sqrt{h_3}}}$$

$$\pm \sqrt{\sqrt{h_1} + \rho^2\sqrt[3]{\sqrt{h_2} + \sqrt{h_3}} + \rho\sqrt[3]{\sqrt{h_2} - \sqrt{h_3}}}$$

where h_1, h_2, h_3 are polynomials in p, q, r.

After these results the main interest was in solving the fifth degree equation by radicals. The first complete proof that this was impossible in general was found by Abel in 1826 (*Beweis der Unmöglichkeit, algebraische Gleichungen von höhern Graden als dem vierten allgemein aufzulösen, J. reine angew. Math.* 1 (1826)). The reasons for it became especially clear in the Galois theory of equations, which is the main subject of this chapter. We first mention a few preliminaries.

7.2 Solution of equations by radicals

We first want to clarify the concept of "*solving an equation by radicals*", and we do so by means of the field concept, which is first found implicitly in Abel, when he speaks of *rational functions of given quantities* x′, x″,... (*Mémoire sur une classe particulière d'equations resolubles algébriquement, J. reine angew. Math.* 4, 1829).

Let $f(x)$ be a polynomial of nth degree with coefficients in a field K of characteristic 0, and let $\alpha_1,\dots,\alpha_n$ be the zeros of $f(x)$ in a field L which includes K. Then the equation $f(x) = 0$ (over K) is solvable by radicals when there is a sequence of fields

$$K_1 = K, K_2,\dots,K_s \tag{6}$$

which satisfies the following conditions:

1. *For each* $i = 1,\dots,s-1$ *there is a prime number* p_i *and a* $\beta_i \in K_{i+1}$ *with* $K_{i+1} = K_i(\beta_i)$, *where* $\gamma_i = \beta_i^{p_i} \in K_i$ *and the polynomial* $x^{p_i} - \gamma_i$ *is irreducible over* K_i (we call β_i a p_i^{th} radical over K_i).

2. K_s *contains* $\alpha_1,\dots,\alpha_s$.

One writes $\beta_i = \sqrt[p_i]{\gamma_i}$, meaning that β_i is one of the p_i zeros of $x^{p_i} = \gamma_i$.

In particular, if $\gamma \in \mathbb{C}$, then the equation $x^n = \gamma$ has the n solutions corresponding to the *de Moivre formula*:

$$\sqrt[n]{r}\,\cos(\phi/n + 2\pi i/n) + \sqrt{-1}\,\sin(\phi/n + 2\pi i/n),\ i = 1,\dots,n,$$

with $r = |\gamma|$ and $\cos\phi = \operatorname{Re}\gamma/r$, $\sin\phi = \operatorname{Im}\gamma/r$. Thus the solutions of an equation by radicals can in principle be computed numerically.

7.3 The general equation of nth degree and the theory of symmetric functions

We also want to clarify the concept of the *general equation of* n^{th} *degree.*

The obvious thing to do corresponds to the original formulation of the problem, in which the x in the equation

$$x^n + a_1x^{n-1} + \dots + a_n = 0 \tag{7}$$

was understood to be an indeterminate and $a_1,...,a_n$ were arbitrary complex numbers. A solution would then be a complex-valued function of $a_1,...,a_n$. We shall study this concept more closely in Chapter 10 in the framework of function theory. Since the solution of (7) is not unique, this leads to difficulties at first. We therefore want to take the general equation of nth degree to be equation (7) in which $a_1,...,a_n$ themselves are indeterminates. Starting from a field K, we construct the field $K_n = K(a_1,...,a_n)$ of rational functions of $a_1,...,a_n$ and ask for a solution of (7) in an extension field L_n of K_n. We provide ourselves with a field L_n in which (7) splits into the product of linear factors as follows:

Consider indeterminates $x_1,...,x_n$ over K. When these are the n zeros of a polynomial f(x), then f(x) has the form

$$f(x) = (x-x_1)...(x-x_n)$$

and hence it has coefficients

$$a'_1 := -(x_1 + ... + x_2) = -\sum_i x_i,$$

$$a'_2 := \sum_{i<j} x_i x_j$$

$$...$$

$$a'_k := (-1)^k \sum_{i_1<..<i_k} x_{i_1}...x_{i_k}$$

$$...$$

$$a'_n := (-1)^n x_1...x_n.$$

$s_k = (-1)^k a'_k$ is called the kth *elementary symmetric polynomial.*

Theorem 1. *The polynomials* $s_1,...,s_n$ *are algebraically independent over* K, *i.e. there is no non-zero polynomial* $p(y_1,...,y_n)$ *in the* n *indeterminates* $y_1,...,y_n$ *with coefficients in* K *such that* $p(s_1,...,s_n) = 0$.

Proof. Suppose there is such a polynomial p. Let the degree d of p as a polynomial in y_n be minimal. Then p has the form

$$p(y_1,...,y_n) = p_0(y_1,...,y_{n-1})+p_1(y_1,...,y_{n-1})y_n+...+p_d(y_1,...,y_{n-1})y_n^d.$$

The term $p_0(y_1,\ldots,y_{n-1})$ is not identically 0, otherwise $p(y_1,\ldots,y_n)/y_n$ would be a polynomial with $p(s_1,\ldots,s_n)/s_n = 0$, contrary to the minimality of d.

We now consider $p(s_1,\ldots,s_n)$ as a polynomial in $x_1,\ldots,x_n$ and set $x_n = 0$. Then we obtain $p_0(s'_1,\ldots,s'_{n-1}) = 0$, where $s'_1,\ldots,s'_n$ are the elementary symmetric polynomials in $x_1,\ldots,x_{n-1}$. Induction on n now leads to the desired contradiction. □

By Theorem 1, $\phi : f(a_1,\ldots,a_n) \mapsto f(a'_1,\ldots,a'_n)$ defines a ring homomorphism of $K[a_1,\ldots,a_n]$ into $K[x_1,\ldots,x_n]$ which is injective. Therefore ϕ may be extended to an embedding of $K(a_1,\ldots,a_n)$ in $K(x_1,\ldots,x_n)$ (Sect. A1.1).

We can therefore regard $K(x_1,\ldots,x_n)$ as an extension of $K(a_1,\ldots,a_n)$, and the polynomial $x^n + a_1x^{n-1} + \ldots + a_n$ has the n zeros $x_1,\ldots,x_n$ in $K(x_1,\ldots,x_n)$.

There is a beautiful characterisation of the image of ϕ in $K[x_1,\ldots,x_n]$. This is the content of the fundamental theorem of symmetric functions.

A polynomial $f(x_1,\ldots,x_n)$ is called *symmetric* when it is unaltered by arbitrary permutations of the indeterminates $x_1,\ldots,x_n$. This is obviously the case for the elementary symmetric polynomials $s_1,\ldots,s_n$.

Theorem 2. *(Fundamental theorem of symmetric polynomials). Each symmetric polynomial may be expressed as a polynomial in* $s_1,\ldots,s_n$.

Proof. By the degree of the monomial $x_1^{k_1}\ldots x_n^{k_n}$ we mean the sum $k_1 + \ldots + k_n$. Among the monomials which result from $x_1^{k_1}\ldots x_n^{k_n}$ by permutation of the indeterminates $x_1,\ldots,x_n$ there is exactly one monomial $x_1^{k_{\pi(1)}}\ldots x_n^{k_{\pi(n)}}$ with $k_{\pi(1)} \leq \ldots \leq k_{\pi(n)}$, where π is a permutation of 1,...,n. Each symmetric polynomial $t(x_1,\ldots,x_n)$ may therefore be written uniquely in the form $t(x_1,\ldots,x_n) = \sum_{k_1,\ldots,k_n} h_{k_1,\ldots,k_n} t_{k_1,\ldots,k_n}(x_1,\ldots,x_n)$, where $k_1,\ldots,k_n$ are non-negative integers with $k_1 \leq \ldots \leq k_n$, $t_{k_1,\ldots,k_n}(x_1,\ldots,x_2)$ equals the sum of all the different monomials $x_1^{k_{\pi(1)}}\ldots x_n^{k_{\pi(n)}}$ for permutations π of 1,...,n, and the coefficients $h_{k_1,\ldots,k_n} \in K$ almost all vanish.

We take the *weight* of the monomial $a_1^{l_1} \ldots a_n^{l_n}$ to be the sum $l_1 + 2l_2 + \ldots + nl_n$. Under the mapping ϕ, a homogeneous polynomial in $K[a_1,\ldots,a_n]$ of weight g, i.e. a linear combination of monomials of weight g, becomes a homogeneous symmetric polynomial of degree g. Because of Theorem 1, it suffices to show that the number of linearly independent homogeneous polynomials in $K[a_1,\ldots,a_n]$ of weight g equals the number of linearly independent homogeneous symmetric polynomials in $K[x_1,\ldots,x_n]$ of degree g, in order to prove Theorem 2. One proves this with the help of the correspondence

$$t_{k_1,\ldots,k_n}(x_1,\ldots,x_n) \mapsto a_1^{k_n - k_{n-1}} a_2^{k_{n-1} - k_{n-2}} \ldots a_n^{k_1}. \qquad \square$$

In particular, one can express the sums of powers $x_1^g + \ldots x_n^g$, $g = 1,\ldots,n$, in terms of the elementary symmetric functions. When K is of characteristic 0 the converse also holds. One proves this by induction on the degree i of the elementary symmetric function: for $i = 0$ there is nothing to prove. Suppose the assertion is already proved for $s_1,\ldots,s_{i-1}$. We have a representation

$$x_1^i + \ldots + x_n^i = hs_i + p(s_1,\ldots,s_{i-1}) \text{ with } h \in Q, \tag{8}$$

where p is a polynomial in $s_1,\ldots,s_{i-1}$. We have to show $h \neq 0$. To do this we specialise (8) by taking $x_j = \exp(j2\pi\sqrt{-1}/i)$ for $j = 1,\ldots,i$ and $x_{i+1} = \ldots = x_n = 0$. Then $x_1,\ldots,x_n$ are the zeros of the polynomial $x^n - x^{n-i}$, i.e. $s_1 = \ldots = s_{i-1} = 0$, $s_i = (-1)^{i+1}$ and hence $h = (-1)^{i+1}i$.

As an example we consider the polynomial

$$d(x_1,\ldots,x_n) = \prod_{i \leq j} (x_i - x_j), \tag{9}$$

which is sent to itself by even permutations and to $-d(x_1,\ldots,x_n)$ by odd permutations. The even permutations form a normal subgroup A_n of index 2 in the group S_n of all permutations of $1,\ldots,n$. S_n is called the *symmetric group*, and A_n the *alternating group, on* n *symbols*. The polynomial $d(x_1,\ldots,x_n)^2$ is invariant under all permutations and is therefore expressible in terms of elementary symmetric polynomials.

When $\beta_1,\ldots,\beta_n$ are the zeros of the polynomial $f(x) = x^n + b_1x^{n-1} + \ldots + b_n$, then $d(\beta_1,\ldots,\beta_n)^2 = D(f)$ is a polynomial in $b_1,\ldots,b_n$ called the *discriminant* of f. The following theorem is obvious.

Theorem 3. *Let* f *be a polynomial which splits into linear factors in* K. *Then* f *has a double zero, i.e.* f(x) *is divisible by* $(x-\beta_i)^2$ *for a certain zero* β_i *if and only if* $D(f) = 0$.

For $f(x) = x^2 + b_1x + b_2$, $D(f) = b_1^2 - 4b_2$, and for $f(x) = x^3 + b_1x^2 + b_2x + b_3$, $D(f) = -4b_1^3b_3 + b_1^2b_2^2 + 18b_1b_2b_3 - 4b_2^3 - 27b_3^2$ (cf. the Cardano formula in Sect. 7.1).

7.4 The Galois memoir on the theory of equations

Galois presented his ideas on the solution of equations in most detail in the work "*Mémoire sur les conditions de résolubilité des equations par radicaux*", first published by Liouville in 1846 in *J. math. pures appl.* 11 (1846), 417-433.

Galois accompanied this memoir by the following note of 16 January 1831:

"*The enclosed memoir is an abstract of a work which I had the honor to submit to the Academy over a year ago. This work was not understood; the theorems it contained were placed in doubt. I have therefore decided to put the general principles in synthetic form and to give a few applications of my theory. I ask my judges to read at least these few pages carefully. One will find here a general condition satisfied by each equation which is solvable by radicals, and which, conversely, guarantees solvability. We make a few appplications of this to equations of prime degree. Here is the theorem which results from our analysis:*

An irreducible equation of prime degree is solvable by radicals; if and only if all its roots are rational functions of two arbitrary roots."

This second version of the work was not accepted either, and was first published 15 years later. The first version is lost.

Galois based his arguments on results of Lagrange (*Réflexions sur la résolution algébrique des équations, Nouv. Mém. Acad. royale Sci. Belles-Lettres Berlin* 1770/1771), on Gauss's division of the circle (Chapter 3) and on the work of Abel "*Mémoire sur une classe particulière d'équations résolubles algébriquement*" (*J. reine angew. Math.* 4 (1829), 131-156). The latter work anticipated Galois theory for commutative groups (the term "*abelian group*" comes from here).

7.5 The theorem of the primitive element

The following *theorem of the primitive element* was discovered by Abel, but first completely proved by Galois.

Theorem 4. *Let* L/K *be a field extension, and let* $\alpha,\beta \in L$ *be zeros of the polynomials* $f(x)$ *and* $g(x)$ *in* $K[x]$, *which both have no multiple zeros. Then there is a* $\gamma \in L$ *with* $K(\alpha,\beta) = K(\gamma)$.

Proof. We first suppose that K has infinitely many elements. Then there is a $c \in K$ with $\alpha_i + c\beta_j \neq \alpha + c\beta$, where α_i, β_j run through the remaining zeros of f and g. In fact, it follows from $\alpha_i + y\beta_j = \alpha + y\beta$ that $y = (\alpha-\alpha_i)/(\beta_j-\beta)$. This holds for only finitely many y.

Let c be an element with the above property and let $\gamma = \alpha + c\beta$. The polynomials $g(x)$ and $f(\gamma-cx)$ have β as their only common zero. Hence the greatest common divisor of $g(x)$ and $f(\gamma-cx)$ equals $x - \beta$. The latter lies in the polynomial ring over the field with the coefficients of $g(x)$ and $f(\gamma-cx)$, i.e. over $K(\gamma)$ (Sect. A 1.2). This implies $K(\alpha,\beta) = K(\sigma)$.

The case where K has finitely many elements does not occur with Galois, since he always understood a field to have characteristic 0. On the other hand, the first results on general finite fields are due to him (*Sur la théorie des nombres, Bull. Sci. math. M. Férrusac* 13 (1830)).

The proof of Theorem 4 in the case of a finite field K follows immediately from the following theorem, whose proof is similar to the proof of Theorem 5 in Chapter 1.

Theorem 5. *The multiplicative group of a finite field is cyclic.* □

A *primitive element* of a field extension L/K is an element γ of L with $L = K(\gamma)$. With this definition one can also express Theorem 4 in the following form.

Theorem 6. *Let* $f(x) \in K[x]$ *be a polynomial without multiple roots and let* L *be an extension of* K *which results from adjunction of zeros of* $f(x)$. *Then* L *has a primitive element.*

By A1, Theorem 11 and Theorem 13, one can always obtain a finite extension of a field of characteristic 0 by adjoining zeros of a polynomial without multiple zeros. Hence in this case each finite extension has a primitive element.

7.6 The Galois group of a polynomial

Let f, as above, be a polynomial in $K[x]$ without multiple zeros in a splitting field L of f, i.e. an extension L of K, over which f splits into linear factors (Sect. A 1.5). Let n be the degree of f. Galois associates f with a subgroup of the symmetric group S_n of all permutations of the n zeros of f. This subgroup is called the *Galois group* G_f of the polynomial f. G_f is defined as follows (deviating only mildly from the procedure of Galois).

Let $y_1,...,y_n$ be indeterminates and let $\alpha_1,...,\alpha_n$ be the zeros of f in L. Also let H be the ring of polynomials $h(y_1,...,y_n)$ in $K[y_1,...,y_n]$ with $h(\alpha_1,...,\alpha_n) \in K$. By Theorem 2, these include all symmetric polynomials.

We define G_f to be the collection of all permutations π of $\alpha_1,...,\alpha_n$ with

$$h(\pi\alpha_1,...,\pi\alpha_n) = h(\alpha_1,...,\alpha_n)$$

for all $h(y_1,...,y_n)$ in H.

G_f is obviously a group.

Example 1. Let $f(x) = x^n + a_1x^{n-1} + ... + a_n$ be the polynomial which belongs to the general equation in the sense of Section 7.3. As in Section 7.3 we consider the indeterminates $a_1,...,a_n$ as symmetric functions $a_i = (-1)^i s_i$ of the indeterminates

$x_1,...,x_n$ over a field K_0, and therefore take $K = K_0(s_1,...,s_n)$ as the ground field. For a polynomial $h(y_1,...,y_n)$ which lies in H, $h(x_1,...,x_n) \in K_0(x_1,...,x_n)$ is symmetric. The group of f is therefore the symmetric group S_n.

We now state a series of theorems on the group of a polynomial, which will be proved in Chapter 17 within a different construction of the theory.

Theorem 7. *Let* $g(y_1,...,y_n) \in K[y_1,...,y_n]$ *be a polynomial with*

$$g(\pi\alpha_1,...,\pi\alpha_n) = g(\alpha_1,...,\alpha_n) \text{ for all } \pi \in G_f.$$

Then $g(\alpha_1,...,\alpha_n) \in K$. ■

Theorem 8. *Let* β *be a primitive element of* $K(\alpha_1,...,\alpha_n)$, *i.e.* $K(\beta) = K(\alpha_1,...,\alpha_n)$, *and let* $h_i(x) \in K[x]$ *be a polynomial with* $h_i(\beta) = \alpha_i$, $i = 1,...,n$. *Also let* $g(x) \in K[x]$ *be the irreducible polynomial corresponding to* β. *Then all the zeros* $\beta^{(1)} = \beta,...,\beta^{(m)}$ *of* $g(x)$ *lie in* $K(\beta)$, *and*

$$\pi_\nu(\alpha_i) = h_i(\beta^{(\nu)}), \; i = 1,...,n, \tag{10}$$

is a permutation of the roots α_i *which belongs to* G_f *for* $\nu = 1,...,m$. *One obtains each permutation in* G_f *uniquely in this form. In particular, the order of* G_f *is equal to the degree of* $K(\alpha_1,...,\alpha_n)$ *over* K. ■

The element β of $K(\alpha_1,...,\alpha_n)$ has the special property that one can express its conjugates relative to K, i.e. $\beta^{(1)},...,\beta^{(m)}$, rationally in terms of β. The associated polynomial $g(x)$ is called a *normal polynomial*. $K(\beta)/K$ is called a *normal extension.*

Example 2. We consider the cyclotomic polynomial $f(x) = x^{p-1} + ... + 1$ (Chapter 3). This $f(x)$ is irreducible (Chapter 3, Theorem 4), and one obtains all zeros of $f(x)$ from a special ζ in the form ζ^ν, $\nu = 1,...,p-1$. When g is a primitive root mod p, then one can also express the p-1 roots of f in the form $\alpha_i = \zeta^{g^i}$, $i = 1,...,p-1$. The permutations in G_f, described by Theorem 8, are the cyclic permutations of $\alpha_1,...,\alpha_{p-1}$. G is therefore the cyclic group of order p-1.

Theorem 9. *Let* $e(x) \in K[x]$ *be an irreducible polynomial with zeros* $\varepsilon_1,...,\varepsilon_t$ *in* L. *Then the group* H_i *of* f *over* $K(\varepsilon_i)$ *is a subgroup of* G_f, *and the groups* H_i *for* $i = 1,...,t$ *are conjugate in* G_f, *i.e. there are* $\pi_i \in G_f$ *with*

$$H_i = \pi_i H_1 \pi_i^{-1} =: \{\pi_i \pi \pi_i^{-1} \mid \pi \in H_1\},\ i = 1,...,t.$$

When $e(x)$ *splits into linear factors in* L, *for each* $\pi \in G_f$ *there is an* i *with* $H_i = \pi H_1 \pi^{-1}$. ∎

By definition of the group of f it is clear that H_i is a subgroup of G_f. We have taken Theorem 9 almost literally from Galois' work. The proof of Theorem 9 in Galois is incomplete. In a marginal note he wrote that he did not have the time to carry out the proof completely. The first detailed account of Galois theory was found in Jordan's book *Traité des substitutions et des équations algébriques, Paris* 1870.

If in particular $e(x)$ is a normal polynomial then $K(\varepsilon_i) = K(\varepsilon_j)$ and hence $H_i = H_j$. In this case H_1 is a subgroup of G_f which coincides with all its conjugates, i.e. a normal subgroup of G_f.

Theorem 10. *If one adjoins to* K *a polynomial* $h(\alpha_1,...,\alpha_n)$, *then the group of* f *over* $K(h(\alpha_1,...,\alpha_n))$ *equals the subgroup of all permutations* π *of* G_f *such that* $h(\pi\alpha_1,...,\pi\alpha_n) = h(\alpha_1,...,\alpha_n)$. ∎

Theorem 11. *For each subgroup* U *of* G_f *there is a polynomial* $h(\alpha_1,...,\alpha_n)$ *such that* U *equals the group of* f *over* $K(h(\alpha_1,...,\alpha_n))$. ∎

7.7 An irreducibility criterion

The basic idea of Galois is that the algebraic properties of the zeros $\alpha_1,...,\alpha_n$ of the polynomial f can be derived from the group G_f of f. This holds in particular for the main question, the representability of $\alpha_1,...,\alpha_n$ by radicals, as we shall see below. In the present section we prove the following irreducibility criterion.

Theorem 12. *A polynomial* f *is irreducible if and only if* G_f *is transitive, i.e. when any two zeros of* f *can be sent to each other by a* $\pi \in G_f$.

Proof. Suppose G_f is transitive and f_1 is the irreducible normalised polynomial associated with α_1. Then $f_1(y_1)$ belongs to H, and by definition of G_f we have $f_1(\pi\alpha_1) = f_1(\alpha_1) = 0$ for all $\pi \in G_f$. Therefore $f = f_1$.

Now suppose on the other hand that G_f is intransitive, but permutes the zeros $\alpha_1,...,\alpha_s$ for $s < n$. Then the elementary symmetric functions of $\alpha_1,...,\alpha_s$ are sent to themselves by all $\pi \in G_f$. Then by Theorem 7 the polynomial $\prod_{i=1}^{s} (x-\alpha_i)$ has coefficients in K. Consequently, f(x) is reducible. □

7.8 Irreducible equations of prime degree with cyclic group

In this section p denotes a prime number, K is a field of characteristic different from p which contains the pth roots of unity, and ζ is a primitive p^{th} root of unity.

Theorem 13. *Let* f *be a polynomial of degree* p, *irreducible over* K *and with cyclic group* G. *Then* f *is a normal polynomial.*

Let α *be a zero of* f *in an extension* L *of* K *and let* σ *be a generator of* G. *Also let*

$$(\zeta,\alpha) := \sum_{i=0}^{p-1} \zeta^{-i}\sigma^{i}\alpha.$$

Then $(\zeta,\alpha)^p \in K$ *and, for a certain primitive* p^{th} *root of unity* ζ, $K(\alpha) = K((\zeta,\alpha))$.

Remark. The equation $f(x) = 0$ is therefore solvable by radicals. (ζ,α) is called the *Lagrange resolvent.*

Proof of Theorem 13. Since f is irreducible, G is transitive. Since G is also cyclic, the zeros of f must be cyclically permuted. G therefore has order p, and by Theorem 8 $K(\alpha,\sigma\alpha,...,\sigma^{p-1}\alpha) = K(\alpha)$, i.e. f is a normal polynomial. Also

$$(\zeta,\sigma\alpha) = \sum_{i=0}^{p-1} \zeta^{-i}\sigma^{i+1}\alpha = \sum_{i=0}^{p-1} \zeta^{-i+1}\sigma^{i}\alpha = \zeta(\zeta,\alpha).$$

Therefore $(\zeta,\alpha)^p$ is unchanged under all permutations in G and by Theorem 7 it lies in K.

If the Lagrange resolvent (ζ,α) were 0 for all primitive p^{th} roots of unity ζ then we should also have

$$(1,\alpha) = \sum_{i=0}^{p-1} (\zeta^i,\alpha) = p\alpha.$$

This equation is impossible, since $(1,\alpha)$ lies in K. Thus there is a ζ with $(\zeta,\alpha) \neq 0$. Since $(\zeta,\sigma\alpha) \neq (\zeta,\alpha)$, (ζ,α) does not belong to K and hence it generates $K(\alpha)$. □

Theorem 14. *Suppose* $a \in K$ *and the polynomial* $f(x) = x^p - a$ *is irreducible. Then the group of* f *is cyclic of order* p.

Proof. Let α be a zero of f. Then the other zeros have the form $\zeta^i\alpha$, $i = 1,\dots,p-1$. The assertion now follows from Theorem 8. □

7.9 The cyclotomic equation

Let p be a prime number. We have already convinced ourselves that the cyclotomic polynomial $\Phi_p(x) = x^{p-1} + \dots + 1$ is irreducible, and that the corresponding group is cyclic of order p-1 (Example 2). We now show

Theorem 15. *The equation* $\Phi_p(x) = 0$ *is solvable by radicals.*

Proof. For $p = 2$ the assertion is trivial. We suppose that Theorem 15 is already proved for prime numbers $p' < p$. Let $K_1 = \mathbb{Q} \subset K_2 \subset \dots \subset K_r$ be a solvable chain of fields for $\prod_{p'<p} \Phi_{p'}(x)$ and let $p-1 = p_1 \dots p_s$. In Section 3.4 we have seen that there are cyclotomic periods $\beta_0 = 1, \beta_1,\dots,\beta_s = \zeta$ such that ζ is a primitive p^{th} root of unity and β_{i+1} satisfies a normal equation f_i of degree p_{i+1} over $\mathbb{Q}(\beta_i)$, $i = 0,\dots,s-1$. Since $\Phi_p(x)$ is irreducible, $\mathbb{Q}(\beta_{i+1})/\mathbb{Q}(\beta_i)$ has degree p_{i+1} (Section A 1.5).

The group G_i corresponding to f_i therefore has order p_{i+1}. The subgroup U_i of G_i corresponding to f_i over $K_r(\beta_i)$ consequently has order 1 or p_{i+1} (in Chapter 25 we shall see that $U_i = G_i$). When $U_i = \{1\}$, $K_r(\beta_{i+1}) = K_r(\beta_i)$. When $U_i = G_i$ we have the case of Section 7.8. □

Theorem 15 is found in Gauss's *Disquisitiones arithmeticae*, Section 7. The special case $p = 11$ was already known to Vandermonde (*Mémoire sur la résolution des équations, Histoires de l'Académie Royale des Sciences de Paris* 1771).

7.10 The fundamental theorem on the solution of equations by radicals

In this section and the next K is a field of characteristic 0.

One of the main results of Galois was a criterion which permits the solvability of an equation by radicals to be recognised from the group of the equation. This is where we meet the concept of solvable group.

A finite group G is called *solvable* when there is a sequence of subgroups

$$U_1 = G \supset U_2 \supset \ldots \supset U_s = \{1\} \tag{11}$$

with the property that U_{i+1} is a normal subgroup of U_i with prime index for $i = 1,\ldots,s-1$. We call (11) a *composition series* of G. As one easily sees, subgroups and quotient groups of solvable groups are again solvable.

Theorem 16. *Let* f *be a polynomial with coefficients in a field* K *of characteristic* 0. *Suppose* f *has no multiple zeros (in a splitting field* L*).*

Then $f(x) = 0$ *is solvable by radicals if and only if* G_f *is solvable.*

Proof. By Theorem 15 we can assume without loss of generality that the roots of unity of prime order mentioned below are in K. Let $\alpha_1,\ldots,\alpha_n$ be the zeros of f.

When $f(x) = 0$ is solvable by radicals, let $K_1 = K \subset K_2 \subset \ldots \subset K_s$ be a solving chain of fields and let $K_{i+1} = K_i(\beta_i)$ with $\beta_i^{p_i} \in K_i$. Then by Theorem 14, K_2/K_1 is a normal extension of degree p_1. Hence by Theorem 9 the group of f over K_2 is a normal subgroup N of G_f. By Theorem 8

$$|G_f| = [K(\alpha_1,...,\alpha_n):K] | [K_2(\alpha_1,...,\alpha_n):K] = |N| [K_2:K].$$

The index of N in G_f therefore equals 1 or p_1. In the latter case we set $U_2 = N$. The same considerations apply to the extensions K_3/K_2,... which follow. Since f has a trivial group over K_s by hypothesis, it follows that G_f is solvable.

Now suppose that G_f is solvable, and let

$$U_1 = G_f \supset U_2 \supset ... \supset U_s = \{1\}$$

be a sequence of subgroups such that U_{i+1} is a normal subgroup of U_i with prime index p_i for $i = 1,...,s-1$. By Theorem 11 there is a polynomial $\theta = h(\alpha_1,...,\alpha_n)$ such that U_2 is the group of f over $K(\theta)$. Let f_1 be the minimal polynomial of θ over K. By Theorem 10 the remaining zeros of f_1 are of the form $\pi\theta := h(\pi\alpha_1,...,\pi\alpha_n)$ with $\pi \in G_f$. Since U_2 is a normal subgroup of G_f, f likewise has group U_2 over $K(\pi\theta)$, by Theorem 9. Hence, by Theorem 7, $\pi\theta$ lies in $K(\theta)$. Thus by Theorem 8 the group of f_1 over K has order p_1. We find ourselves in the situation of Theorem 13, i.e. there is a β_1 with $\beta_1^{p_1} \in K$ and $K(\theta) = K(\beta_1)$. We set $K_2 := K(\beta_1)$. Now we can apply the same argument to $U_3 \subset U_2$. Proceeding in this way, we obtain the desired chain of fields $K_1 = K \subset K_2 \subset ... \subset K_s = K(\alpha_1,...,\alpha_n)$. □

7.11 Permutation groups

To apply the general theory of Galois just described, it is necessary to become acquainted with permutations and permutation groups.

Since the objects which are permuted (in Galois theory the roots of the equation) do not matter, it is usual to take them to be the numbers 1 to n.

A transparent representation of a permutation π of n numbers is obtained with the cycle decomposition: one begins with an arbitrary number a_{11} from $\{1,...,n\}$, say 1. After it one writes the numbers $a_{12} = \pi a_{11}$,..., until one comes back to the first number $a_{11} = \pi^{n_1} a_{11}$. This gives the first cycle $(a_{11} a_{12} ... a_{1n_1})$. Now one chooses an arbitrary

a_{21} from the remaining numbers and continues the process. One obtains the cycle $(a_{21}a_{22}...a_{2n_2})$, etc.. Finally one comes to a representation

$$\pi = (a_{11}a_{12}...a_{1n_1})...(a_{s1}a_{s2}...a_{sn_s}). \tag{12}$$

The cycles of length 1 are usually omitted. The identity permutation is simply written (1).

The representation (12) yields the following:

1. The order of π is the lowest common multiple of the cycle lengths $n_1,...,n_s$.

2. For any $\varepsilon \in S_n$,

$$\varepsilon\pi\varepsilon^{-1} = (\varepsilon a_{11}\varepsilon a_{12}...\varepsilon a_{1n_1})...(\varepsilon a_{s1}\varepsilon a_{s2}...\varepsilon a_{sn_s}).$$

In particular, conjugation does not alter the type $\{n_1,...,n_s\}$ of π.

We now consider the subgroups of S_n. By Example 1 and Theorem 16, the general equation of nth degree is solvable if and only if S_n is solvable.

S_3 consists of six permutations (1), (12), (13), (23), (123), (132). The permutations (1), (123), (132) constitute the alternating group A_3. The sequence $S_3 \supset A_3 \supset \{(1)\}$ is a composition series for S_3. It is clear from this, in conjunction with Section 7.10, that the general equation of third degree is solvable by adjunction of the cube roots of unity, using the extraction of a square root and finally a cube root. This corresponds to the Cardano formula (cf. Section 7.1).

In S_4 we have the normal subgroup $K_4 = \{(1),(12)(34),(13)(24),(14)(23)\}$ and hence the composition series $S_4 \supset A_4 \supset K_4 \supset \{(1),(12)(34)\} \supset \{(1)\}$. This explains the solution given above for the general equation of fourth degree.

In general one goes from S_n to A_n by adjoining the square root of the discriminant of the equation (see (9)).

A_5 consists of 20 permutations of type $\{3\}$, 15 permutations of type $\{2,3\}$, 24 permutations of type $\{5\}$. and the identity permutation. We want to show that, apart from itself, A_5 contains only $\{(1)\}$ as a normal subgroup. A group with this property is called *simple*. Suppose N is a normal subgroup of A_5 different from $\{(1)\}$. Any element of N can be carried to any other of the same type by conjugation with elements of A_5 and raising to a power. Since the order of N is a divisor of 60, it follows from this that $N = A_5$. More generally, one can easily show that the groups A_n are simple for $n \geq 5$. These considerations yield the theorem of Abel:

Theorem 17. *The general equation of* n^{th} *degree is not solvable by radicals for* $n \geq 5$.

Proof. If a group G is solvable then so is each subgroup of G. Since A_5 can be embedded in S_n for $n \geq 5$, S_n cannot be solvable. □

To conclude this section we prove the following theorem, which was first proved by Cauchy in his work "*Mémoire sur le nombre des valeurs qu'une fonction peut acquérir lorsqu'on y permute de toutes les manières possibles les quantités qu'elle renferme*". (*J. Ec. Polyt*. 17 (1815)), and used by Galois.

Theorem 18. *If the order of a finite group* G *is divisible by a prime number* p, *then* G *contains an element of order* p.

Proof. The theorem is proved by induction on the order of the group. For $|G| = 1$ there is nothing to prove. Suppose that the theorem is already proved for all proper subgroups of G. When one of these subgroups has an order divisible by p then we are finished. Suppose this is not the case and let $g \in G$. The group $H_g = \{h \in G \mid hgh^{-1} = g\}$ is called the *centraliser* of g. The elements g with $H_g = G$ form an abelian subgroup Z of G, the *centre*. If $H_g \neq G$, then by hypothesis p is a divisor of $[G:H_g]$, and the class $K_g = \{g'gg'^{-1} \mid g' \in G\}$ of elements conjugate to g has $[G:H_g]$ members. Now since G decomposes into disjoint classes of conjugate elements, p is a divisor of $|Z|$. Being an abelian group, Z therefore has an element of order p. □

7.12 On irreducible equations of prime degree

We want to prove the following theorem of Galois, which was mentioned in Section 7.4

Theorem 19. *An irreducible equation of prime degree* p *is solvable by radicals if and only if all its solutions* $\alpha_1,\ldots,\alpha_p$ *are rationally expressible in terms of an arbitrary pair of solutions.*

Proof. By Theorem 12 and Theorem 16, Theorem 19 is equivalent to the following purely group-theoretic assertion:

Theorem 20. *A transitive permutation group* G *on* p *symbols is solvable if and only if* G *contains no permutation, apart from the identity, which leaves two symbols fixed.*

To prove Theorem 20 we first prove two further theorems.

Theorem 21. *A permutation group* G *on* p *symbols is transitive if and only if* p *divides the order of* G.

Proof. If p divides the order of G, then G contains a permutation π of order p by Theorem 18. This permutation is necessarily of the form $(a_1a_2\ldots a_p)$, and any given symbol can be carried to any other by a power of π, i.e. G is transitive.

Conversely, suppose that G is transitive and that G_1 is the subgroup of all permutations in G which leave 1 fixed. Also, let π_a be a permutation in G which carries 1 to a, for $a = 1,\ldots,p$. Such a permutation exists because G is transitive. Then

$$G = \bigcup_{a=1}^{p} \pi_a G_1$$

is a decomposition of G into disjoint cosets of G modulo G_1. It follows that $|G|/|G_1| = p$. □

We introduce the following notations. The symbols $1,\ldots,p$ are identified with their classes in $\mathbb{Z}/p\mathbb{Z}$. The *one-dimensional affine group* $A_1(\mathbb{Z}/p\mathbb{Z})$ *over* $\mathbb{Z}/p\mathbb{Z}$ is the group of all permutations in S_p with

$$\pi(k) = a + bk \quad \text{for} \quad k \in \mathbb{Z}/p\mathbb{Z},$$

where a,b are elements of $\mathbb{Z}/p\mathbb{Z}$ with $b \neq 0$.

One easily checks that the order of π is p for $b = 1$, and for $b \neq 1$ it equals the order of b in the multiplicative group $(\mathbb{Z}/p\mathbb{Z})^{\times}$ of $\mathbb{Z}/p\mathbb{Z}$.

Theorem 22. *A transitive permutation group on* p *symbols is solvable if and only if it is conjugate to a subgroup of* $A_1(\mathbb{Z}/p\mathbb{Z})$.

Proof. The group $A_1(\mathbb{Z}/p\mathbb{Z})$ is solvable, because it contains the cyclic normal subgroup generated by (12...p), and the quotient by this normal subgroup is isomorphic to $(\mathbb{Z}/p\mathbb{Z})^{\times}$.

Conversely, suppose G is a solvable transitive permutation group on $\mathbb{Z}/p\mathbb{Z}$. By Theorem 18 and Theorem 21, G contains an element of order p. Let

$$N_0 = \{(1)\} \subset N_1 \subset \ldots \subset N_s = G$$

be a composition series for G and let N_i be the smallest group in this series such that $p \,\big|\, |N_i|$. Then N_i contains a permutation of the form $(a_1 \ldots a_p)$. Since N_{i-1} is a normal subgroup of N_i, it follows that $N_{i-1} = \{(1)\}$ or N_{i-1} is transitive. But in the second case p would be a divisor of $|N_{i-1}|$ by Theorem 21, contrary to hypothesis. This implies $i = 1$. Hence $|N_1| = p$. We can assume without loss of generality that N_1 is generated by $\pi = (12\ldots p)$.

Now suppose $\pi \in N_2$. Then

$$\pi'\pi = \pi^b\pi' \tag{13}$$

for some number b, i.e.

$$\pi'(k+1) = \pi'(k) + b \quad \text{for } k \in \mathbb{Z}/p\mathbb{Z},$$

so $\pi'(k) = \pi'(0) + kb$, i.e. $N_2 \subset A_1(\mathbb{Z}/p\mathbb{Z})$.

As remarked above, the only elements of order p in $A_1(\mathbb{Z}/p\mathbb{Z})$ are $\pi, \pi^2, \ldots, \pi^{p-1}$. Hence for $\pi' \in N_3$ there is an exchange relation with π analogous to (13). From this one gets $N_3 \subset A_1(\mathbb{Z}/p\mathbb{Z})$. By continuing this process one finally obtains $G \subset A_1(\mathbb{Z}/p\mathbb{Z})$. □

We now come to the proof of Theorem 20. Suppose G is solvable. Then by Theorem 22, G is conjugate to a subgroup of $A_1(\mathbb{Z}/p\mathbb{Z})$. The group $A_1(\mathbb{Z}/p\mathbb{Z})$ contains no element, except the identity, which fixes two symbols. Suppose conversely that G is a transitive permutation group on p symbols which contains no element, apart from the identity, which fixes two symbols i, j. We let H denote the group of all permutations in S_p which fix i and j. H has order $(p-2)!$. Obviously G can contain at most one element from each coset of H in S_p. Therefore $|G| \leq p(p-1)$. By Theorem 18 and Theorem 21, G contains an element π of order p. Without loss of generality we can assume that $\pi = (12...p)$. Apart from the powers of π there can be no further elements of order p in G, otherwise G would have order at least p^2. If $\pi' \in G$ then $\pi'\pi\pi'^{-1}$ is a cycle of length p, and hence a power of π. From this one concludes as above that $\pi' \in A_1(\mathbb{Z}/p\mathbb{Z})$. □

The results of Galois expressed in the latter few theorems can be regarded as an interesting extension of the problem of solving the general n^{th} degree equation by radicals in the case where n is a prime p. As follows from the structure of $A_1(\mathbb{Z}/p\mathbb{Z})$, the solutions of an irreducible equation of prime degree, if it is solvable by radicals at all, may be expressed by a formula analogous to the Cardano formula.

7.13 Equations of fifth degree with symmetric group

We now consider the case of the fifth degree equation in more detail.

Theorem 23. *A nonsolvable transitive subgroup of* S_5 *is equal to* S_5 *or* A_5.

Proof. Let U be such a subgroup. Without loss of generality we can assume that U contains the permutation $\pi = (12345)$. By Theorem 20, U contains an element $\eta \neq (1)$ which fixes two symbols. Thus η is a cycle of length 2 or 3. In the first case one easily shows that U contains all cycles of length 2 and hence equals S_5. In the second case let $\pi' = \eta\pi\eta^{-1}$. Then π' cannot be a power of π, otherwise the group generated by π and η would be solvable, and thus η would not be able to fix two symbols. Therefore, one obtains 25 different elements in U in the form $\pi^i\pi'^j$ for $i,j = 0,1,2,3,4$. Thus U contains A_5, or a subgroup of A_5 of order 30. The latter subgroup would be a normal subgroup of A_5, which is impossible since A_5 is simple. □

One can show, in fact, that each group G with $|G| < 60$ is solvable.

We want to construct fifth degree polynomials over $\mathbb{Q}$ whose groups equal S_5. By Theorem 19, the group of an irreducible polynomial f of degree 5 with rational coefficients is certainly nonsolvable when f has three real and two complex zeros. By Theorem 23, f then has the group A_5 or S_5. One can decide between these two possibilities with the help of the discriminant of f. The first case occurs precisely when this is a square in $\mathbb{Q}$. As one easily sees, f has three real and two complex zeros precisely when the discriminant of f is negative (cf. (9)). Thus polynomials with group S_5 are always available. The discriminant of $f(x) = x^5 + ax + b$ is $4^4a^5 + 5^5b^4$. If one makes sure that f is an Eisenstein polynomial (Chapter 3, Theorem 3), then f is irreducible. In this way one easily finds a polynomial with group S_5 over $\mathbb{Q}$, e.g. $x^5 - 6x + 2$.

Exercises

7.1 Compute the discriminant of general equation of third and fourth degree.

7.2 Let $\alpha_1, \alpha_2, \ldots, \alpha_n$ be the zeros of the polynomial f. Show that the discriminant $D(f)$ of f has the form

$$D(f) = (-1)^{n(n-1)/2} \prod_{i=1}^{n} f'(\alpha_i).$$

7.3 Show that the polynomial $x^n + ax + b$ has the discriminant

$$(-1)^{(n-1)(n-2)/2}(n-1)^{n-1}a^n + (-1)^{n(n-1)/2}n^nb^{n-1}.$$

(Hint: First consider the polynomials $x^n - 1$ and $x^n - x$.)

7.4 Clarify the connection between the discriminant of equation (4) and the polynomial $x^3 + b_1x^2 + b_2x + b_3$. Compute h_1, h_2, h_3.

7.5 Express $\cos 3\alpha$ as a polynomial in $\cos \alpha$ and hence derive the solution of the equation $f(x) := x^3 + px + q = 0$ by angle trisection for the case where $f(x)$ has three real zeros. Also derive this solution from the Cardano formula with the help of the de Moivre formula.

7.6 Let p be a prime number and let K be a field with characteristic different from p. Show that the polynoial $x^p - a \in K[x]$ is reducible if and only if a is a pth power in K.

7.7 Let $x_1,\dots,x_n$ be indeterminates, let $s_1,\dots,s_n$ be the elementary symmetric functions of $x_1,\dots,x_n$ and let

$$p_i = x_1^i + \dots + x_n^i, \; i = 1,2,\dots .$$

Prove the Newton formulae

$$p_i - p_{i-1}s_1 + p_{i-2}s_2 - \dots + (-1)^{i-1}p_1 s_{i-1} + (-1)^i i s_i = 0 \quad \text{for} \quad i \le n$$

and

$$p - p_{i-1}s_1 + p_{i-2}s_2 - \dots + (-1)^n p_{i-n} s_n = 0 \quad \text{for} \quad i > n.$$

7.8 Let p be a prime number and let K be a field with characteristic different from p. Compute the Galois group of the polynomial $x^p - a \in K[x]$.

7.9 Let $f(x) \in \mathbb{Q}[x]$ be an irreducible polynomial of third degree with three real zeros. Show that $f(x) = 0$ is not solvable by radicals; with a sequence (6) of fields $K_1 = K, K_2, \dots, K_s$ which all lie in $\mathbb{R}$.

7.10 Use Theorem 7 to prove Theorem 2 in the framework of Galois theory.

7.11 Show that the Galois group of the polynomial $x^3 + x^2 - 2x - 1 \in \mathbb{Q}[x]$ is cyclic of order 3.

7.12 Show that the Galois group of the polynomial $x^5 + x^4 - 4x^3 - 3x^2 + 3x + 1 \in \mathbb{Q}[x]$ is cyclic of order 5.
(Hint: Substitute $x = y + 1/y$.)

8. The beginnings of complex function theory

8.1 Introduction: from a letter of Gauss to Bessel

The beginning of the 19th century marked the transition from the study of real functions to the study of complex functions. In a letter to Bessel on 18 December 1811 Gauss wrote:

"*I would first like to ask anyone who wants to introduce a new function into analysis to explain whether he wishes it to be applied merely to real quantities (real values of the argument of the function), and regards imaginary values of the argument only as an appendage, or whether he agrees with my thesis that in the realm of quantities the imaginaries* $a + b\sqrt{-1} = a + bi$ *have to be accorded equal rights with the reals. Here it is not a question of practical value; analysis is for me an independent science, which would suffer serious loss of beauty and completeness, and would have constantly to impose very tiresome restrictions on truths which would hold generally otherwise, if these imaginary quantities were to be neglected .... Now what is one to think of* $\int \phi x.dx$ *for* $x = a + bi$? *Obviously, if one wants to start from clear concepts, then one must assume that* x *changes by infinitely small increments (each of the form* $\alpha + \beta i$*) from that value for which the integral is to be* 0 *to* $x = a + bi$, *and then sum all the* $\phi x\, dx$. *This makes the meaning completely definite. However, the transition can take place in infinitely many ways: just as one can think of the realm of all real quantities as (represented by) an infinite straight line, so one can make sense of the realm of all quantities, real and imaginary, as an infinite plane in which each point, determined by abscissa* a *and ordinate* b, *represents the quantity* $a + bi$. *The continuous transition from one value of* x *to another* $a + ib$ *accordingly takes place along a curve, and hence is possible in infinitely many ways.*

I now claim that the integral $\int \phi x.dx$ *always has the same value for two different transitions if* ϕx *never becomes* ∞ *in the region enclosed between the two curves representing the transitions. This is a beautiful theorem (strictly speaking, one needs the additional assumption that* ϕx *is a single-valued function of* x, *or at least one whose value within each region is one continuously varying member of a system of values), whose proof is not difficult, as I shall show in a suitable occasion. It is connected with other beautiful truths concerning series expansions. The transition to each point may always be carried out without touching a place where* $\phi x = \infty$. *I therefore demand that such points be avoided, since the original concept of* $\int\phi x.dx$ *obviously loses clarity at such points, and easily leads to contradictions. Incidentally, this also shows clearly how a function defined by* $\int\phi x.dx$ *can have many values for a single argument* x, *since in making a transition one can wind around a point where* $\phi x = \infty$, *once, several times, or not at all. E.g. if one defines* log x *by* $\int\frac{1}{x}\,dx$ *from* $x = 1$, *then one comes to* log x *either without winding around the point* 0, *or by winding one or more times, each time adding the constant* $+ 2\pi i$ *or* $- 2\pi i$. *In this way the multiple logarithms of each number are completely clear. If* ϕx *cannot become infinite for a finite value of* x, *then the integral is always a single-valued function."*

Gauss never carried out this program for complex function theory. Rather, it was Cauchy who, beginning in the year 1814, developed the beginnings of complex function theory in over 200 works.

In his appreciation of the mathematical works of Weierstrass (*Acta Math.* 22 (1899)), Poincaré named Gauss, Cauchy, Riemann and Weierstrass as the founders of the "*modern*" theory of analytic functions.

Gauss did not publish his results, however he significantly influenced Riemann. Cauchy's theory already contains the germ of the geometric conception of Riemann (Chapters 10, 11) and the arithmetic conception of Weierstrass. In contrast to Riemann, Weierstrass wished to do without all geometric intuition, and based his theory on the concept of power series alone. However, his method is often arduous, and we prefer, following the textbooks of function theory, to prove all the results of this chapter by Cauchy's method.

8.2 Basic concepts

Let $z = x + iy$ be a complex number with real part $\text{Re } z := x$ and imaginary part $\text{Im } z := y$. We interpret z as the point (x,y) in the euclidean plane. The quantity $|z| = \sqrt{x^2+y^2}$ is called the *absolute value* of z. Each complex number $z \neq 0$ has a unique representation $z = re^{i\phi}$, where $r = |z|$ is positive and $0 \leq \phi < 2\pi$. The angle ϕ is called the *argument* of z. The field of all complex numbers is denoted by $\mathbb{C}$.

Suppose $z_1, z_2 \in \mathbb{C}$. A *(directed) simple smooth curve* C from initial point z_1 to final point z_2 is given by a continuously differentiable function z(t) for $t \in [a,b]$, with complex values and $z(a) = z_1$, $z(b) = z_2$, which is one-to-one in the open interval (a,b). We call t the *parameter* of C. When $t(t')$ is a continuously differentiable one-to-one mapping of the interval $[a',b']$ onto the interval $[a,b]$ with $dt/dt' > 0$ the curves given by $z(t)$ and $z_1(t') := z(t(t'))$ are considered to be essentially the same. By a (piecewise smooth) curve C we mean a sequence of smooth simple curves $C_1,\dots,C_n$ where the final point of C_i equals the initial point of C_{i+1} for $i = 1,\dots,n-1$. When these curves have no other points in common, except the final point z_{n+1} of C_n and the initial point z_1 of C_1, C is called *simple.* When $z_{n+1} = z_1$, C is called a *closed* curve.

Let ε be a positive real number and let $z_0 \in \mathbb{C}$. Then $K_\varepsilon(z_0) = \{z \in \mathbb{C} \mid |z-z_0| < \varepsilon\}$ is called the ε-neighbourhood of z_0. $K_\varepsilon(z_0)$ is bounded by the curve $C_\varepsilon(z_0)$ given parametrically by $z(t) = z_0 + \varepsilon\rho^{it}$ for $t \in [0,2\pi]$. A subset U of $\mathbb{C}$ is called *open* when it contains an ε-neighbourhood of each of its members z_0. $U \subset \mathbb{C}$ is called *(arcwise) connected* when any two points $z_1,z_2 \in U$ can be connected by a curve in U. Finally, U is called *simply connected* when U is connected and each closed curve in U is contractible in U to a point. A connected open set in $\mathbb{C}$ is called a *region.*

A simple closed curve divides the plane into two regions. This Jordan curve theorem will not be proved here. In all applications the curve will be of a form for which the theorem is obvious.

A complex-valued function $f(z)$ which is defined in a region U of the plane is called *differentiable* at $z_0 \in U$ when, for all sequences $z_1,z_2,\dots$ converging to z_0, the limit value

$$\lim_{z_n \to z_0} \frac{f(z_n)-f(z_0)}{z_n-z_0}$$

exists and has the same value, which is denoted by $f'(z_0)$ or $\frac{df}{dz}(z_0)$.
When $f(z)$ is continuously differentiable over all of U then $f(z)$ is called *regular* or *holomorphic* in U.

Goursat showed in 1884 that differentiability of $f(z)$ in U implies the continuity of $f'(z)$ in U (*Acta Math.* 4 (1884)).

Let $u(x,y)$ and $v(x,y)$ be the real and imaginary parts of $f(z)$, i.e. $f(x+iy) = u(x,y) + iv(x,y)$.

Theorem 1. *The following conditions are equivalent:*

a) $f(z)$ *is regular in* U.

b) *The first order partial derivatives of* u *and* v *exist and are continuous in* U. *They satisfy the Cauchy-Riemann differential equations*

$$\frac{\partial u}{\partial x} = \frac{\partial v}{\partial y}, \quad \frac{\partial u}{\partial y} = -\frac{\partial v}{\partial x} \tag{1}$$

Proof. a) ⇒ b) follows immediately by passing to real and imaginary parts. Suppose that b) holds and that $z = x + iy$, $z_0 = x_0 + iy_0 \in U$. Then by Taylor's theorem

$$f(z) - f(z_0) = \left[\frac{\partial u}{\partial x}(z_0) + r_1(z)\right](x-x_0) + \left[\frac{\partial u}{\partial y}(z_0) + r_2(z)\right](y-y_0)$$

$$+ i\left[\frac{\partial v}{\partial x}(z_0) + r_3(z)\right](x-x_0) + i\left[\frac{\partial v}{\partial y}(z_0) + r_4(z)\right](y-y_0)$$

where the $r_i(z)$ are continuous functions which $\to 0$ as $z \to z_0$. Because of (1),

$$\frac{f(z)-f(z_0)}{z-z_0} = \frac{\partial u}{\partial x}(z_0) + i\frac{\partial v}{\partial x}(z_0) + \frac{(r_1(z)+ir_3(z))(x-x_0)}{z-z_0} + \frac{(r_2(z)+ir_4(z))(y-y_0)}{z-z_0}.$$

Since

$$\left|\frac{x-x_0}{z-z_0}\right| \le 1, \quad \left|\frac{y-y_0}{z-z_0}\right| \le 1$$

it follows that

$$\lim_{z\to z_0} \frac{f(z)-f(z_0)}{z-z_0} = \frac{\partial u}{\partial x}(z_0) + i\frac{\partial v}{\partial x}(z_0). \qquad \square$$

The equations (1) already occur in the mid-18th century with d'Alembert and Euler.

8.3 The line integral

A concept of decisive importance for Cauchy's function theory is that of the line integral (Section 8.1): let C be a curve in a region U, given by $z(t)$ for $t \in [a,b]$, and let f be a continuous function on U. Then we define

$$\int_C f(z)dx := \int_a^b f(z)\frac{dz}{dt}dt = \int_a^b \left[u\frac{dx}{dt}-v\frac{dy}{dt}\right]dt + i\int_a^b \left[u\frac{dy}{dt}+v\frac{dx}{dt}\right]dt. \tag{2}$$

This definition is independent of the choice of parameter t.

When a function $F(z)$ with $F'(z) = f(z)$ exists in the region U we have

$$\int_C f(z)dz = F(z(b)) - F(z(a)). \tag{3}$$

This follows easily from Theorem 1. In this case $\int_C f(z)dz$ is therefore independent of the choice of path from $z(a)$ to $z(b)$. An equivalent statement is that the integral of $f(z)$ over a closed curve vanishes. Conversely, the following Cauchy integral theorem, from which the remainder of Cauchy's function theory follows as a corollary, says something about the existence of an *antiderivative* $F(x)$.

Theorem 2. *Let* U *be a simply connected region in* $\mathbb{C}$ *and let* $f(z)$ *be a regular function in* U. *Also let* C *be a closed curve in* U. *Then* $\int_C f(z)dz = 0$.

Proof. First suppose C is simple. By the Gauss (or Green's) theorem (A.3)

$$\mathrm{Re}\int_C f(z)dz = \int_C (udx-vdy) = -\iint_{F(C)} \left[\frac{\partial u}{\partial y} + \frac{\partial v}{\partial x}\right]dx\, dy,$$

where $F(C)$ is the region enclosed by C. The integrand in the integral on the right vanishes by Theorem 1. Similarly one finds $\mathrm{Im}\int_C f(z)dz = 0$.

Now let C be an arbitrary closed curve, given by the parametric representation $z(t)$ for $t \in [a,b]$. Since C is piecewise smooth by definition, each $t \in [a,b]$ has an ε-neighbourhood, in the closure of which C can be replaced by a finite polygon in U, without altering the value of the integral. Since $[a,b]$ can be covered by finitely many such ε-neighbourhoods, it follows that C can be replaced by a polygon without altering the value of the integral.

By finite subdivision of the polygon edges one can also arrange that two edges either consist of the same points or else have at most an initial or final point in common. If C contains two successive edges which differ only in orientation, then one can omit them without altering the value of the integral. The same is true of any simple closed subcurve of C. The assertion of Theorem 2 now follows by induction on the number of edges in C.

We now give some consequences of the Cauchy integral theorem.

8.4 The Cauchy integral formula

Theorem 3. *(Cauchy integral formula). Let* f(z) *be regular in the simply connected region* U *and let* C *be a positively traversed circle in* U *with centre* z. *Then*

$$f(z) = \frac{1}{2\pi i}\int_C \frac{f(\zeta)d\zeta}{\zeta - z}, \tag{4}$$

and the nth *derivative* $f^{(n)}(z)$ *of* f(z) *satisfies*

$$f^{(n)}(z) = \frac{n!}{2\pi i}\int_C \frac{f(\zeta)d\zeta}{(\zeta - z)^{n+1}}. \tag{5}$$

In particular, f(z) *has derivatives of arbitrarily high order.*

Proof. Let C_ρ be a positively traversed circle of sufficiently small radius ρ and centre z. We reduce the annulus between C and C_ρ to a simply-connected region by a cut A. The latter region and its boundary may be embedded in a region in which $f(\zeta)/(\zeta - z)$ is regular as a function of ζ (Figure 2). One can therefore apply Theorem 2 to the

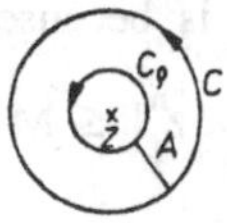

Fig. 2

boundary and obtain

$$\int_C \frac{f(\zeta)d\zeta}{\zeta - z} = \int_{C_\rho} \frac{f(\zeta)d\zeta}{\zeta - z} = \int_{C_\rho} \frac{f(z)d\zeta}{\zeta - z} + \int_{C_\rho} \left[\frac{f(\zeta)-f(z)}{\zeta - z}\right] d\zeta.$$

For any $\varepsilon > 0$, $|f(\zeta)-f(z)| < \varepsilon$ for all ζ on C_ρ when ρ is sufficiently small. With $\zeta = z + \rho e^{it}$ we therefore have

$$\left|\int_{C_\rho}\left[\frac{f(\zeta)-f(z)}{\zeta-z}\right]d\zeta\right| = \left|\int_0^{2\pi}(f(\zeta)-f(z))dt\right| \le 2\pi\varepsilon.$$

Also

$$\int_{C_\rho}\frac{f(z)}{\zeta-z}d\zeta = \int_0^{2\pi} f(z)i\ dt = 2\pi i f(z).$$

This implies (4).

One obtains (5) from (4) by differentiation with respect to z under the integral sign. This procedure is justified by reference to the interchangeability of limit and integral for a uniformly continuous integrand. □

8.5 Power series expansion

We first consider the convergence behaviour of the power series

$$\sum_{n=0}^{\infty} a_n z^n, \tag{6}$$

where $a_0, a_1, \ldots$ are arbitrary complex numbers.

Suppose (6) is convergent for $z = z_1$. Then (6) is absolutely and uniformly convergent for $|z| \le r < |z_1|$. This is because $\lim_{n\to\infty} a_n z_1^n = 0$ by hypothesis. Hence there is a positive number M with $|a_n z_1^n| \le M$ for all n, and for natural numbers h

$$\sum_{n=h}^{\infty} |a_n||z|^n \le \sum_{n=h}^{\infty} M\left|\frac{z}{z_1}\right|^n \le \sum_{n=h}^{\infty} Mr_1^n = M\frac{r_1^h}{(1-r_1)}$$

with $r_1 = r/|z_1| < 1$. This yields the assertion.

The supremum R of all $|z_1|$ for which $\sum_{n=0}^{\infty} a_n z_1^n$ is convergent is called the *radius of convergence* of (6). Thus (6) is convergent for $|z| < R$ and divergent for $|z| > R$.

Theorem 4. *(Power series expansion). Let* f(z) *be regular in* $K_\rho(z_0)$. *Then the power series*

$$\sum_{n=0}^{\infty} \frac{f^{(n)}(z_0)}{n!}(z-z_0)^n \tag{7}$$

is convergent for $z \in K_\rho(z_0)$ *and equals* f(z).

Proof. Assume without loss of generality that $z_0 = 0$. Let ζ be a point on a circle C about 0 with radius ρ', with $|z| < \rho' < \rho$. We have

$$\frac{f(\zeta)}{\zeta-z} = f(\zeta)\frac{1}{\zeta}(1-\frac{z}{\zeta})^{-1} = \sum_{n=0}^{\infty} f(\zeta)\frac{z^n}{\zeta^{n+1}}.$$

The series on the right-hand side is uniformly convergent in $\zeta \in C$. Hence by Theorem 3

$$2\pi i f(z) = \int_C \frac{f(\zeta)}{\zeta-z}d\zeta = \sum_{n=0}^{\infty} z^n \int_C \frac{f(\zeta)}{\zeta^{n+1}} = 2\pi i \sum_{n=0}^{\infty} \frac{f^{(n)}(0)}{n!}z^n. \qquad \square$$

By Theorem 4 a regular function in $K_\rho(z_0)$ is uniquely determined by its values in an arbitrarily small neighbourhood of z_0. More generally, we have

Theorem 5. *Let* $f_1(z)$ *and* $f_2(z)$ *be functions, regular in a region* U, *and let* $f_1(z) = f_2(z)$ *over a subregion* U′. *Then* $f_1(z) = f_2(z)$ *over the whole of* U.

Proof. Let z_1 be an arbitrary point of U and let $z_0 \in U'$. Also, let z(t) for $t \in [a,b]$ with $z(a) = z_0$, $z(b) = z_1$ define a curve in U which connects z_0 to z_1. We set $v(z) := f_1(z) - f_2(z)$. Let T be the set of $t \in [a,b]$ with $v(z) = 0$ for all z in a neighbourhood of z(t), let $t_1 = \sup T$ and let U_ε be an ε-neighbourhood of $z(t_1)$ which lies in U. We choose $t_2 \in T$ so near to t_1 that $|z(t_1)-z(t_2)| < \varepsilon/2$. Then the circle K about $z(t_2)$ with radius $\varepsilon/2$ is contained in U, and $z(t_1)$ lies inside K. Hence it follows from Theorem 4 that v(z) vanishes in K. That means $t_1 = b$ and $v(z_1) = 0$. $\square$

8.6 Coefficient estimates

Theorem 6. *(Cauchy's inequality). Let* $f(z) = \sum_{n=0}^{\infty} a_n z^n$ *be a function, regular for* $z \in C_\rho = C_\rho(0)$, *and let*

$$M(r) = \max_{|z|=r} |f(z)|$$

for positive $r < \rho$. *Then*

$$|a_n| \leq \frac{M(r)}{r^n}.$$

Proof. By Theorems 3 and 4 with $\zeta = re^{it}$ we have

$$|a_n| = \left|\frac{f^{(n)}(0)}{n!}\right| = \frac{1}{2\pi}\left|\int_{C_r} \frac{f(\zeta)d\zeta}{\zeta^{n+1}}\right| = \frac{1}{2\pi}\left|\int_0^{2\pi} \frac{f(\zeta)dt}{\zeta^n}\right| \leq \frac{1}{r^n}M(r). \qquad \square$$

From Theorem 6 one easily obtains the absolute convergence of (7) and the following theorem of Liouville.

Theorem 7. *Let* $f(z)$ *be regular and bounded over the whole plane. Then* $f(z)$ *is constant.*

Proof. Let $|f(z)| \leq M$. Then $|a_n| \leq M/r^n$ for all $r > 0$ and hence $a_n = 0$ for $n = 1,2,\ldots$. □

Liouville's theorem yields the *fundamental theorem of algebra* (Section 7.1).

Theorem 8. *Let* $f(z)$ *be a non-constant polynomial. Then* $f(z)$ *has a zero.*

Proof. If $f(z)$ has no zero, then $1/f(z)$ is regular and bounded over the whole plane, hence constant by Theorem 7. □

8.7 The maximum-modulus principle

Theorem 9. *Let* $f(z)$ *be regular in the region* U, *and let*

$$M := \sup_{z \in U} |f(z)| < \infty.$$

Then f(z) *is constant, or else*

$$|f(z)| < M \quad \textit{for all} \quad z \in U.$$

Proof. Let $z_0 \in U$ with $|f(z_0)| = M$, let ρ be the radius of a circle about z_0 which lies in U, and let $0 < r \le \rho$. Then by Theorem 6

$$M = |f(z_0)| \le \max_{|z-z_0|=r} |f(z)| \le M,$$

hence $M = |f(z)|$ for all z in $K_\rho(z_0)$. One concludes from this that $|f(z)|$ is constant. Suppose $M > 0$. Then we consider log f(z) in a sufficiently small neighbourhood of z_0. This function has a constant real part, and hence by the Cauchy-Riemann equations (1) it is itself constant. Hence f(z) is constant in a sufficiently small neighbourhood of z_0, and thus by Theorem 5 it is constant over the whole of U. □

8.8 The Laurent expansion

A series of the form

$$\sum_{n=-\infty}^{\infty} a_n z^n := \sum_{n=0}^{\infty} a_n z^n + \sum_{n=1}^{\infty} a_{-n} z^{-n} \tag{8}$$

is called a *Laurent series.* When (8) converges for z_1, z_2 with $|z_1| < |z_2|$ then it converges absolutely and uniformly for all z with $|z_1| < r_1 \le |z| \le r_2 < |z_2|$ (cf. Section 8.5).

Theorem 10. *(Laurent expansion). Let* f(z) *be regular in an open annulus* K_{12} *with midpoint* z_0 *and radii* $r_1 < r_2$. *Then* f(z) *is uniquely representable in the form*

$$f(z) = \sum_{n=-\infty}^{\infty} a_n (z-z_0)^n,$$

where the coefficients a_n *are given by*

$$a_n = \frac{1}{2\pi i}\int_C \frac{f(\zeta)d\zeta}{(\zeta - z_0)^{n+1}} \tag{9}$$

and C *is a positively traversed circle in* K_{12} *with centre* z_0.

Proof. Without loss of generality suppose $z_0 = 0$. By Theorem 2, and with the notation of Fig. 3 we have

$$\begin{aligned} 2\pi i f(z) &= \int_{B_0} \frac{f(\zeta)}{\zeta - z}d\zeta \\ &= \int_{B_1} \frac{f(\zeta)}{\zeta - z}d\zeta - \int_{B_2} \frac{f(\zeta)}{\zeta - z}d\zeta \\ &= \int_{B_1} \sum_{n=0}^{\infty} f(\zeta)\frac{z^n}{\zeta^{n+1}}d\zeta + \int_{B_2} \sum_{n=-1}^{\infty} f(\zeta)\frac{z^n}{\zeta^{n+1}}d\zeta \\ &= \sum_{n=-1}^{\infty} z^n \int_C f(\zeta)\frac{d\zeta}{\zeta^{n+1}} . \end{aligned}$$

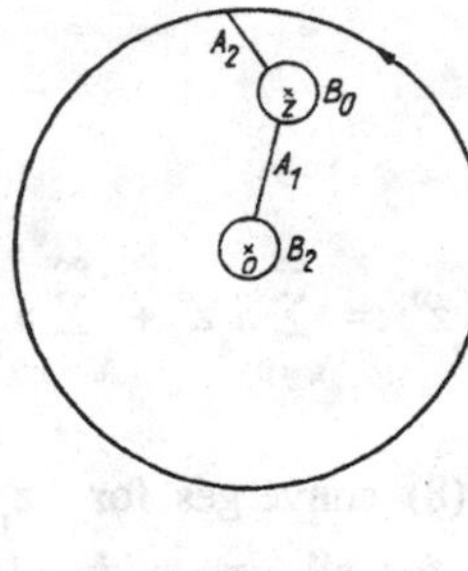

Fig. 3

Conversely, it follows from

$$f(z) = \sum_{n=-\infty}^{\infty} b_n z^n,$$

by the uniform convergence of this series, that

$$\int_C \frac{f(\zeta}{\zeta^{m+1}}d\zeta = \int_C \sum_{n=-\infty}^{\infty} b_n \zeta^{n-m-1}d\zeta = b_m \int_C \frac{d\zeta}{\zeta} = 2\pi i b_m . \qquad \square$$

Now suppose $f(z)$ is bounded in a neighbourhood U of the origin, and regular in $U - \{0\}$. Then $\lim_{z\to 0} f(z)$ exists by Theorem 10, and by taking $f(0) = \lim_{z\to 0} f(z)$ one obtains a function regular in U.

On the other hand, suppose $f(z)$ is regular in $U - \{0\}$ and has a Laurent expansion with infinitely many negative powers with non-zero coefficients. Then $f(z)$ comes arbitrarily close to each complex number in every neighbourhood of 0 (Casorati-Weierstrass theorem). If this were not the case for a $w \in \mathbb{C}$, then $1/(f(z)-w)$ would be bounded in a neighbourhood of 0 and regular outside $z = 0$, hence also regularly definable at 0. Consequently, we should have

$$\frac{1}{f(z)-w} = \sum_{n=n_0}^{\infty} a_n z^n \quad \text{with} \quad a_{n_0} \neq 0$$

and

$$f(z)-w = \sum_{n=-n_0}^{\infty} b_n z^n,$$

contrary to hypothesis.

A function $f(z)$ which is representable as

$$f(z) = \sum_{n=n_0}^{\infty} a_n (z-z_0)^n \quad \text{with} \quad a_{n_0} \neq 0$$

in the neighbourhood of a point z_0 is called *meromorphic* at z_0 of order $n_0 =: \nu_{z_0}(f)$. When n_0 is positive, $f(z)$ has an n_0*-tuple zero*, when n_0 is negative a $(-n_0)$*-tuple pole*. In the latter case one sets $f(z_0) = \infty$.

Theorem 11. *Let* $f(z)$ *be meromorphic in a region* U, *let* $z_1, z_2, \ldots$ *be a sequence of points in* U *which have an accumulation point in* U, *and let* $f(z_m) = 0$ *for* $m = 1,2,\ldots$. *Then* $f(z) = 0$ *in* U.

Proof. Assume without loss of generality that the accumulation point is 0 and that $z_1, z_2, \ldots \neq 0$. Suppose we had $f(z) = \sum_{n=n_0}^{\infty} a_n z^n$ with $a_{n_0} \neq 0$. Then $f_1(z) = f(z)z^{-n_0}$ would be regular and $\neq 0$ in a neighbourhood of 0, yet

$f_1(z_1) = f_1(z_2) = ... = 0$. This is a contradiction. Hence $f(z) = 0$ in a neighbourhood of 0. The assertion then follows by Theorem 5. □

It follows from Theorem 11 that a meromorphic function is uniquely determined by its values on a sequence of points which accumulates in the interior of its region of definition.

8.9 Residues

Suppose $f(z)$ is meromorphic in the neighbourhood of the point z_0 and let

$$f(z) = \sum_{n=n_0}^{\infty} a_n(z-z_0)^n.$$

Then a_{-1} is called the *residue* of $f(z)$ at the point z_0.

Theorem 12. *(Residue theorem). Let* $f(z)$ *be a meromorphic function in a region* U, *let* C *be a positively oriented closed curve in* U *which bounds a simply connected region* $\mathcal{F}(C)$, *and let* $f(z)$ *be regular in* U *except at points* $z_1,...,z_s$ *in the interior of* $\mathcal{F}(C)$ *where the residues are* $c_1,...,c_s$. *Then*

$$\int_C f(z)dz = \sum_{m=1}^{s} 2\pi i c_m.$$

Proof. This results from the Cauchy integral theorem (Theorem 2) and the Laurent expansion (see Figure 4 and the proof of Theorem 10). □

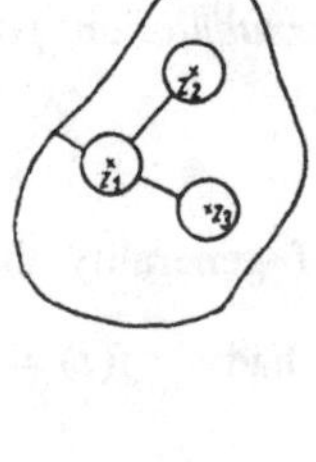

Fig. 4

Theorem 13. *With notation as in Theorem 12, suppose that* f(z) *is nonzero on* C. *Then*

$$\int_C (f'(z)/f(z))dz = 2\pi i \Sigma \nu_{z_0}(f), \quad \int_C (f'(z)/f(z))z dz = 2\pi i \Sigma z_0 \nu_{z_0}(f), \tag{10}$$

where the sums are taken over all zeros and poles z_0 *of* f *in* $\mathcal{F}$(C).

Proof. To prove (10) we simply observe that the functions $f'(z)/f(z)$ and $(f'(z)/f(z))z$ are regular in $\mathcal{F}$(C) except at the zeros and poles of f(z). At the latter they have simple poles with residues $\nu_{z_0}(f)$ and $z_0\nu_{z_0}(f)$ respectively. □

We use Theorem 13 to show the existence and uniqueness of implicit functions.

Theorem 14. *Let* U *and* V *be regions in* ℂ *and let* g(z,w) *be a function which is defined for all pairs* z,w *in* U × V *and regular as a function of* z *and* w. *Also let* z_0 *be a point in* U *for which* $g(z_0,w)$ *has a simple zero* w_0.

Then there are neighbourhoods K_δ *of* z_0 *and* K_ε *of* w_0 *such that the equation* g(z,w) = 0 *for* $z \in K_\delta$ *has a unique solution* w = f(z) *in* K_ε. *The function* f(z) *is regular in* K_ε.

Proof. There is an $\varepsilon > 0$ such that $g(z_0,w)$ has the single zero w_0 in $K_{2\varepsilon}(w_0)$ (Theorem 11). Then by the continuity of g(z,w) there is a $\delta > 0$ such that $g(z,w) \neq 0$ for $z \in K_\delta(z_0) = K_\delta$ and $w \in C_\varepsilon(w_0) = C_\varepsilon$. Let g_w be the partial derivative of g with respect to w. By Theorem 13,

$$n(z) = \frac{1}{2\pi i}\int_{C_\varepsilon} \frac{g_w(z,w)}{g(z,w)}dw$$

is the number of zeros of g(z,w) for fixed $z \in K_\delta$ in K_ε. Since $n(z_0) = 1$ and n(z) is a continuous function in K_δ, n(z) = 1 for all $z \in K_\delta$. This implies the first part of the assertion of Theorem 14. The unique solution of g(z,w) = 0 has the form

$$f(z) = \frac{1}{2\pi i}\int_C \frac{g_w(z,w)}{g(z,w)}w dw.$$

The integrand is a regular function of z. Therefore $f(z)$ is also regular for $z \in K_\delta$. □

8.10 The double series theorem

In dealings with series of regular functions the following *double series theorem* of Weierstrass is of great importance.

Theorem 15. *Let* $f_1(z), f_2(z), \ldots$ *be regular functions in a region* U, *and suppose*

$$f(z) = \sum_{n=1}^{\infty} f_n(z)$$

converges uniformly on each bounded closed subset of U. *Then* $f(z)$ *is a regular function in* U, *and one obtains the derivatives of* $f(z)$ *by termwise differentiation.*

Proof. By (5) we have

$$f_n^{(k)}(z) = \frac{k!}{2\pi i}\int_C \frac{f_n(\zeta)d\zeta}{(\zeta-z)^{k+1}}$$

for a circle C about z in U. By hypothesis

$$\sum_{n=1}^{\infty} f_n(\zeta)\frac{1}{(\zeta-z)^{k+1}} = f(\zeta)\frac{1}{(\zeta-z)^{k+1}}$$

is uniformly convergent for $\zeta \in C$. Therefore

$$\frac{k!}{2\pi i}\int_C \frac{f(\zeta)d\zeta}{(\zeta-z)^{k+1}} = \frac{k!}{2\pi i}\sum_{n=1}^{\infty}\int_C \frac{f_n(\zeta)d\zeta}{(\zeta-z)^{k+1}} = \sum_{n=1}^{\infty} f_n^{(k)}(z).$$

The left side of this equation is a regular function, the kth derivative of

$$\frac{1}{2\pi i}\int_C \frac{f(\zeta)d\zeta}{(\zeta-z)} = \sum_{n=1}^{\infty} f_n(z) = f(z). \qquad \square$$

One can apply Theorem 15 to a power series $f(z) = \sum_{n=0}^{\infty} a_n z^n$ with radius of convergence $\rho > 0$. It follows that $f(z)$ is regular for $|z| < \rho$, with derivative

$f'(z) = \sum_{n=0}^{\infty} na_n z^{n-1}$. Since, by Theorem 4, one has conversely that each function regular for $|z| < \rho$ admits a power series representation, the study of regular functions in the neighbourhood of a point z_0 is the same as the study of power series $\sum_{n=0}^{\infty} a_n(z-z_0)^n$ which converge in a neighbourhood of z_0. This was the starting point of Weierstrass (Section 8.1), and it explains the name "double series theorem" for Theorem 15.

8.11 Analytic continuation

Let U be a region of the complex plane and let $f_1(z)$ be a function regular in some subregion U_1 of U. Let $z_1 \in U_1$, $z_2 \in U - U_1$ and let C be a simple curve from z_1 to z_2. One says that $f_1(z)$ may be *analytically continued along* C when there is a subregion U_2 of U, containing C, and a function $f_2(z)$ regular in U_2 which agrees with f_1 in $U_1 \cap U_2$. By Theorem 5, $f_2(z)$ is uniquely determined.

Theorem 16. *(Monodromy theorem). Let* U *be a simply connected region, let* z_1 *be a point of* U *and let* $f_1(z)$ *be a function regular in a neighbourhood* U_1 *of* z_1. *Suppose the function* $f_1(z)$ *can be analytically continued along each simple curve which runs from* z_1 *to a point* z_2 *in* $U - U_1$. *Then there is a function* $f(z)$, *regular in* U, *which agrees with* $f_1(z)$ *in* U_1.

Proof. It suffices to construct $f(z)$ for all simply connected, bounded, closed subsets G of U which are squares. Let M be such a square with vertices v_1, v_2, v_3, v_4, let C_M be a curve from z_1 to v_1 (which is determined more precisely below), and let $v \in M$. Then we define $f_M(v)$ to be the continuation of $f_1(z)$ along the curve consisting of C_M and the line joining v_1 to v (Fig. 5). Since this

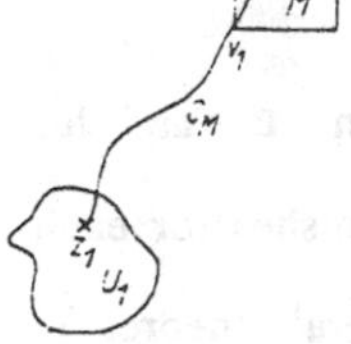

Fig. 5

continuation defines a unique regular function in a neighbourhood of v (Theorem 5), one obtains in this way a function f_M regular in M.

It remains to show that for any two squares M and M' the functions f_M and $f_{M'}$ agree on $M \cap M'$. Suppose this is already proved for a set $\mathfrak{M}$ of squares whose union G' is a connected subset of G, and let $f_{G'}$ be the resulting regular function on G'. Since G is simply connected, if $G' \neq G$ there is a further square M'' such that $G \cap M''$ is connected and not empty. We choose $C_{M''}$ so that $C_{M''}$ has a non-empty intersection with an $M \in \mathfrak{M}$ bordering M'' and so that the continuation of $f_1(z)$ along $C_{M''}$ in M agrees with $f_{G'}$. The continuation of $f_{G'}|_M$ along $M'' \cap G'$ must agree with $f_{G'}$ on the one hand, and with $f_{M''}$ on the other, whence it follows that $f_{G'}$ may be continued to $M'' \cup G'$. □

Exercises

8.1 Let $a_1, a_2, \ldots$ be a sequence of complex numbers. We set

$$l := \limsup \sqrt[n]{|a_n|}.$$

a) Let $l = \infty$. Show that the power series $a_0 + a_1 z + \ldots$ converges only for $z = 0$.

b) Let $l = 0$. Show that the power series $a_0 + a_1 z + \ldots$ converges for all complex numbers z.

c) Let l be finite and different from 0. Show that the power series $a_0 + a_1 z + \ldots$ has radius of convergence $1/l$.

8.2 Determine the radius of convergence of the power series $a_0 + a_1 z + \ldots$ when $a_n = n!/n^n$ and $a_n = n^{\log n}$.

8.3 Let U be a region in $\mathbb{C}$ and let $f(z)$ be a function continuous in U. Suppose the integral $\int_C f(z)dz$ vanishes for each closed curve C in U. Show that $f(z)$ is a regular function in U (Morera's theorem).

8.4 Compute the following integrals with the help of the residue theorem (Theorem 12):

$$\int_{-\infty}^{\infty} \frac{dx}{(1+x^2)^n}, \quad \int_0^{2\pi} \cos^{2n} x \, dx.$$

(Hint: In the second integral change the variable to $\zeta := e^{ix}$).

8.5 Show that the power series $\sum_{n=0}^{\infty} z^{2^n}$ cannot be analytically continued past the unit circle.

9. Entire functions

9.1 Functions with finitely many singular points

In his work "*Zur Theorie der eindeutigen analytischen Funktionen*" (*Abh. Preuss. Akad. Wiss.* 1876), Weierstrass investigated the class of functions which are regular over the whole plane except at finitely many points. When such a function has no singular points at all, it is called *entire*.

Theorem 1. *Let* $f(z)$ *be a function which is regular over the whole plane except at the points* $a_1,...,a_m$. *Then there are* $m+1$ *entire functions* $f_0(z), f_1(z),...,f_m(z)$ *with*

$$f(z) = f_0(z) + f_1\left(\frac{1}{z-a_1}\right) + ... + f_m\left(\frac{1}{z-a_m}\right).$$

Proof. By Section 8.8, $f(z)$ has the form $f(z) = \sum_{n=-\infty}^{\infty} b_n(z-a_m)^n$ in a neighbourhood of a_m. Then $f_m\left(\frac{1}{z-a_m}\right) = \sum_{n=1}^{\infty} b_{-n}(z-a_m)^{-n}$ is a function which is defined over the whole plane except at $z = a_m$, and regular. Consequently $f(z) - f_m\left(\frac{1}{z-a_m}\right)$ is regular over the whole plane except at $a_1,...,a_{m-1}$. Theorem 1 now follows by induction. □

9.2 The Weierstrass product theorem

Theorem 1 underlines the importance of entire functions. As his main result of the above-mentioned work, Weierstrass gave a generalisation of the representation of a polynomial as the product of linear factors to arbitrary entire functions.

We derive the *Weierstrass product theorem* in this section, and begin with a few preliminaries on infinite products.

Let $p_1, p_2, \ldots$ be a sequence of complex numbers. $\prod_{n=1}^{\infty} p_n$ is called *convergent* when the partial products $\prod_{n=1}^{n} p_n$ converge to a value $\neq 0$ as h ∞, or when $p_n = 0$ for some n. The former is the case if and only if the series $\sum_{n=1}^{\infty} \log p_n$ converges, taking the principal value of the argument, i.e. $-\pi < \arg p_n \leq \pi$. The same applies for uniform convergence. Also, $\sum_{n=1}^{\infty} \log p_n(z)$ is absolutely and uniformly convergent if and only if $\sum_{n=1}^{\infty} (p_n(z)-1)$ is.

We construct an entire function with given zeros $a_1, a_2, \ldots \in \mathbb{C} - \{0\}$, where the set $\{a_1, a_2, \ldots\}$ accumulates only at infinity (Chapter 8, Theorem 11). Multiple zeros are accommodated by including them the corresponding number of times in the sequence $a_1, a_2, \ldots$.

Let $e(z,k) := (1-z)\exp \sum_{n=1}^{k} z^n/n$. For $|z| < 1$, $e(z,k) = \exp \sum_{n=k+1}^{\infty} (-z^n/n)$ and for $|z| \leq \frac{1}{2}$

$$|\log e(z,k)| \leq \sum_{n=k+1}^{\infty} |z|^n/(k+1) \leq 2|z|^{k+1}/(k+1).$$

Hence for any given $r > 0$ the product $\prod_{n=1}^{\infty} e(z/a_n, k_n)$ is absolutely and uniformly convergent for $|z| < r$ when

$$\sum_{n=1}^{\infty} \frac{1}{k_n+1} \left|\frac{r}{a_n}\right|^{k_n+1} \tag{1}$$

converges. This is the case, e.g., when one sets $k_n = n$. (For almost all a_n it is true that $r \leq |a_n|/2$.)

Thus for suitable k_n, $\prod_{n=1}^{\infty} e(z/a_n, k_n)$ is an entire function with zeros $a_1, a_2, \ldots$.

If $f_1(z)$ and $f_2(z)$ are entire functions which have the same zeros, then $g(z) := \frac{f_1(z)}{f_2(z)}$ is an entire function without zeros. The function $g(z)$ may be represented in the form $\exp h(z)$ for an entire function $h(z)$. In fact, $g'(z)/g(z)$ is an entire function by hypothesis, and hence representable as a power series $\sum_{n=0}^{\infty} c_n z^n$ for all $z \in \mathbb{C}$. Then $h(z) := \log g(0) + \sum_{n=0}^{\infty} (c_n/(n+1))z^{n+1}$ does what we require.

In this way we have proved the following Weierstrass product theorem.

Theorem 2. *Let* $f(z)$ *be an entire function, with zeros* $a_1, a_2, \ldots$ *other than* 0. *Also, let* $k_1, k_2, \ldots$ *be positive integers such that* (1) *converges for all* $r > 0$. *Then* $f(z)$ *may be represented in the form*

$$f(z) = e^{h(z)} z^h \prod_{n=1}^{\infty} e(z/a_n, k_n), \tag{2}$$

where $h(z)$ *is an entire function and* $h = \nu_0(f)$ (Section 8.8).

Conversely, (2) *is an entire function with zeros* $a_1, a_2, \ldots$ *and the* h*-tuple zero* 0 *for any sequence* $a_1, a_2, \ldots \in \mathbb{C} - \{0\}$ *with no finite accumulation point.* □

Now let $f(z)$ be a function meromorphic over the whole of $\mathbb{C}$, with zeros $a_1, a_2, \ldots$ and poles $b_1, b_2, \ldots$. Then $f(z)$ is the quotient of two entire functions $f_1(z)$ and $f_2(z)$ with zeros $a_1, a_2, \ldots$ and $b_1, b_2, \ldots$ respectively.

Example. (Euler). We consider the function $\sin \pi z$. It has simple zeros for $z = n \in \mathbb{Z}$. (1) is convergent for $k_n = 1$. Thus

$$\sin \pi z = e^{h(z)} z \prod_{n=1}^{\infty} \left(1 - \frac{z}{n}\right) e^{z/n} \prod_{n=1}^{\infty} \left(1 + \frac{z}{n}\right) e^{-z/n}$$

$$= e^{h(z)} z \prod_{n=1}^{\infty} \left(1 - \frac{z^2}{n^2}\right). \tag{3}$$

To compute $h(z)$ we compare the logarithmic derivatives of the two sides of (3) :

$$\pi \cot \pi z = h'(z) + \frac{1}{z} + \sum_{n=1}^{\infty} \frac{-2z}{n^2 - z^2}.$$

Not only is $h'(z)$ an entire function, it is also bounded: in the first place $h'(z) = h'(z+1)$, in the second place, for $z = x + iy$, $|y| \geq 1$, $0 \leq x < 1$ we have the bounds

$$|\cot z| = \left|\frac{e^{iz}+e^{-iz}}{e^{iz}-e^{-iz}}\right| \le \frac{e^{|y|}+e^{-|y|}}{e^{|y|}-e^{-|y|}}$$

and $|n^2-z^2| = |n-z|\cdot|n+z| \ge |n-1-iy|\cdot|n-1+iy|$, hence

$$\left|\frac{1}{z} + \sum_{n=1}^{\infty} \frac{-2z}{n^2-z^2}\right| \le \frac{1}{|y|} + \sum_{n=1}^{\infty} \frac{2|y|+2}{(n-1)^2+y^2}$$

$$\le \frac{1}{|y|} + \int_1^{\infty} \frac{2|y|+2}{(u-1)^2+y^2}du + \frac{2|y|+2}{y^2}.$$

The expression on the right is bounded because

$$\int_1^{\infty} \frac{du}{(u-1)^2+y^2} = \int_0^{\infty} \frac{dv}{v^2+y^2} = \frac{1}{|y|}\int_0^{\infty} \frac{dw}{w^2+1}.$$

Therefore $h'(z)$ is a constant (Chapter 8, Theorem 7). This constant equals 0, because $\pi \cot \pi z$ and $\frac{1}{z} + \sum_{n=1}^{\infty} \frac{-2z}{n^2-z^2}$ are both odd functions. Thus $h(z)$ is a constant. If one divides (3) by z and lets z tend to 0, one finds $h(z) = \pi$. Thus we have the following result:

$$\sin \pi z = \pi z \prod_{n=1}^{\infty} \left(1-\frac{z^2}{n^2}\right). \tag{4}$$

9.3 The Γ-function

As a further example we consider the product

$$f(z) = e^{\gamma z} z \prod_{n=1}^{\infty} \left(1+\frac{z}{n}\right)e^{-z/n}, \tag{5}$$

where

$$\gamma := \lim_{m\to\infty}\left(1 + \frac{1}{2} + \ldots + \frac{1}{m} - \log m\right) = \int_1^{\infty}\left(\frac{1}{[u]} - \frac{1}{u}\right)du = 0.5772\ \ldots \tag{6}$$

is *Euler's constant*.

By Theorem 2, $f(z)$ is an entire function with simple zeros $0, -1, -2, \ldots$. The meromorphic function $\Gamma(z) = 1/f(z)$ is called the *Γ-function.* This function was first studied by Euler. The discovery of the expansion (5) gave Weierstrass the impetus to find Theorem 2. We shall see that $\Gamma(z)$ interpolates the factorial: $\Phi(n) = (n-1)!$ for $n = 1,2,\ldots$. The Γ-function also plays an important rôle in other connections (Chapter 15, Chapter 26).

We can also write $1/\Gamma(z)$ in the form

$$1/\Gamma(z) = \lim_{m\to\infty} z(1+z)\left(1+\frac{z}{2}\right) \ldots \left(1+\frac{z}{m}\right)m^{-z}, \tag{7}$$

which is called the *Gauss product representation.* From this it easily follows that

$$\Gamma(1) = 1, \; \Gamma(z+1) = z\Gamma(z) \tag{8}$$

and hence $\Gamma(n) = (n-1)!$.

Euler proceeded from the following expression for $\Gamma(z)$:

Theorem 3. *For* $\operatorname{Re} z > 0$,

$$\Gamma(z) = \int_0^\infty e^{-t} t^{z-1} dt. \tag{9}$$

Proof. The integral on the right is uniformly convergent for $x_1 \geq \operatorname{Re} z \geq x_0 > 0$ and hence represents a regular function for $\operatorname{Re} z > 0$ (Section 8.10). In fact,

$$\lim_{h\to\infty}\left|\int_h^1 e^{-t} t^{z-1} dt\right| \leq \lim_{h\to\infty}\int_h^1 e^{-t} t^{x_0-1} dt = \frac{e^{-1}}{x_0} + \int_h^1 (e^{-t})' \frac{t^{x_0}}{x_0} dt,$$

$$\lim_{h\to\infty}\left|\int_1^h e^{-t} t^{z-1} dt\right| \leq \lim_{h\to\infty}\int_1^h e^{-t} t^{x_1-1} dt \leq c \lim_{h\to\infty}\int_1^h e^{-t/2} dt = 2ce^{-1/2}$$

where c is a constant independent of x with $t^{x_1-1} \leq ce^{t/2}$.

To prove Theorem 3 we can therefore assume that z is real and $0 < z < 1$ (Chapter 8, Theorem 11).

For positive integers m we have

$$\int_0^m (1-\frac{t}{m})^m t^{z-1}dt = m^z\int_0^1 (1-t)^m t^{z-1}dt$$

$$= m^z\frac{m}{z}\int_0^1 (1-t)^{m-1}t^z dt$$

$$\ldots \qquad = m^z\frac{m(m-1)\ldots 1}{z(z+1)\ldots(z+m-1)}\int_0^1 t^{z+m-1}dt$$

$$= m^z\frac{m!}{z(z+1)\ldots(z+m)}.$$

Hence by (7)

$$\Gamma(z) = \lim_{m\to\infty}\int_0^m (1-\frac{t}{m})^m t^{z-1}dt = \lim_{m\to\infty}\int_0^m e^{m\log(1-t/m)}t^{z-1}dt.$$

Also, for $0 \leq n \leq m$,

$$\Gamma(z) = \lim_{m\to\infty}\int_0^n e^{m\log(1-t/m)}t^{z-1}dt + \lim_{m\to\infty}\int_n^m (1-\frac{t}{m})^m t^{z-1}dt$$

$$= \int_0^n e^{-t}t^{z-1}dt + \lim_{m\to\infty}\int_n^m (1-\frac{t}{m})^m t^{z-1}dt. \tag{10}$$

and

$$\int_n^m (1-\frac{t}{m})^m t^{z-1}dt \leq n^{z-1}\int_n^m (1-\frac{t}{m})^m dt = n^{z-1}\frac{m}{m+1}(1-\frac{n}{m})^{m+1} < n^{z-1}.$$

Thus as n increases the second integral in (10) becomes arbitrarily small. □

9.4 Stirling's formula

Now we shall estimate the Γ-function for large arguments n :

Theorem 4 *(Stirling's formula). For natural numbers* n,

$$\log \Gamma(n) = (n-\tfrac{1}{2})\log n - n + \log\sqrt{2\pi} + O(\tfrac{1}{n}).$$

Proof: This is based on the following identity $(h > \frac{1}{2})$:

$$\int_{h-1/2}^{h+1/2} \log t \, dt = \int_0^{1/2} \log(h^2-t^2)dt = \log h + \int_0^{1/2} \log(1-\frac{t^2}{h^2})dt.$$

With $c_h := \int_0^{1/2} \log(1-\frac{t^2}{h^2})dt = -\sum_{k=1}^{\infty} \frac{1}{(2k+1)2k(2k)^{2k}}$ we get

$$|c_h| \le \frac{1}{16h^2} \sum_{k=1}^{\infty} \frac{1}{k^2} < \frac{1}{h^2}.$$

It follows that

$$\begin{aligned}\log \Gamma(n) &= \sum_{h=1}^{n-1} \log h \\ &= \int_{1/2}^{n-1/2} \log t \, dt - \sum_{h=1}^{n-1} c_h \\ &= (n-\tfrac{1}{2})\log(n-\tfrac{1}{2})-(n-\tfrac{1}{2})- \tfrac{1}{2} \log \tfrac{1}{2} + \tfrac{1}{2} - \sum_{h=1}^{\infty} c_h + \sum_{h=n}^{\infty} c_h.\end{aligned}$$

Since

$$\sum_{h=n}^{\infty} c_h < \sum_{h=n}^{\infty} \frac{1}{h^2} \le \int_{n-1}^{\infty} \frac{dt}{t^2} = \frac{1}{n-1} = O(\frac{1}{n})$$

and

$$\log(n-\tfrac{1}{2}) - \log n = \log(1-\tfrac{1}{2n}) - \frac{1}{2n} + O(\frac{1}{n^2})$$

we get

$$\log \Gamma(n) = (n-\tfrac{1}{2})\log n - n + c + O(\tfrac{1}{n}) \tag{11}$$

with a constant c, which we find to be $\log\sqrt{2\pi}$ in the proof of the next theorem. □

Stirling's formula may be extended to an estimate of $\Gamma(z)$ in a sector which omits the negative real axis, on which the poles of $\Gamma(z)$ lie :

Theorem 5. *For* $\alpha > 0$ *and* $|\arg z| \leq \pi-\alpha$

$$\log \Gamma(z) = (z-\tfrac{1}{2})\log z - z + \log\sqrt{2\pi} + O\left(\frac{1}{|z|\sin\alpha}\right).$$

Here $\log z$ *and* $\log \Gamma(z)$ *are the functions in the sector* $|\arg z| \leq \pi-\alpha$ *determined by* $\log 1 = 0$ (Chapter 8, Theorem 16).

Proof. By taking the logarithm of (5) one gets

$$\log \Gamma(z) = -\gamma z - \log z + \sum_{n=1}^{\infty} (\tfrac{z}{n} - \log(1+\tfrac{z}{n})).$$

For a natural number N

$$\int_0^N \frac{[u]-u+1/2}{u+z}du = \sum_{n=0}^{N-1}\int_n^{n+1} \frac{n}{u+z}du + \int_0^N\left(\frac{z+1/2}{u+z} - 1\right)du$$

$$= -\sum_{n=1}^{N-1} \log(n+z) + (N-1)\log(N+z) + (\tfrac{1}{2}+z)(\log(N+z) - \log z) - N$$

$$= \sum_{n=1}^{N-1} (\tfrac{z}{n} - \log(1+\tfrac{z}{n})) - \log(N-1)! - z\sum_{n=1}^{N-1}\frac{1}{n}$$

$$+ (N-\tfrac{1}{2}+z)\log(N+z) - (\tfrac{1}{2}+z)\log z - N.$$

By use of (11),

$$\sum_{n=1}^{N-1} \frac{1}{n} - \log N = \gamma - \int_N^{\infty}\left(\frac{1}{[u]} - \frac{1}{u}\right)du = \gamma + O(\tfrac{1}{N}),$$

and the fact that

$$\log(N+z) = \log N + \log(1+\tfrac{z}{N}) = \log N + \frac{z}{N} + O(\tfrac{1}{N^2}),$$

one also finds that

$$\int_0^N \frac{[u]-u+1/2}{u+z}du = \sum_{n=1}^{N-1}\left[\frac{z}{n} - \log(1+\tfrac{z}{n})\right] - z\left[\sum_{n=1}^{N-1}\frac{1}{n} - \log N\right] - c$$

$$+ (N-\tfrac{1}{2}+z)\frac{z}{N} - (\tfrac{1}{2}+z)\log z + O(\tfrac{1}{N}).$$

As $N \to \infty$ this yields

$$\log \Gamma(z) = \int_0^\infty \frac{[u]-u+1/2}{u+z}du - z + (z-\tfrac{1}{2})\log z + c. \tag{12}$$

We set $\phi(u) := \int_0^u ([v] - v + \frac{1}{2})dv$. Then $|\phi(u)| < 1$ and

$$\left|\int_0^\infty \frac{[u]-u+1/2}{u+z}du\right| = \left|\int_0^\infty \frac{\phi'(u)}{u+z}du\right| = \left|\int_0^\infty \frac{\phi(u)}{(u+z)^2}\,du\right| \le \int_0^\infty \frac{du}{|u+z|^2}.$$

Let $z = |z|e^{i\delta}$. Since

$$|u+z|^2 = u^2 + 2|z|u \cos \delta + |z|^2 \ge u^2 - 2|z|u \cos \alpha + |z|^2,$$

$w := \frac{u}{|z| \sin \alpha}$ gives

$$\int_0^\infty \frac{du}{|u+z|^2} \le \int_0^\infty \frac{du}{u^2-2|z|u \cos \alpha+|z|^2}$$

$$= \frac{1}{|z|\sin \alpha}\int_0^\infty \frac{dw}{(w-\cot\alpha)^2+1}$$

$$= 0\left(\frac{1}{|z|\sin \alpha}\right).$$

It remains to determine the constant c. To do that we first note that, by (4) and (5),

$$\Gamma(z)\Gamma(1-z) = \frac{\pi}{\sin \pi z} \tag{13}$$

and hence $\Gamma(\frac{1}{2}) = \sqrt{\pi}$. Also

$$\Gamma(z)\Gamma(z+\tfrac{1}{2}) = \sqrt{\pi}\ 2^{1-2z}\Gamma(2z). \tag{14}$$

(*Legendre's duplication formula*) One proves (14) by proceeding from (7) to show that $\Gamma(z)\Gamma(z+\frac{1}{2})2^{2z-1}\Gamma(2z)^{-1}$ is constant. The assertion now follows by substituting $z = \frac{1}{2}$.

If one now substitutes the values n, $n + \frac{1}{2}$, $2n$ for z in (12), and recalls (14), then one finds the value $c = \log\sqrt{2\pi}$ by letting $n \to \infty$. □

A more delicate estimation of $\int_0^\infty \frac{[u]-u+1/2}{n+z}du$ enables one to show that

$$\int_0^\infty \frac{[u]-u+1/2}{u+z}du = \sum_{k=1}^{n} \frac{B_{2k}}{2k(2k-1)2^{2k-1}} + O\left(\frac{1}{(|z|\sin\alpha)^{2n+1}}\right) \qquad (15)$$

for $n = 1,2,\ldots$, where $B_2, B_4, \ldots$ are the Bernoulli numbers (Section 5.6). The series (15) is *divergent*, i.e. $\sum_{k=1}^{\infty} \frac{B_{2k}}{2k(2k-1)z^{2k-1}}$ diverges for all $z \in \mathbb{C} - \{0\}$.

Exercises

9.1 Show $\frac{2}{1}\cdot\frac{2}{3}\cdot\frac{4}{3}\cdot\frac{4}{5}\cdot\ldots = \frac{\pi}{2}$ (Wallis' product).

9.2 Give the product representation of the function $e^{2\pi z}-1$.

9.3 Show that there are entire functions which take arbitrarily prescribed values $w_1, w_2, \ldots$ at given points $z_1, z_2, \ldots$ with no finite accumulation point.

9.4 Prove the Gauss multiplication formula

$$\prod_{r=0}^{m-1} \Gamma\left(z+\frac{r}{m}\right) = (2\pi)^{(m-1)/2} m^{1/2-mz}\Gamma(mz).$$

(Hint: Proceed as in the proof of (14) and use (13).)

9.5 Prove the formula (15).

10. Riemann surfaces

10.1 The 19th century view of the problems of function theory

One of the main questions of function theory, already in the 18th century and above all in the 19th, was the study of integrals of algebraic functions. By an *integral* we initially mean an antiderivative, i.e. a function whose derivative equals the given function.

A rational function $f(x,y)$, where $y = \sqrt{g(x)}$ and $g(x)$ is a quadratic or linear polynomial, was known to be integrable by rational functions in x, y and elementary functions $\log r(x,y)$ and (if one admits only real functions) $\arcsin r(x,y)$, $\arctan r(x,y)$, where $r(x,y)$ is a rational function of x, y.

In the complex numbers these functions admittedly have only a formal existence at present. If we consider the integrals as line integrals (Section 8.2), then they depend on the path to an unpleasant extent. E.g. the function $\int \left(\frac{1}{t+1} + \frac{\sqrt{2}}{t-1}\right) dt$ is many-valued with periods $2\pi i$ and $\sqrt{2}\cdot 2\pi i$. I.e. the values of the function for a fixed argument lie densely on a line parallel to the imaginary axis, and in the same way one constructs integrals whose values for a fixed argument lie densely over the whole plane. In order to obtain uniqueness of the integral, one must, in view of the Cauchy integral theorem (Chapter 8, Theorem 2), restrict the domain of the function to a simply connected region in which it is regular.

For the case where $g(x)$ is a polynomial of third or fourth degree without multiple zeros, integration by elementary functions is no longer possible. New functions arise. The study of such functions is one of the main themes of 19th century mathematics. Gauss wrote, in a letter of 1808 to the astronomer Schumacher:

"*Integral calculus is of far less interest to me when it is matter of skilfully operating the machinery of substitutions, transformations, etc. to reduce integrals to algebraic, logarithmic or circular functions, rather than the deeper considerations involving those transcendental functions which are not so reducible. We are now as familiar with circular functions and logarithms as with 1 times 1, but the marvellous goldmine of higher functions is still virtually Terra Incognita. I have investigated this extensively in the past, and one day will produce an appropriately large work on the subject, a hint of which was given in my Disquis. arithm. p. 593, Art. 335. You would be astounded at the overflowing wealth of new and highly interesting truths and relationships exhibited by these functions (including, among others, those associated with the rectification of the ellipse and the hyperbola)."*

In the *Disquisitiones arithmeticae,* Art. 335, Gauss remarks that his theory of division of the circle (Chapter 3) applies not only to circular functions but also to other transcendents, such as those which depend on the integral $\int \frac{dx}{\sqrt{1-x^4}}$.

Gauss never wrote the "*large work*" of which he spoke here. But his remark in the *Disquisitiones arithmeticae* acted as a stimulus for Abel and Jacobi in their competition to investigate the properties of integrals; of the form

$$\int \frac{dx}{\sqrt{(1-x^2)(1-k^2x^2)}} . \tag{1}$$

Abel's decisive discovery was that the inverse function of (1) was a single-valued function, meromorphic over the whole plane, with two periods linearly independent over $\mathbb{R}$. Such functions are called *elliptic functions.* We consider them in Chapter 13, after we have introduced the general foundations for the study of integrals of algebraic functions, due to Riemann, in the present chapter.

Riemann succeeded in overcoming the difficulties associated with many-valued functions by viewing the argument of the function, not as confined to the plane, but as ranging over a more complicated structure, which was later called a *Riemann surface.* Weierstrass overcame the same difficulties by passing from analytic functions to analytic configurations consisting of all the power series resulting from analytic continuation of a given one. After Riemann's death in 1866, Weierstrass won the day with his conception for a number of reasons. It was only at the beginning of the 20th century, when Riemann's principles could be put on a sound basis, that their significance, extending far beyond function theory, became clear. We return to this in Chap. 29.

Riemann published his ideas on function theory principally in the following two works:

Grundlagen für eine allgemeine Theorie der Funktionen einer veränderlichen komplexen Grösse (Dissertation, Göttingen 1851);

Theorie der Abelschen Funktionen (J. reine angew. Math. 54 (1857)).

10.2 The extended complex plane

First we extend the complex plane by a point ∞. A neighbourhood U of ∞ is of the form $U = U_1 \cup \{\infty\}$, where U_1 is an open subset of the plane complementary to a closed disc. A complex valued function $f(z)$ on U is called *regular at* ∞ when $f(1/t)$ is regular at $t = 0$. The function $t = 1/z$ of $z \in U$ is called a *local uniformising variable at* ∞. At a finite point a of the plane we call $t = z-a$ the local uniformising variable. We set $\psi_\infty(t) = 1/t, \psi_a(t) = t+a$.

Now suppose $a \in \mathbb{C} \cup \{\infty\}$ and let U be a region in $\mathbb{C} \cup \{\infty\}$ which contains a. A function $f(z)$ which is defined in $U - \{a\}$ is called *meromorphic at* a when $f(\psi_a(t))$ has a power series expansion

$$f(\psi_a(t)) = \sum_{n=h}^{\infty} a_n t^n$$

in a neighbourhood of 0, where h is an integer and $a_h \neq 0$. The number $h =: \nu_a(f)$ is called the *order* of f at a. When h is positive (resp. negative), f has an *h-tuple zero* (resp. *h-tuple pole*) at a. These terms generalise those of Section 8.8.

Theorem 1. *A function* $f(z)$ *which is meromorphic on the extended plane is rational.*

Proof. Without loss of generality assume $f(z)$ is not identically zero. A function meromorphic at $a \in \mathbb{C} \cup \{\infty\}$ is regular in a neighbourhood of a with the possible exception of a itself. Hence there is no accumulation point of poles or zeros (Section 8.8). By definition of the neighbourhoods of ∞, $f(z)$ therefore has only finitely many poles and zeros. By multiplication by a suitable rational function $f(z)$ can then be converted to a function $f_1(z)$ which has no zeros or poles for finite z. Since $f_1(z)$ is either convergent or divergent as $z \to \infty$, either $|f_1(z)|$ or $\left|\frac{1}{f_1(z)}\right|$ is bounded. Hence, by Chapter 8, Theorem 7, $f_1(z)$ is constant. □

By Theorem 1 we can characterise the rational functions as those which are meromorphic over the extended plane.

10.3 The nth root

Before passing from the extended plane to Riemann surfaces, we consider two examples.

Example 1. $f(z) = \sqrt[n]{z-a},\ a \in \mathbb{C}$.

This function can be defined as single-valued when one cuts the plane, e.g. along the semi-axis $A = \{a - r \mid r \in \mathbb{R}_+\}$, and defines the function with the help of polar coordinates as follows:

$$z = r\cdot e^{i\phi} + a,\ \sqrt[n]{z-a} = \sqrt[n]{r}\cdot e^{i\phi/n} \quad \text{for} \quad -\pi < \phi \leq \pi.$$

The function becomes discontinuous as z crosses A. Riemann's idea was to think of several sheets lying over the plane (n in our example). Starting on the first sheet, one arrives on the second sheet after crossing A, and after a further circuit around the point a one reaches the third sheet, and so on, returning to the first sheet after the nth circuit around a.

Over each point of the extended plane except a and ∞ there are n points of the n-sheeted surface $\mathcal{F}_{a,n}$ thus defined. There is only a single point over each of a and ∞; a and ∞ are the *branch points* of the surface. The function $\sqrt[n]{z-a}$ (and in general each rational function of $\sqrt[n]{z-a}$) is a single-valued function on $\mathcal{F}_{a,n}$. At the n points over $z \in \mathbb{C}$ it takes the n values $\sqrt[n]{r}\cdot e^{i(\phi+2(k-1)\pi)/n}$ for $k = 1,\dots,n$, where $z = a + r\cdot e^{i\phi}$ and $-\pi < \phi \leq \pi$. We denote this function by $\phi_{a,n}$. $\mathcal{F}_{\infty,n}$ and $\phi_{\infty,n}$ are defined correspondingly for the function $1/\sqrt[n]{z}$.

Example 2. A more interesting example is the Riemann surface for the function $\sqrt{(z^2-1)(z^2-4)}$.

Corresponding to the two-valuedness of the function we have two sheets. There are four branch points $z = \pm 1, \pm 2$. We cut the real axis between -2 and -1 and between 1 and 2, and pass from one sheet to the other on crossing the cuts (Figure 6).

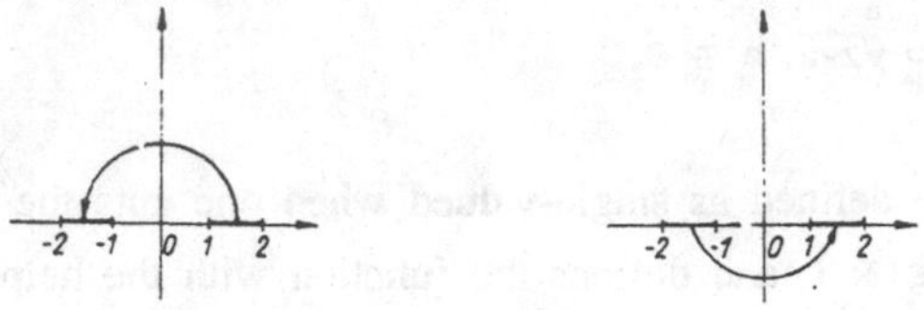

sheet 1 sheet 2 Fig. 6

The surface $\mathcal{F}$ defined in this way cannot be mapped homeomorphically onto the extended plane (in contrast to $\mathcal{F}_{a,n}$), the latter surface being the same as the sphere. One sees this, e.g., because there is a closed curve on $\mathcal{F}$ which does not separate $\mathcal{F}$ into two parts (Figure 6). $\mathcal{F}$ is homeomorphic to the surface which results from joining two doubly-slit spheres, i.e. $\mathcal{F}$ is homeomorphic to the torus.

For the sake of a later generalisation we conceive a somewhat different definition of $\mathcal{F}$. Let I_1 resp. I_2 be the closed interval [-2,-1] resp. [1,2], and let K_1 resp. K_2 be the open disc in $\mathbb{C}$ with midpoint $-\frac{3}{2}$ resp. $\frac{3}{2}$ and radius $\frac{1}{2}$. Then we assemble $\mathcal{F}$ from six regions $U_1,\ldots,U_6$ and the four branch points ± 1, ± 2. U_1 and U_2 are two copies of the region $(\mathbb{C} \cup \{\infty\}) - I_1 - I_2$ in the extended plane, U_3 and U_4 are two copies of K_1, and U_5, U_6 are two copies of K_2. $\mathcal{F}$ results by joining these together as follows: as subsets of $\mathbb{C} \cup \{\infty\}$, $U_j \cap U_k$ for $j = 1,2$, $k = 3,4$ separate into two open half discs K_1^o, K_1^u. We join U_1 to U_3 and U_2 to U_4 with K_1^o, and U_1 to U_4 and U_2 to U_3 with K_1^u. U_1, U_2 are joined to U_5, U_6 similarly.

Along with $\mathcal{F}$, a projection $\pi : \mathcal{F} \to \mathbb{C} \cup \{\infty\}$ is defined. One obtains a neighbourhood U of the branch point $P_a \in \mathcal{F}$ for $a = \pm 1, \pm 2$, as the preimage under π of a neighbourhood U_a of a in $\mathbb{C} \cup \{\infty\}$. When U_a is sufficiently small we can identify U with the subset of $\mathcal{F}_{a,2}$ lying over U_a.

We now consider how the concept of a uniformising variable can be transported to $\mathcal{F}$. As the uniformising variable of a non-branch point $P \in \mathcal{F}$ we take the uniformising variable of πP in a sufficiently small neighbourhood of πP. As the uniformising variable of a branch point P_a we take the function $\phi_{a,2}$. Here it must be noted that the identification of U with a subset of $\mathcal{F}_{a,2}$ can be carried out in two ways. Passing from one to the other corresponds to exchanging the sheets of $\mathcal{F}_{a,2}$, which corresponds to passing from $\phi_{a,2}$ to $-\phi_{a,2}$. Thus the uniformising variable of a branch point of $\mathcal{F}$ is determined only up to sign.

We can now define the concept of a meromorphic function on $\mathcal{F}$ as in Section 10.2. One easily checks that $\sqrt{(z^2-1)(z^2-4)}$ is a meromorphic function on $\mathcal{F}$, with a simple zero at each of the four branch points, and a double pole at each of the two points lying over ∞.

10.4 Definition of a Riemann surface

We now come to the concept of a Riemann surface. Riemann wrote the following on this in his work on abelian functions:

"For many investigations, such as the investigation of algebraic and abelian functions, it is convenient to represent the mode of branching of a many-valued function in the following geometric way. One thinks of the (x,y)*-plane being covered by another surface coincident with it (or lying at an infinitesimal distance above it) for as far as the function is defined. On continuation of the function the surface is extended correspondingly. In a part of the plane where the function has two or more continuations the surface is double or multiple; there it consists of two or more sheets, each of which represents a branch of the function. Around a branch point of the function one sheet of the surface continues into another, so that in the neighbourhood of such a point the surface can be regarded as a helicoidal surface with axis through the point and perpendicular to the* (x,y)*-plane, and infinitesimal pitch. When the function returns to its previous value after a number of circuits of* z *around the branch point (as, e.g., with* $(z-a)^{m/n}$*, when* m,n *are relatively prime, after* n *circuits of* z *around* a*), then of course one has to assume that the topmost sheet passes back through the others to join the lowest.*

The many-valued function has only one value for each point on this surface which represents its branching, and hence can be regarded as a completely determinate function on the surface."

We make this definition precise by extending Section 10.3 as follows.

An unbranched Riemann surface $\mathcal{F}$ is given by a family $\mathfrak{U} = \{U_j \mid j \in J\}$ of regions of the extended complex plane, together with a rule Φ for joining the regions U_j, U_k along a connected component U_{jk} of $U_j \cap U_k$, for any $j,k \in J$. Here we require that a point of U_j is joined at most once, i.e. that $U_{jk} \cap U_{j\ell} = \emptyset$ for all $\ell \neq j,k$.

The Riemann surface $\mathcal{F} = (\mathfrak{U},\Phi)$ then results from the disjoint union $\amalg\mathfrak{U}$ of the U_j by identification of joined points. Let $\tilde{U}_j$ be the image of $U_j \subset \amalg\mathfrak{U}$ in $\mathcal{F}$. Along with $\mathcal{F}$ we also have defined the projection π of F into $\mathbb{C} \cup \{\infty\}$.

The concept of neighbourhood of a point, and hence every other topological concept, can be carried over immediately from $\mathbb{C} \cup \{\infty\}$ to $\mathcal{F}$. The surface $\mathcal{F}$ is called an *unbranched covering* of $\mathbb{C} \cup \{\infty\}$, because for each point P of $\mathcal{F}$ there is a neighbourhood mapped one-to-one by π onto a neighbourhood of πP. Such a neighbourhood is called *schlicht*.

Two coverings $\pi_1 : \mathcal{F}_1 \to \mathbb{C} \cup \{\infty\}$ and $\pi_2 : \mathcal{F}_2 \to \mathbb{C} \cup \{\infty\}$ are called *isomorphic* when there is a homeomorphism ϕ of $\mathcal{F}_1$ onto $\mathcal{F}_2$ compatible with the projections, i.e. $\pi_2\phi = \pi_1$.

In particular, consider $\mathcal{F}_{a,n}$ in a neighbourhood $\mathcal{F}_{a,n}(U_a)$ of P over the neighbourhood U of a. The punctured surface $\mathcal{F}_{a,n}(U) - \{P_a\}$ is an unbranched covering. As one easily sees, there are exactly n isomorphisms of $\mathcal{F}_{a,n}(U) - \{P_a\}$ onto itself, resulting from cyclic permutation of the sheets of $\mathcal{F}_{a,n}(U) - \{P_a\}$.

We shall now define the concept of a *branch point* of a Riemann surface $\mathcal{F}$.

Let $\pi : \mathcal{F}_0 \to \mathbb{C} \cup \{\infty\}$ be an unbranched covering and let $a \in \mathbb{C} \cup \{\infty\}$. Also let U be a sufficiently small neighbourhood of a and let U_1 be a connected component of $\pi^{-1}U$. When U_1 contains a point P with $\pi P = a$, then U_1 is homeomorphic to U. This case does not interest us here. Suppose then that U_1 contains no point over a. Then U_1 is called *branched* over a when U_1 is isomorphic, as a covering, to $\mathcal{F}_{a,n}(U) - \{P_a\}$ for some $n > 1$. In this case we add a point P to $\mathcal{F}_0$ which lies over a. P is called a *branch point* with branching index n. By definition, $U_P := U \cup \{P\}$ is a neighbourhood of P. The case $n = 1$ can also occur. Then π is a homeomorphism from $U_1 \cup \{P\}$ onto U, and P is unbranched. Thus a branched covering surface $\mathcal{F}$ of $\mathbb{C} \cup \{\infty\}$ results from an unbranched covering $\mathcal{F}_0$ by attaching branch points.

Just as we represent regions of the earth by overlapping charts in an atlas, a Riemann surface $\mathcal{F}$ is represented by a covering by the open sets $\tilde{U}_j$, $j \in J$ and the U_P for the branch points P. One therefore calls these open sets *charts of* $\mathcal{F}$ and their totality an *atlas of* $\mathcal{F}$.

We now define a uniformising variable ϕ_P for each point P of $\mathcal{F}$ as a function which maps a sufficiently small neighbourhood U_P of P homeomorphically onto a neighbourhood of 0 in $\mathbb{C}$, where $\phi_P(P) = 0$. As uniformising variable of an unbranched point P we take the uniformising variable of πP. For a branch point P we set

$$\phi_P : U_1 \to \mathcal{F}_{a,n}(U) \xrightarrow[\phi_{a,n}]{} \mathbb{C},$$

in the notation from above. Here a is the image πP of P. There are exactly n possibilities for the choice of ϕ_P, which differ only by a constant factor, which is an nth root of unity. In each case $\pi\phi_P^{-1}(t) = t^n + a$.

The definition of Riemann surface as a covering of the extended plane was for Riemann only a crutch in the construction of the theory. In 1913 Weyl was able to throw this crutch away (Chapter 29). A *holomorphic* (or *regular* or *conformal*) mapping of the Riemann surface $\mathcal{F}_1$ into the Riemann surface $\mathcal{F}_2$ is defined to be a continuous mapping χ which is holomorphic locally; i.e. for each point $P \in \mathcal{F}_1$ there is a neighbourhood U_P such that $\phi_{\chi P}\chi\phi_P^{-1}$ is a holomorphic function on $\phi_P(U_P)$. (The independence of the definition from the choice of uniformising variable is obvious here, as also in what follows.)

Two Riemann surfaces $\mathcal{F}_1$, $\mathcal{F}_2$ are called *isomorphic* when there are inverse holomorphic mappings $\chi_1 : \mathcal{F}_1 \to \mathcal{F}_2$, $\chi_1^{-1} : \mathcal{F}_2 \to \mathcal{F}_1$.

A function f which maps a region of a Riemann surface $\mathcal{F}$ into $\mathbb{C} \cup \{\infty\}$ is called *meromorphic* at point P when $f\phi_P^{-1}$ is meromorphic at 0. The order of f at P, zeros and poles are now defined as in Section 8.8.

The concept of a (piecewise smooth) curve introduced in Section 8.2 carries over immediately to Riemann surfaces. With it we also get the concept of an (arcwise) connected open set. Let C_1 and C_2 be curves for which the final point of C_1 coincides with the initial point of C_2. Then C_1C_2 denotes the curve which one obtains by traversing C_1, then C_2. C_1^{-1} denotes the curve C_1 with the opposite orientation, i.e. traversed from final point to initial point.

10.5 The Riemann surface of an algebraic function

Suppose we have a monic polynomial

$$g(w) = g(z,w) = w^m + a_1(z)w^{m-1} + \dots + a_m(z) \tag{2}$$

whose coefficients are in $\mathbb{C}(z)$, i.e. rational functions. We want to associate g with a Riemann surface, on which there is a meromorphic function f with $g(f) = 0$.

We first exclude the finitely many points $z_1,\dots,z_s$ at which one of the functions $a_1(z),\dots,a_m(z)$ has a pole. For the remaining $z \in \mathbb{C}$ the discriminant $D(z)$ of g is defined, and it vanishes if and only if g has z as a multiple zero (Section 7.3). Since $D(z)$ is a rational function, this happens for only finitely many points which, together with ∞ and $z_1,\dots,z_s$, constitute the set M_g of *critical points* of g. For the remaining points z' the implicit function theorem (Chapter 8, Theorem 14) gives exactly m distinct regular functions $w_1(z),\dots,w_m(z)$, defined in a neighbourhood of z', with $g(z,w_j(z)) = 0$ for $j = 1,\dots,m$. By the monodromy theorem (Chapter 8, Theorem 10), $w_1(z),\dots,w_m(z)$ are then uniquely defined as regular functions in each simply connected region G of the complex plane such that $G \cap M_g = \emptyset$.

We now come to the definition of the Riemann surface $\mathcal{F}_g$ of g. We connect the points of M_g by a simple curve C. Let $z'_1,\dots,z'_n = \infty$ be the points of M_g in the order in which they occur on C. Then $G = \mathbb{C} - C$ is a simply connected region of the complex plane, and hence there are functions $w_1(z),\dots,w_m(z)$ regular in G with

$$g(w) = (w-w_1(z))\dots(w-w_m(z)).$$

We set $U_1 = \dots = U_m = G$. Let $C^k \subset C$ be the arc which connects the point z'_k to z'_{k+1} in M_g. We choose simply connected, pairwise disjoint regions G^k with $G^k \cap C = C^k - \{z'_k,z'_{k+1}\}$ and $G^k \cap G = G^k(r) \cup G^k(\ell)$, where $G^k(r)$ is the right connected component of $G^k \cap G = G^k - C$, in the sense of traversal of C, and $G^k(\ell)$ is the left. We set $U^k_j = G^k$ for $k = 1,\dots,h - 1;\ j = 1,\dots,m$ and

$$\mathfrak{U} = \{U_j,U^k_j \mid j = 1,\dots,m; k = 1,\dots,h-1\}.$$

In this system of sets $U_1,\dots,U_m$ resp. $U^k_1,\dots,U^k_m$ are to be regarded as m different copies of G resp. G^k. To establish the joining rule for $\mathfrak{U}$ we consider $w_j(z)$ on

$G^k(r)$ for $j = 1,2,...,m$. By continuation of this function on G^k we arrive on $G^k(\ell)$ at another function, which is a zero of $g(w)$ and hence equal to one of the functions $w_1(z),...,w_m(z)$. Suppose it is $w_{p_k(j)}(z)$. Then p_k is a permutation of $1,...,m$. We join U_j to U_j^k in $G^k(r)$ and U_j^k to $U_{p_k(j)}$ in $G^k(\ell)$ for $j = 1,...,m$; $k = 1,...,h-1$. In this way we obtain an m-sheeted Riemann surface $\mathcal{F}_0$ which covers $\mathbb{C} - M_g$ without branching. The joining rules enable the regular functions $w_j(z)$ on U_j to be continued to a regular function f on $\mathcal{F}_0$.

We now extend $\mathcal{F}_0$ by points which lie over M_g. We consider $\mathcal{F}_0$ over a sufficiently small circle $K_\varepsilon(z_k')$ around z_k'. If we start with a point P_1 in sheet $\tilde{U}_1$ of $\mathcal{F}_0$ with $\pi P = z \in K_\varepsilon(z_k')$, then after traversing a uniquely defined curve over the circle around z_k' we come to a point P_2 over z on one of the sheets $\tilde{U}_1,...,\tilde{U}_m$. After at most m circuits we must come back to P_1. Suppose this happens for the first time after the eth circuit. Then we extend $\mathcal{F}_0$ by a branch point over z_k' with branching index e. When $e < m$ we repeat the process and obtain further branch points over z_k'. If we count each branch point with its multiplicity e, then we can say that over each point of M_g there are also m points of the Riemann surface $\mathcal{F}_g$ just constructed. As a covering surface of $\mathbb{C} \cup \{\infty\}$, $\mathcal{F}_g$ is determined up to isomorphism by g.

We still have to show that f can be extended to a meromorphic function at the points of M_g. Consider f in the neighbourhood of a point Q over z_k' with branching index e. Let $t = \phi_Q(P)$ be the uniformising variable of $\mathcal{F}_g$ at Q. Then $f_u(t) = f(\phi_Q^{-1}(t))$ is a function regular in a neighbourhood 0 with the exception of 0, and $z = \pi P = t^e + z_k'$ for $z_k' \neq \infty$ while $z = \pi P = t^{-e}$ for $z_k' = \infty$. The function f_n satisfies the equation $g(f_u(t)) = 0$ whose coefficients are rational functions of t. By multiplying f_u by a suitable polynomial in t one arrives at a function $f_v(t)$ which satisfies an equation

$$f_v(t)^m + b_1(t)f_v(t)^{m-1} + ... + b_m(t) = 0, \tag{3}$$

where $b_1(t),...,b_m(t)$ are polynomials. If $f_u(t)$, and with it $f_v(t)$, was not definable as a meromorphic function for $t = 0$, then the Casorati-Weierstrass theorem would give, for each number $c \in \mathbb{C}$, a sequence $t_1,t_2,...$ converging to 0 with $\lim_{n\to\infty} f_v(t_n) = c$ (Section 8.8), in contradiction to (2). Thus f can be defined as a meromorphic function at the point Q. The definition of $\mathcal{F}_g$ easily yields the following

Theorem 2. *The polynomial* (1) *is irreducible if and only if the associated Riemann surface is connected.* □

In what follows we confine ourselves to connected Riemann surfaces.

10.6 The topology of closed Riemann surfaces

In the second section of the introduction to his work on abelian functions Riemann sketched the proof of a few theorems from a mathematical discipline he called *analysis situs*, and for which the accepted term later became *topology*. Riemann's remarks represent the germ of today's *algebraic topology*. By way of introduction he writes:

"In the investigation of functions which result from the integration of complete differentials, some theorems of analysis situs are almost indispensable. This term of Leibniz, though perhaps with not quite his meaning, covers a part of the theory of continuous quantities in which the quantities are not considered to exist independently of their position, or to be measurable by each other, but to be something independent of magnitude, with only their positional properties coming under investigation."

We first define a few basic concepts in the topology of Riemann surfaces.

A Riemann surface $\mathcal{F}$ is called *closed* when each infinite point set in $\mathcal{F}$ has a point of accumulation. It then follows from the compactness of closed, bounded sets in $\mathbb{C}$ that a Riemann surface is closed if and only if it is compact (A.2).

As one easily sees, the Riemann surface $\mathcal{F}_g$ of a polynomial g is closed. It is a fundamental insight of Riemann's that, conversely, each closed Riemann surface is isomorphic to $\mathcal{F}_g$ for a certain polynomial g. Admittedly, this theorem was not proved rigorously until later (see Chapter 29). However, it is not difficult to prove the following partial result.

Theorem 3. *A closed Riemann surface* $\mathcal{F}$ *has only finitely many branch points. Let* a *be a point of* $\mathbb{C} \cup \{\infty\}$ *over which no branch points of* $\mathcal{F}$ *lie. Then the number* n *of points* P *of* $\mathcal{F}$ *which lie over* a *is finite and independent of the choice of* a.

The number n is called the *sheet number* of $\mathcal{F}$, and $\mathcal{F}$ is called an *n-sheeted covering* of $\mathbb{C} \cup \{\infty\}$.

Proof. The set of branch points of $\mathcal{F}$ has no accumulation point, since each point P of $\mathcal{F}$ has a neighbourhood containing no branch points apart from P. Let π be the projection of $\mathcal{F}$ into $\mathbb{C} \cup \{\infty\}$. (Since $\pi\mathcal{F}$ is open and compact, $\pi\mathcal{F}$ in fact equals $\mathbb{C} \cup \{\infty\}$. The set $\pi^{-1}a$ has no accumulation point and is therefore finite. Let a' be another point of $\mathbb{C} \cup \{\infty\}$ over which $\mathcal{F}$ is unbranched, let $P \in \pi^{-1}a$ and $P' \in \pi^{-1}a'$, and let $\mathcal{F}_0$ be the set of non-branch points of $\mathcal{F}$. Since $\mathcal{F}_0$ is connected, there is a curve C in $\mathcal{F}_0$ which connects P to P'. We now construct a one-to-one mapping of $\pi^{-1}a$ onto $\pi^{-1}a'$ in the following way. For $P_1 \in \pi^{-1}a$ there is a unique curve which begins at P_1 and whose projection by π equals πC. We set $\phi(P_1)$ equal to the endpoint of this curve. □

10.7 Polygon complexes

With the help of Theorem 3 it is easy to decompose a closed Riemann surface $\mathcal{F}$ into easily understood parts, called *topological polygons*. We now define the basic concepts of *surface topology* needed in this connection.

A *(topological) polygon* in a topological space T is the continuous image of a disc, whose circumference is divided by r $(r \geq 1)$ points into r segments. The images of these points, which need not be all different, are called the *vertices*, and the images of the segments the *sides*, of the polygon. The interior of the disc, as well as the interior of each segment, must be mapped homeomorphically. However, different segments can be mapped to the same side. We speak of a polygon with r vertices and sides, where vertices as well as sides can appear multiply. An *arc* in T is a homeomorphic image of the closed unit interval.

A polygon complex $\mathfrak{K}$ in a topological space T is a finite set of polygons, arcs and points in T. The arcs and points in $\mathfrak{K}$ are called the *edges* and *vertices*, respectively, of $\mathfrak{K}$. In order to have a common term for polygons, edges and vertices we speak of *cells* (of dimension 2, 1, 0). The cells of a polygon complex have to satisfy the following conditions:

(i) *Along with each polygon* G, *the sides of* G *also belong to* $\mathfrak{K}$. *Along with each edge* A, *the vertices of* A *also belong to* $\mathfrak{K}$.

(ii) *The intersection of two distinct cells* X, Y *of* $\mathfrak{K}$ *is a union of cells in* $\mathfrak{K}$ *whose dimensions are smaller than the maximum of the dimensions of* X *and* Y.

The union $\cup\mathfrak{K}$ of a polygon complex $\mathfrak{K}$ is called a *surface* when the following conditions are satisfied.

(iii) $\mathfrak{K}$ *consists of polygons and their sides and vertices.*

(iv) *An edge* A *of* $\mathfrak{K}$ *appears at most twice as a side in the polygons of* $\mathfrak{K}$. *More precisely: either there is a polygon in* $\mathfrak{K}$ *in which* A *appears twice as a side, in which case* A *appears as a side in no other polygon; or this is not so, in which case* A *appears in no polygon twice, and in at most two polygons once.*

When each edge appears exactly twice as a side of a polygon in $\mathfrak{K}$ the surface is called *closed*, otherwise it is said to be *with boundary.*

We also define the concept of *orientability* of a polygon complex. A polygon of $\mathfrak{K}$ can be oriented in two ways, corresponding to the two directions of traversal of the circumference. The sides of the polygon are thereby oriented so that each vertex is initial point of one side and final point of another.

We write the sides of the polygon in this order and give each of them an exponent $+1$ or -1 according as the given orientation of edges coincides with that induced by that of the circumference or not. For example the torus corresponding to Figure 7 is given by the edge sequence $ABA^{-1}B^{-1}$

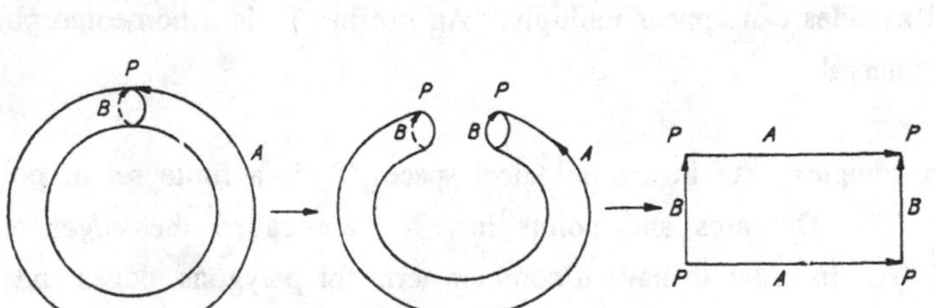

Figure 7

$\mathfrak{K}$ is called orientable when its polygons can be oriented in such a way that each edge in $\mathfrak{K}$ appearing twice as a polygon side occurs with opposite orientations.

A closed Riemann surface $\mathcal{F}$ with projection ϕ onto $\mathbb{C} \cup \{\infty\}$ may be decomposed into a polygon complex.

In fact, by Theorem 3 $\mathcal{F}$ has only finitely many branch points. We first decompose the extended plane into a polygon complex $\mathfrak{K}$ so that all images of branch points appear as vertices. All polygons are given the counterclockwise orientation. The extended plane is therefore an orientable closed surface in the sense of the above definition. Figure 8 represents a decomposition of $\mathbb{C} \cup \{\infty\}$ into one hexagon, where each side (necessarily) appears twice. The side sequence of the polygon is $ABCC^{-1}B^{-1}A^{-1}$.

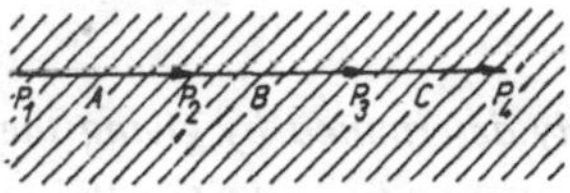

Figure 8

Let $\mathcal{F}$ be an n-sheeted covering of $\mathbb{C} \cup \{\infty\}$, and let G be a polygon of $\mathfrak{K}$. Then $\phi^{-1}(G)$ consists of n polygons, with at most vertices, which are branch points, in common, and no common edges at all. From this it is easy to see that the totality of polygons one obtains in this way, as G runs through $\mathfrak{K}$, decomposes $\mathcal{F}$ into a polygon complex in which each edge appears exactly twice, and with opposite orientations, i.e. $\mathcal{F}$ is a closed orientable surface in the sense of the above definition.

An example of a closed non-orientable surface is the projective plane $\mathbb{P}^2_{\mathbb{R}}$ (A.5). This is homeomorphic to a sphere with antipodal points identified. An associated polygon complex is therefore a 2-gon with both sides the same and with the same orientation. The polygon complex has a single vertex.

The phenomenon of non-orientable surfaces (with boundary) was discovered by Möbius. In a prize submission to the Paris Academy in 1861 he described the band named after him, which one obtains by joining the ends of a rectangular strip after a twist of 180°. If one cuts the Möbius band along the centre line, one obtains a connected surface. The decomposition of the Möbius band into a 4-gon is shown in Figure 9.

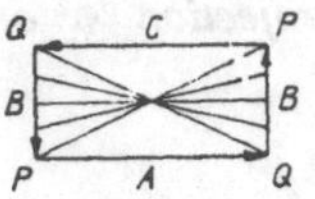

Figure 9

10.8 Classification of closed orientable surfaces

We want to classify the closed orientable surfaces up to homeomorphism. First a remark:

A polygon complex is completely determined by giving the preimages of the polygons in the euclidean plane, together with the pairing which tells which sides are mapped to the same edge in the topological space T : A homeomorphism of the circle onto itself may obviously be extended to a homeomorphism of the disc onto itself. Let K_1 and K_2 be discs, which are mapped onto polygons G_1 and G_2 by continuous mappings ϕ_1 and ϕ_2. Suppose the side A_1 resp. A_2 of K_1 resp. K_2 is mapped onto the edge A, where the initial and final points of A_1 resp. A_2 are denoted by P_1, Q_2. This gives a homeomorphism ψ of A_1 onto A_2, under which P_1 is mapped onto P_2 and Q_1 onto Q_2, or else P_1 onto Q_1 and Q_1 onto P_2. By composing with a suitable homeomorphism of K_1 onto itself one can convert ψ into a prescribed homeomorphism. It therefore suffices to give the mapping of vertices $\{P_1,Q_1\} \to \{P_2,Q_2\}$.

We now define *elementary transformations*, for which it is easy to show that they transform a polygon complex for a surface into a polygon complex for the same surface.

A *dimension* 1 *subdivision* consists of the subdivision of an edge of the complex into two edges by distinguishing an interior point of the edge as a new vertex. The converse process is called a *dimension* 1 *amalgamation.*

A *dimension* 2 *subdivision* consists similarly of the subdivision of a polygon into two polygons by introduction of a new edge which connects two non-adjacent vertices of the polygon. The converse process is a *dimension* 2 *amalgamation.*

Polygon complexes whose surfaces are homeomorphic to the sphere were already considered by Euler, who gave the theorem that the number e_2 of polygons minus the number e_1 of edges plus the number e_0 of vertices of the complex always equals 2. For an arbitrary polygon complex, $e_2 - e_1 + e_0$ is called the *Euler characteristic* of the complex. It is easy to convince oneself that the Euler characteristic and the orientability are invariant under elementary transformations.

Our goal is to use elementary transformations to reduce polygon complexes to normal forms which depend only on the Euler characteristic of the complex.

The passage to normal form occurs in four steps.

Step 1. When the complex consists of at least two polygons one can always find an edge which belongs to two different polygons. By removing this edge (dimension 2 amalgamation), the two polygons become one. By continuation of this process one finally obtains a complex consisting of a single polygon.

Step 2. Suppose the edge sequence of the polygon is of the form $W_1AA^{-1}W_2$, where A is an edge and W_1, W_2 are sequences of edges. When $W_1 = W_2 = 1$, i.e. the polygon is a 2-gon, then we have one of the desired normal forms. In this case the surface is homeomorphic to the extended plane. When W_1 or W_2 are not empty, then we want to go to a polygon with the edge sequence W_1W_2. By hypothesis there is a vertex P of the polygon apart from the initial point P_1 and the final point P_2 of A. By drawing an edge X from P_1 to P one obtains a complex of two polygons W_1AX, $X^{-1}A^{-1}W_2$ (dimension 2 subdivision). We now convert AX to a single edge Y (dimension 1 amalgamation) and then remove the latter edge (dimension 2 amalgamation). As a result we obtain the polygon W_1W_2. By continuing this process one comes either to the form AA^{-1} or else to a polygon in which no sequence of the form AA^{-1} appears. In what follows we assume that the polygon to be transformed is not of type AA^{-1}.

Step 3. Suppose the polygon has at least two different vertices. We want to go to a polygon with one vertex less. Let P and Q be different vertices, connected by an edge B from P to Q. Suppose B is followed by the edge A with endpoint R. We draw an edge D from R to P in our polygon (dimension 2 subdivision) and obtain two polygons BAD and $D^{-1}W$, where W is a sequence of edges. By hypothesis $B \neq A^{-1}$. If we had $B = A$ we should have $P = Q$. Thus the edge A occurs in the polygon $D^{-1}W$ and can be removed by dimension 2 amalgamation. In the resulting polygon the vertex P occurs once more and the vertex Q once less. We now remove edge sequences AA^{-1} again and obtain a polygon in which Q occurs at least once less often. By repetition of the process one arrives at a polygon with only one vertex.

Step 4. The edge sequence must be of the form $W_1CD^{-1}W_2C^{-1}W_3DW_4$ where C, D are edges. We draw an edge A connecting the initial points of the two copies of C, and remove the edge D. Then we obtain $W_1A^{-1}W_3W_2C^{-1}ACW_4$. Next we draw an edge B connecting the initial points of the two copies of A, and remove the edge C. Then we obtain $A^{-1}B^{-1}ABW_3W_2W_4W_1$. By continuing the process we arrive at a polygon whose edge sequence has the form

$$A_1^{-1}B_1^{-1}A_1B_1A_2^{-1}B_2^{-1}A_2B_2...A_g^{-1}B_g^{-1}A_gB_g.$$

This is the desired normal form.

The decomposition of the surface $\mathscr{F}$ into a 4-gon obtained above is called the *canonical dissection* of $\mathscr{F}$ by $A_1,B_1,...,A_g,B_g$. The corresponding Euler characteristic is 2-2g. The normal form AA^{-1} has Euler characteristic 2. We therefore set $g = 0$ in this case. The number g is called the *genus* of the surface; g is invariant under homeomorphisms, since a surface of genus g is homeomorphic to a sphere with g handles, which are preserved by homeomorphisms.

Thus we have proved the following theorem.

Theorem 4. *Let $\mathscr{F}$ be a closed orientable surface. Then either $\mathscr{F}$ is homeomorphic to the extended plane, or else there is a natural number g such that $\mathscr{F}$ is homeomorphic to a 4g-gon with edge sequence $A_1^{-1}B_1^{-1}A_1B_1A_2^{-1}B_2^{-1}A_2B_2...A_g^{-1}B_g^{-1}A_gB_g$. g is invariant under homeomorphisms of $\mathscr{F}$.* □

As a byproduct we have proved the following generalisation of the Euler polyhedron formula.

Theorem 5. *Let $\mathscr{F}$ be an orientable closed surface of genus g and let $\mathfrak{K}$ be a decomposition of $\mathscr{F}$ into a polygon complex. Then the Euler characteristic of $\mathfrak{K}$ is 2-2g.* □

It is only in the case $g = 0$ that two Riemann surfaces of the same genus are necessarily isomorphic (Chapter 11, Theorem 8). The surfaces of genus 1 may be characterised up to isomorphism by a complex number, the associated value of the *elliptic modular function* (Section 13.3).

For $g > 1$ the isomorphism classes of Riemann surfaces of genus g constitute a complex manifold of dimension $3g - 3$, as was already known to Riemann. The precise working out of this fact and its further development has been one of the main themes of mathematics in the 20th century.

Exercises

10.1 Define a mapping ϕ of the extended plane onto $P^1_{\mathbb{C}}$ by $\phi(z) = (1:z)$ for $z \in \mathbb{C}$, $\phi(\infty) = (0:1)$. Show that ϕ is a homeomorphism of $\mathbb{C} \cup \{\infty\}$ onto $P^1_{\mathbb{C}}$.

10.2 Let $g(z,w)$ be the polynomial of degree m considered in Section 10.5 and let $\mathcal{F}_g$ be the associated Riemann surface. Also let $a_i(z) = b_i(z)/c(z)$ for $i = 1,\dots,m$ be the representation of the coefficients of g as fractions with $(b_1(z),\dots,b_m(z),c(z)) = 1$. Suppose the polynomial $c(z)g(z,w)$ has total degree h in z and w. Then

$$g_1(u_0,u_1,u_2) := u_0^h c(u_1/u_0) g(u_1/u_0,u_2/u_0)$$

is a homogeneous polynomial of degree h.

With the notation of Section 10.5 we define a mapping ϕ of $\mathcal{F}_g - \pi^{-1}C$ into the projective curve associated with g_1,

$$\mathcal{K}_{g_1} := \{(x_0:x_1:x_2) \in P^2_{\mathbb{C}} \mid g_1(x_0,x_1,x_2) = 0\},$$

by letting the point P of U_i with $\pi P = z$ correspond to the point $(1:z:w_i(z))$.

a) Show that ϕ is a continuous monomorphism.

b) Show that ϕ extends uniquely to a continuous mapping of the whole of $\mathcal{F}_g$.

c) A point Q of $\mathcal{K}_{g_1}$ is called *singular* when $\phi^{-1}Q$ consists of at least two points. Show that $\mathcal{K}_{g_1}$ is singular at most at the image points of the critical set of g.

10.3 Let $h_3(z)$ be a third degree polynomial without multiple zeros. Show that the projective curve associated with $u_2^2u_0 - u_0^3h_3(u_1/u_0)$ has no singular points.

10.4 Let $h_4(z)$ be a fourth degree polynomial without multiple zeros. Show that the projective curve associated with $u_2^2u_0^2 - u_0^4h_4(u_1/u_0)$ has the one singular point (0:0:1).

10.5 Show that the *Fermat curve* of points $(x_0:x_1:x_2) \in \mathbb{P}_{\mathbb{C}}^2$ with $u_1^n + u_2^n = u_0^n$ has no singular points.

10.6 Let the notation be as in Exercise 2. Let Q be a point of $\mathcal{H}_{g_1}$ whose first coordinate is not 0. Show that Q is singular if and only if the two partial derivatives $\partial g_1/\partial u_1$ and $\partial g_1/\partial u_2$ equal 0 at Q. The corresponding result holds for a point whose second or third coordinate is not 0.

10.7 (Riemann). Let $\mathcal{F}$ be a closed Riemann surface which is an n-sheeted covering of the extended plane with branch points $P_1,\ldots,P_s$ and corresponding branching indices $e_1,\ldots,e_s$. Show that $\mathcal{F}$ has the Euler characteristic

$$2n - \sum_{j=1}^{s}(e_j-1).$$

10.8 Compute the genera of the Riemann surfaces which appear in Exercises 10.3 to 10.5.

10.9 Let $\mathcal{F}$ be a non-orientable closed surface. Show that $\mathcal{F}$ may be decomposed into a polygon complex consisting of a single polygon with edge sequence of the form $A_1A_1A_2A_2\ldots A_gA_g$.

11. Meromorphic differentials and functions on closed Riemann surfaces

11.1 Differentials and integrals on Riemann surfaces

The main goal of this chapter is to present the basic properties of meromorphic functions on closed Riemann surfaces. To do this we must first carry over the principles of Cauchy's function theory from the plane to Riemann surfaces.

We begin with the definition of a meromorphic differential on a Riemann surface $\mathcal{F}$.

Let U be an open subset of $\mathcal{F}$ on which meromorphic functions f_1, f_2 are defined, where f_2 is not constant. Also let Q be a point of U and let $t = \phi_Q(P)$ be a uniformising variable for Q, defined at points P in a neighbourhood $U_Q \subset U$ of Q. Then the *differential* $\omega = f_1 df_2$ on U_Q is given by the function $f_1 df_2/dt$. Let

$$f_1 df_2/dt = \sum_{n=h}^{\infty} c_n t^n. \qquad (1)$$

The concepts of *order, zero* and *pole* are carried over to ω from the meromorphic function (1). The coefficient c_{-1} is called the *residue* of ω at the point Q and is denoted by $\mathrm{res}_Q\omega$. All these concepts are independent of the choice of uniformising variable. On the other hand, the *principal part* $\sum_{n=h}^{-1} c_n t^n$ of ω at the point Q does depend on the choice of ϕ_Q (for $h \geq 0$ the residue and principal part at Q are set equal to 0).

Let $\mathcal{F}$ be covered by open sets U_j, $j \in J$: $\mathcal{F} = \bigcup_{j\in J} U_j$. Then a differential ω on $\mathcal{F}$ is given by differentials ω_j on U_j which are compatible with each other, i.e. for each point Q in $U_j \cap U_k$, where $j,k \in J$, we must have $\omega_j/dt = \omega_k/dt$. This definition is independent of the choice of uniformising variable t. Two differentials on $\mathcal{F}$ are equal when they are equal locally, i.e. in the neighbourhood of each point.

As an example we consider the Riemann surface $\mathcal{F}_g$ associated with $g = w^2 - (z^2-1)(z^2-4)$. Let $f = \sqrt{(z^2-1)(z^2-4)}$. Then $\omega = dz/f$ is a differential on $\mathcal{F}_g$ with no pole or zero anywhere: at a non-branch point, lying over $z_0 \in \mathbb{C}$, we have $t = z-z_0$, hence $\omega/dt = f^{-1} = c_0 + c_1 t + ...$ with $c_0 \neq 0$. For a point over ∞, $t = z^{-1}$ and

$$\omega/dt = -t^{-2}\left[\sqrt{(t^{-2}-1)(t^{-2}-4)}\right]^{-1}$$
$$= -\left[\sqrt{(1-t^2)(1-4t^2)}\right]^{-1}$$
$$= \mp (1 + c_1 t + \ldots).$$

For a branch point z_0, $t = \sqrt{z-z_0}$, hence

$$\omega/dt = 2t\left[\sqrt{((t^2+z_0)^2-1)((t^2+z_0)^2-4)}\right]^{-1}$$

$$= c_0 + c_1 t + \ldots \quad \text{with} \quad c_0 \neq 0.$$

Let C be a curve on $\mathcal{F}$ and let ω be a differential on $\mathcal{F}$ which is regular on C. Let C be given by the function $P(s)$ for $s \in [a,b]$.

Then the integral $\int_C \omega$ is defined by

$$\int_C \omega = \int_a^b (\omega/ds)ds, \tag{2}$$

where ω/ds is given in the neighbourhood of the point Q with uniformising variable $t = \phi_Q(P)$ by

$$\omega/ds = (\omega/dt)dt/ds, \; t = \phi_Q(P(s)). \tag{3}$$

For a smooth curve ω/ds represents a continuous function of s, independent of the choice of uniformising variable. Hence the integral (2) is well defined for a (piecewise smooth) curve C.

As one easily sees, the Cauchy integral theorem (Chapter 8, Theorem 2) carries over easily to simply connected regions on Riemann surfaces. One merely has to subdivide G enough so that each subregion lies in a single chart of $\mathcal{F}$.

11.2 The Riemann existence theorem for differentials

In the remaining sections of this chapter $\mathcal{F}$ will always denote a closed Riemann surface. It is customary to use the following terminology for differentials ω on $\mathcal{F}$: ω

is called a *differential of the first kind* when ω is everywhere regular, of the *second kind* when all residues of ω equal 0, and of the *third kind* when ω has only simple poles.

Thus the integral of a differential of the first kind is defined for each curve on $\mathcal{F}$. It depends only on the initial and final point of C, for all curves C on $\mathcal{F}$, if and only if it vanishes for each closed curve.

Let ω be a differential of the second kind and let C be any closed curve which begins (and ends) at a fixed point P_1 and contains no pole of ω. In what follows we shall call such a curve *admissible* (for ω). The number $\int_C \omega$ is called a *period* of ω. The set $\Lambda(\omega)$ of all periods of ω is a subgroup of the additive group of $\mathbb{C}$. Because for two admissible curves C_1, C_2 we have

$$\int_{C_1} \omega + \int_{C_2} \omega = \int_{C_3} \omega\,, \quad -\int_{C_1} \omega = \int_{C_1^{-1}} \omega$$

with $C_3 = C_1 D C_2 D^{-1}$, where D is a curve which begins at a point P_1 of C_1 and ends at a point P_2 of C_2.

In Section 10.7 we have decomposed $\mathcal{F}$ into a polygon complex. We can assume that the edges of the complex are curves on $\mathcal{F}$ in the sense of Section 10.4.

Theorem 1. *Let* ω *be a differential of the second kind and let* $A_1, B_1, \ldots, A_g, B_g$ *be a canonical dissection of* $\mathcal{F}$ *into a polygon* G (Section 10.8) *which is admissible for* ω. *Then the group* $\Lambda(\omega)$ *is generated by the fundamental periods;*

$$a_k = \int_{A_k} \omega,\ b_k = \int_{B_k} \omega \quad \text{for } k = 1, \ldots, g.$$

Proof. Because of the Cauchy integral theorem, which also holds for differentials of the second kind, it suffices to prove the theorem for crosscuts of G which begin and end at the same point P of $\mathcal{F}$. If, for example, P lies on A_j, then $\int_C \omega = \int_{B_j} \omega$ by the Cauchy integral theorem (Figure 10). □

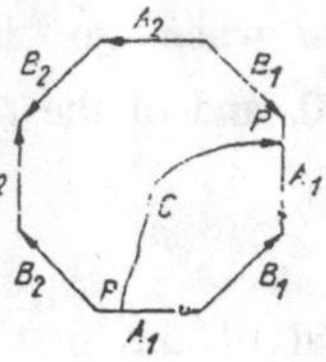

Fig. 10

By Section 8.8 a meromorphic differential on $\mathcal{F}$ can have only finitely many zeros and poles. But more is true:

Theorem 2. *Let* ω *be a meromorphic differential on* $\mathcal{F}$. *Then the sum of the residues of* ω *equals* 0.

Proof. First suppose $g \geq 1$ and let $\mathcal{F}$ be dissected by canonical cuts $A_1, B_1, \ldots, A_g, B_g$ into a polygon with interior $\mathcal{F}_0$ in such a way that no poles of ω lie on $A_1, B_1, \ldots, A_g, B_g$. By an easy generalisation of the residue theorem (Chapter 8, Theorem 12), the sum of the residues equals $\frac{1}{2\pi i}\int_C \omega$ where C, the boundary of $\mathcal{F}_0$, consists of the curve $A_1^{-1}B_1^{-1}A_1B_1 \ldots A_g^{-1}B_g^{-1}A_gB_g$. This integral vanishes, since along with each curve A_j, B_j in C the curve A_j^{-1}, B_j^{-1} also appears.

When $g = 0$ we choose a closed curve C with no poles of ω on it, and apply the residue theorem to each of the two simply connected regions into which $\mathcal{F}$ is divided by C. □

The crux of Riemann's function theory on closed surfaces $\mathcal{F}$ is the following existence theorem for differentials.

Theorem 3. *Let* $\mathcal{F}$ *be dissected by canonical cuts* $A_1, B_1, \ldots, A_g, B_g$ *into a polygon with interior* $\mathcal{F}_0$. *Let* $P_1, \ldots, P_s$ *be points of* $\mathcal{F}_0$ *with uniformising variables* ϕ_{P_j} *for* $j = 1, \ldots, s$, *and let* $f_1(z), \ldots, f_s(z)$ *be polynomials of the form* $f_j(z) = \sum_{j=1}^{h_j} c_{nj} z^n$ *with*

$$\sum_{j=1}^{s} c_{1j} = 0. \tag{4}$$

Also let $\hat{a}_1,\hat{b}_1,\dots,\hat{a}_g,\hat{b}_g$ *be arbitrary real numbers.*

Then there is exactly one meromorphic differential ω *on* $\mathcal{F}$ *with the following properties:*

1. ω *is regular, except at the points* $P_1,\dots,P_s$.
2. ω *has principal part* $f_j(t^{-1})$ *at* P_j *for* $t = \phi_{P_j}(P)$ *and* $j = 1,\dots,s$.
3. $\mathrm{Re}\int_{A_k} \omega = \hat{a}_k$, $\mathrm{Re}\int_{B_k} \omega = \hat{b}_k$ *for* $k = 1,\dots,g$. ∎

The restriction (4) is necessary by Theorem 2. Theorem 3 therefore represents a parametrisation of all meromorphic differentials on $\mathcal{F}$. Riemann sketched a proof of Theorem 3 with the help of what he called *Dirichlet's principle.* However, this proof contained a gap, which Hilbert was first able to fill in 1901 in his work "*Über das Dirichletsche Prinzip*" (Math. Ann. 59 (1901)). We shall prove Theorem 3 in Chapter 29.

As an immediate consequence of Theorem 3 one obtains

Theorem 4. *The differentials of the first kind on a closed Riemann surface of genus* g *form a vector space over* $\mathbb{C}$ *of dimension* g. □

11.3 The Riemann period relations

Let $\mathcal{F}$ be a closed Riemann surface which is dissected by canonical cuts $A_1,B_1,\dots,A_g,B_g$ into a polygon with interior $\mathcal{F}_0$. Theorem 3 shows that one cannot arbitrarily prescribe the periods $a_k := \int_{A_k} \omega$, $b_k := \int_{B_k} \omega$, for $k = 1,\dots,g$, of a differential ω of the first kind. The relations between the periods of differentials include the *Riemann period relations,* which we derive among the results which follow.

Let O be a fixed point on $\mathcal{F}_0$. Then integration of the differential ω from O to a variable point P in $\mathcal{F}_0$ along a curve in $\mathcal{F}_0$ yields a single-valued regular function f(P). We continue f to the boundary of $\mathcal{F}_0$; this is possible in a unique manner if we distinguish the two copies $A_k,\hat{A}_k$ and $B_k,\hat{B}_k$ of the boundary curves. At corresponding points P and $\tilde{P}$ on A_k and $\tilde{A}_k$ resp. B_k and $\tilde{B}_k$ we have $f(\tilde{P}) = f(P) + b_k$ resp. $f(\tilde{P}) = f(P) - a_k$ for $k = 1,...,g$ (Figure 11).

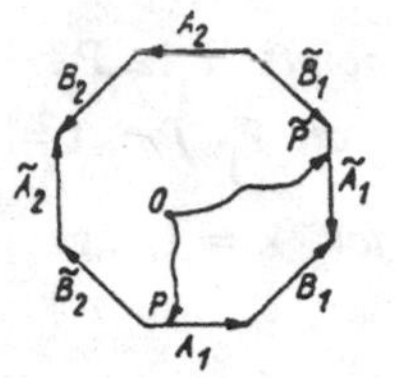

Fig. 11

Let C be the closed curve $A_1^{-1}B_1^{-1}A_1B_1,\ldots,A_g^{-1}B_g^{-1}A_gB_g$, and let ω' be a meromorphic differential of $\mathscr{F}$ which is regular on C, with periods $a_1', b_1', \ldots, a_g', b_g'$. Then

$$\int_{A_k} f\omega' - \int_{\tilde{A}_k} f\omega' = \int_{A_k} f\omega' - \int_{A_k} (f+b_k)\omega' = -b_k a_k'$$

and

$$\int_{B_k} f\omega' - \int_{\tilde{B}_k} f\omega' = a_k b_k',$$

hence

$$\int_C f\omega' = \sum_{k=1}^{g} (a_k b_k' - b_k a_k'). \tag{5}$$

On the other hand, by the residue theorem

$$\int_C f\omega' = 2\pi i \sum_{P\in F_0} \mathrm{res}_P f\omega'.$$

This yields the formula

$$2\pi i \sum_{P\in F_0} \mathrm{res}_P f\omega' = \sum_{k=1}^{g} (a_k b_k' - b_k a_k'). \tag{6}$$

When ω' is a differential of the first kind, $f\omega'$ is a regular differential in $\mathscr{F}_0 \cup C$ and hence

$$\sum_{k=1}^{g} (a_k b_k' - b_k a_k') = 0. \tag{7}$$

This is the *first Riemann period relation.* The second concerns the periods only of a differential $\omega \neq 0$ of the first kind:

$$i \sum_{k=1}^{g} (a_k \bar{b}_k - b_k \bar{a}_k) > 0. \tag{8}$$

Riemann proves (8) by proceeding from the following formula analogous to (5):

$$\int_C \bar{f}\omega = \sum_{k=1}^{g} (\bar{a}_k b_k - \bar{b}_k a_k).$$

By applying Green's theorem (A.3) he converts $\int_C \bar{f}\omega$ to a surface integral: with $f = u + iv$ and $\omega = du + idv$ we get

$$\int_C \bar{f}\omega = \int_C (u-iv)(du+idv)$$

$$= \int_C (udu+vdv) + i\int_C (udv+vdu)$$

$$= 2i\int_{F_0} du\, dv.$$

Using the uniformising variable $t = x + iy$, and bearing in mind the Cauchy-Riemann equations (Chapter 8, (1)), we get

$$du\, dv = \left(\left(\frac{\partial u}{\partial x}\right)^2 + \left(\frac{\partial u}{\partial y}\right)^2\right) dx\, dy$$

and hence $-i\int_C \bar{f}\omega > 0$.

The justification for this procedure comes from dividing the region of integration into sufficiently small subregions, each of which lies in a single chart of $\mathcal{F}$. □

11.4 Meromorphic functions

In this section we derive the simplest properties of meromorphic functions on $\mathcal{F}$.

Theorem 5. *Let* f *be a nonconstant meromorphic function on* $\mathcal{F}$. *Then* f *takes each value* c *(including* ∞*) only finitely often.*

Proof. Let $c \in \mathbb{C}$. Suppose that there is a sequence $P_1, P_2, \ldots$ of points of $\mathcal{F}$ with $f(P_1) = f(P_2) = \ldots = c$. Since $\mathcal{F}$ is closed, the sequence has an accumulation point Q. This implies $f(P) = c$ for all $P \in \mathcal{F}$ by Chapter 8, Theorem 11. For $c = \infty$ one considers $1/f$. □

One says that f *takes the value* $c \in \mathbb{C}$ *with multiplicity* n *at point* P when f-c has an n-tuple zero at P ; f *takes the value* ∞ *with multiplicity* n *at point* P when f has an n-tuple pole at P.

Theorem 6. *Let* f *be a non-constant meromorphic function on* $\mathcal{F}$. *Then* f *takes each value equally often (counted according to multiplicity).*

Proof. Let $c \in \mathbb{C}$. We apply Theorem 2 to the differential $df/(f-c)$. It follows that f takes the values c and ∞ equally often (cf. Chapter 8, Theorem 13). □

Theorem 7. *An everywhere regular function* f *on* $\mathcal{F}$ *is a constant.*

Proof. If the function f were not constant, then it would have to take every value equally often, and hence would have a pole. □

By Theorem 7, a meromorphic function is determined up to a constant factor by giving its zeros and poles with the associated multiplicities. This leads to the idea of describing a function by these data.

We formalise this concept by defining the *divisors of* f. In general a divisor D of $\mathcal{F}$ is a formal linear combination $D := n_1P_1 + \ldots + n_sP_s$ of points with integer coefficients (i.e. D is a mapping of $\mathcal{F}$ into $\mathbb{Z}$ which sends almost all $P \in \mathcal{F}$ to the number 0). The set of divisors of $\mathcal{F}$ forms an abelian group $\mathfrak{D}$ under addition of coefficients. The sum $n_1 + \ldots + n_s$ of coefficients of D is called the *degree*, deg D, of D. A divisor is called *effective* when all its coefficients are non-negative. We associate with a meromorphic function f the divisor

$$(f) = \sum_{P \in F} v_P(f)P \qquad \text{(cf. Section 8.8).}$$

By Theorem 5 this definition is meaningful, and by Theorem 6

$$\deg(f) = 0. \tag{9}$$

A divisor of the form (f) is called a *principal divisor.*

Let $\mathfrak{D}_0$ be the group of divisors D with $\deg D = 0$. The group $\mathfrak{D}_H$ of principal divisors is therefore a subgroup of $\mathfrak{D}_0$. The elements of the quotient group $\mathfrak{D}/\mathfrak{D}_H$ are called *divisor classes.* We have an exact sequence

$$0 \to \mathfrak{D}_0/\mathfrak{D}_H \to \mathfrak{D}/\mathfrak{D}_h \xrightarrow[\deg]{} \mathbb{Z} \to 0.$$

$\mathfrak{D}/\mathfrak{D}_H$ is called the *Picard group* of $\mathcal{F}$. We study this group in more detail in Chapter 12.

11.5 Riemann surfaces of genus 0

Let $\mathcal{F}$ be a surface of genus 0. Then $\deg D = 0$ is the only condition a divisor D has to satisfy in order to be the divisor of a meromorphic function on $\mathcal{F}$. In order to prove this, it suffices to construct a function f_Q which has a simple pole at a given point Q and which is regular at all other points of $\mathcal{F}$. Then $\sum_{j=1}^{s} (f_{Q_j} - f_{Q_j}(P_j))$ has the divisor $\sum_{j=1}^{s} (P_j - Q_j)$ (by Theorem 6). One obtains f_Q from the Riemann existence theorem for differentials; (Theorem 3). According to that theorem, there is a differential ω_Q which has principal part $1/t^2$ for Q and which is otherwise regular. Then the integral $f_Q(P)$ of ω_Q from a fixed point P_0 to a variable point P is independent of the path, and hence a meromorphic function of P, which has a simple pole at Q and is otherwise regular.

Also, f_Q is an isomorphism of $\mathcal{F}$ onto the extended plane. Hence we have

Theorem 8. *A Riemann surface of genus* 0 *is isomorphic to the extended plane.* □

11.6 The Riemann-Roch theorem

The considerations of Section 11.5 show that a surface $\mathcal{F}$ on which there is a meromorphic function f with (f) = D corresponding to any given divisor D with deg D = 0 has genus 0. Investigation of the conditions a divisor on a surface of genus g > 0 has to satisfy in order to be the divisor of a function is a central problem of the function theory on such surfaces. The general constraints occurring in this situation are described by the Riemann-Roch theorem, to which we now turn.

Since, on a closed surface $\mathcal{F}$, there are no everywhere regular functions apart from constants, it is an interesting question how many poles have to be admitted. Accordingly, we associate with a divisor $D = \sum_{P \in F} n_P P$ the vector space V(D) of all meromorphic functions f on $\mathcal{F}$ with

$$\nu_P(f) \geq -n_P \quad \text{for all} \quad P \in \mathcal{F}. \tag{10}$$

Here, following Riemann, we have effective divisors D particularly in mind. For the latter, (10) means there is a bound given by D on the number of possible poles.

By passing from f to df, (10) gives restrictions on the principal part of df, from which it easily follows by Theorem 3 that V(D) is a finite dimensional vector space over $\mathbb{C}$, whose dimension we denote by $l(D)$.

A differential ω is associated with a divisor

$$(\omega) = \sum_{P \in F} \nu_P(\omega) P. \tag{11}$$

With the terminology just introduced, the *Riemann-Roch theorem* can now be formulated as follows.

Theorem 9. *Let* D *be an arbitrary divisor and let* ω *be a non-zero meromorphic differential on a Riemann surface* $\mathcal{F}$ *of genus* g. *Then*

$$l(D) = \deg D - g + 1 + l((\omega) - D). \tag{12}$$

Remark. For a non-zero meromorphic function f_0,

$$l(D+(f_0)) = l(D),\ \deg(D+(f_0)) = \deg D. \qquad (13)$$

The first of these equations follows from the isomorphism $\phi : V(D) \to V(D+(f_0))$ given by $\phi(f) = (ff_0^{-1})$; the second follows from $\deg(f_0) = 0$.

The equations (13) show that Theorem 9 is a theorem about divisor classes. The non-zero divisors of differentials all lie in one class, called the *canonical class* of $\mathcal{F}$. In fact, for two non-zero meromorphic differentials ω_1, ω_2, ω_1/ω_2 is a non-zero meromorphic function.

As for the proof of Theorem 9, we first remark that the theorem is correct for $g = 0$ by Section 11.5. We can therefore assume from now on that $g \geq 1$. We first want to prove Theorem 9 for effective divisors $D = \sum_{P \in \mathcal{F}} n_P P$, and we introduce the following notation:

Ω_1 := $\mathbb{C}$-vector space of differentials of the first kind,

$\Omega_1(D) := \{\omega \in \Omega_1 \mid \nu_P(\omega) \geq n_P \text{ for } P \in \mathcal{F}\}$,

Ω_{Re} := $\mathbb{R}$-vector space of differentials of the second kind whose periods have vanishing real part,

$\Omega_{Re}(D) := \{\omega \in \Omega_{Re} \mid \nu_P(\omega) \geq -n_P - 1 \text{ for } P \in \mathcal{F} \text{ with } n_P > 0,$ and $\nu_P(\omega) \geq 0$ otherwise$\}$,

$\Omega_0(D) := \{\omega \in \Omega_{Re}(D) \mid \omega \text{ exact}\}$.

A differential ω is called *exact* when it is of the second kind and its periods vanish. This is equivalent to the integral of ω being independent of the path and hence defining a single-valued meromorphic function on $\mathcal{F}$. It follows that

$$\dim_{\mathbb{R}} \Omega_0(D) = 2l(D) - 2. \qquad (14)$$

Let ω_0 be an arbitrary non-zero meromorphic differential on $\mathcal{F}$. Then

$$\dim_{\mathbb{R}} \Omega_1(D) = \dim_{\mathbb{R}} V((\omega_0)-D). \qquad (15)$$

In fact, $\omega \mapsto \omega/\omega_0$ defines an isomorphism of $\Omega_1(D)$ onto $V((\omega_0)-D)$.

We can regard (5) as a bilinear form on the vector spaces Ω_1 and $\Omega_{Re}(D)$. We want to investigate the behaviour of this bilinear form β.

Proposition 1. $\beta(\omega,\omega') = 0$ *for all* $\omega' \in \Omega_{Re}(D)$ *if and only if* $\omega \in \Omega_1(D)$.

Proof. For $P \in \mathcal{F}$ with $n_P > 0$ and uniformising variable t of P, ω and ω' have the power series expansions

$$\omega/dt = c_0 + c_1 t + \ldots,\ \omega'/dt = d_n t^n + \ldots + d_{-2}t^{-2} + d_0 + \ldots$$

with $n = -n_P - 1$. Then $\beta(\omega,\omega') = 2\pi i \sum_{P\in F} \mathrm{res}_P f\omega' = 0$ for all $\omega' \in \Omega_{Re}(D)$ means that $c_0 = c_1 = \ldots = c_{n_P-1} = 0$ for all P with $n_P > 0$. □

Proposition 2. $\beta(\omega,\omega') = 0$ *for all* $\omega \in \Omega_1$ *if and only if* $\omega' \in \Omega_0(D)$.

Proof. Let $a_1',b_1',\ldots,a_g',b_g'$ be the periods of ω', purely imaginary by hypothesis, and let $a_1,b_1,\ldots,a_g,b_g$ be the periods of ω. It follows from $\beta(\omega,\omega') = \sum_{k=1}^{g} (a_k b_k' - b_k a_k') = 0$ for $\omega \in \Omega_1$ that

$$\sum_{k=1}^{g} b_k' \mathrm{Re}\, a_k - \sum_{k=1}^{g} a_k' \mathrm{Re}\, b_k = 0 \text{ for } \omega \in \Omega_1.$$

Since we can prescribe $\mathrm{Re}\, a_k$ and $\mathrm{Re}\, b_k$ arbitrarily for $k = 1,\ldots,g$ by Theorem 3, it follows that $a_k' = b_k' = 0$ for $k = 1,\ldots,g$ and hence that $\omega' \in \Omega_0(D)$. Conversely, it obviously follows from $\omega' \in \Omega_0(D)$ that $\beta(\omega,\omega') = 0$ for all $\omega \in \Omega_1$. □

It follows from Propositions 1 and 2 that the $\mathbb{R}$-vector spaces $\Omega_1/\Omega_1(D)$ and $\Omega_{Re}(D)/\Omega_0(D)$ have the same dimension:

$$\dim_{\mathbb{R}}\Omega_1 - \dim_{\mathbb{R}}\Omega_1(D) = \dim_{\mathbb{R}}\Omega_{Re}(D) - \dim_{\mathbb{R}}\Omega_0(D). \tag{16}$$

By Theorem 3

$$\dim_{\mathbb{R}}\Omega_1 = 2g\ ,\ \dim_{\mathbb{R}}\Omega_{Re}(D) = 2 \deg D. \tag{17}$$

Theorem 9 follows, for effective D, from (14) to (17).

In order to prove Theorem 9 for arbitrary divisors D, we first remark that

$$l((\omega)) = g\ ,\ \deg((\omega)) = 2g - 2 \tag{18}$$

holds for any non-zero meromorphic differential ω. It suffices to show (18) for a single differential ω. Since $g \geq 1$ we can assume that ω is a differential of the first kind. We set $D = 0$ in (15) and obtain $l((\omega)) = g$. Applying Theorem 9 to the effective divisor (ω) and to $D = 0$, we then get $\deg((\omega)) = 2g - 2$.

We can now write Theorem 9 in the symmetric form

$$l(D) - \frac{1}{2} \deg D = l((\omega_o)\text{-}D) - \frac{1}{2} \deg((\omega_o)\text{-}D). \tag{19}$$

We have proved (19) when one of the divisors D, $(\omega_0) - D$ is effective. Since $l(D)$ and $\deg D$ depend only on the class in which these divisors lie, (19) is already proved when an effective divisor lies in the class of D or $(\omega_0) - D$. Suppose this is not the case. Then $l(D) = l((\omega_0)\text{-}D) = 0$. It remains to show that $\deg D = g - 1$ in this situation.

We suppose that $\deg D \geq g$, and decompose D into its positive and negative parts D_1 and D_2 : $D = D_1 - D_2$. Then

$$l(D_1) \geq \deg D_1 - g + 1 = \deg D + \deg D_2 - g + 1 \geq \deg D_2 + 1. \tag{20}$$

Let $D_2 = n_1P_1 + ... + n_sP_s$. For $f \in V(D_1)$ to lie in $V(D)$, f must be of the form $c_0t_j^{n_j} + c_1t_j^{n_j+1} + ...$ for the uniformising variables t_j of P_j, $j = 1,...,s$. This yields $\deg D_2$ vanishing conditions for the coefficients of the power series expansion of f at the points $P_1,...,P_s$. Consequently

$$l(D) \geq l(D_1) - \deg D_2.$$

Hence $l(D) \geq 1$ by (20), contrary to hypothesis. It follows that $\deg D \leq g\text{-}1$. Similarly, one gets $\deg((\omega_0)\text{-}D) \leq g - 1$, and hence $\deg D = g - 1$ by (18). □

Riemann himself proved only the inequality

$$l(D) \geq \deg D - g + 1, \tag{21}$$

for effective divisors D, in his work on abelian functions. It can be proved without using (5) from the existence theorem for differentials. Theorem 9 was proved for effective divisors by Roch (*J. reine angew. Math.* 64 (1865)), and was first proved for arbitrary divisors by Dedekind and Weber (see Chapter 23).

The inequality (21) gives the answer to the question posed at the beginning of this section on the existence of meromorphic functions on Riemann surfaces of genus g. To be precise we have

Theorem 10. *In the notation of Theorem 9:*

1. $l(D) = 0$ *when* $\deg D \leq 0$ *and* D *is not a principal divisor.*

2. $l(D) = \deg D + 1 - g$ *when* $\deg D \geq 2g - 2$ *and* D *does not lie in the canonical class.*

Proof. Let $\deg D \leq 0$ and $f \in V(D)$. Then by definition $D + (f)$ is a positive divisor of degree ≤ 0, i.e. $D + (f) = 0$. This implies 1. Also, 2 follows from 1 and Theorem 9. □

For $0 < \deg D < 2g-2$ one has only the inequality (21).

11.7 The field of meromorphic functions on a closed Riemann surface

The set $K(\mathcal{F})$ of meromorphic functions on a Riemann surface obviously constitutes a field under addition and multiplication of functions. In this section we want to investigate the field $K(\mathcal{F})$ in more detail for closed surfaces $\mathcal{F}$.

In Chapter 10, Theorem 1, we have seen that $K(\mathbb{C} \cup \{\infty\}) = \mathbb{C}(z)$, the field of rational functions in one indeterminate z. Since $\mathcal{F}$ is a covering of $\mathbb{C} \cup \{\infty\}$, each function on $\mathbb{C} \cup \{\infty\}$ can also be regarded as a function on $\mathcal{F}$, i.e. we have $\mathbb{C}(z) \subset K(\mathcal{F})$. Let $\mathcal{F}$ be an n-tuple covering of $\mathbb{C} \cup \{\infty\}$. Then z takes each value n times as a function on $\mathcal{F}$: z is the function which associates with each $P \in \mathcal{F}$ its projection on $\mathbb{C} \cup \{\infty\}$. From now on we assume the $\mathcal{F}$ is closed.

Theorem 11. *Each function* f *in* $K(\mathcal{F})$ *satisfies an equation*

$$f^n + r_1(z)f^{n-1} + \dots + r_n(z) = 0,$$

where $r_1(z),\dots,r_n(z)$ *are in* $\mathbb{C}(z)$.

Proof. Let M be the finite set of points z_0 of $\mathbb{C}\cup\{\infty\}$ over which $\mathcal{F}$ branches or f has a pole. For $z_0 \in \mathbb{C}\cup\{\infty\} - M$ we can construct the elementary symmetric functions (Section 7.3)

$$r_1(z) = -\sum_{j=1}^{n} f(P_j),$$

$$r_2(z) = \sum_{1\le j_1<j_2\le n} f(P_{j_1})f(P_{j_2}),$$

$$\vdots$$

$$r_k(z) = (-1)^k \sum_{1\le j_1<\dots<j_k\le n} f(P_{j_1})\dots f(P_{j_k}),$$

$$\vdots$$

$$r_n(z) = (-1)^n f(P_1)\dots f(P_n),$$

where $P_1,\dots,P_n$ are the points of $\mathcal{F}$ over z. Since $P_1,\dots,P_n$ have the same uniformising variable, $r_k(z)$ is a regular function for $z \in \mathbb{C}\cup\{\infty\} - M$. For $z_0 \in M$, $r_k(z)$ has a Laurent expansion. Since f is meromorphic on $\mathcal{F}$, the latter can have only finitely many terms with negative powers of the uniformising variable (Section 8.8). I.e., $r_k(z)$ is meromorphic over the whole of $\mathbb{C}\cup\{\infty\}$, and hence $r_k(z) \in \mathbb{C}(z)$, for $k = 1,\dots,n$.

Let x be an indeterminate and let

$$\phi_f(x,z) = \phi_f(x) := (x-f(P_1))\dots(x-f(P_n)) = x^n + r_1(z)x^{n-1} + \dots + r_n(z).$$

Then $\phi_f(f(P)) = 0$. □

Theorem 12. *There is a function* f *on* $\mathcal{F}$ *for which the polynomial* $\phi_f(x)$ *constructed in the proof of Theorem 11 is irreducible over the field* $\mathbb{C}(z)$, *i.e.* $\mathbb{C}(z,f)$ *is an extension of degree* n *over* $\mathbb{C}(z)$.

Proof. Let z_0 be a point of $\mathbb{C}$, over which the n different points $P_1,\ldots,P_n$ of $\mathcal{F}$ lie. By Theorem 3 there is a meromorphic differential ω_k on $\mathcal{F}$ with a single pole at P_k and principal part $dz/(z-z_0)^2$. Let $c_1,\ldots,c_n$ be distinct complex numbers. Then

$$f := (z-z_0)^2\left[c_1\frac{\omega_1}{dz} + \ldots + c_n\frac{\omega_n}{dz}\right] \tag{22}$$

is a meromorphic function on $\mathcal{F}$ with $f(P_k) = c_k$ for $k = 1,\ldots,n$.

We want to show that $\phi_f(x)$ is irreducible. Suppose $\phi_f(x) = \phi_1(x)\phi_2(x)$ with $\phi_1(f) = 0$. Then in particular $\phi_1(f(P_k),z_0) = 0$ for $k = 1,\ldots,n$, i.e. $\phi_1(x,z_0)$ has n different zeros. This implies $\phi_f(x) = \phi_1(x)$. □

Theorem 13. *The field* $K(\mathcal{F})$ *is an extension of degree* n *over* $\mathbb{C}(z)$.

Proof. The function (22) generates $K(\mathcal{F})$. For suppose f_1 is an arbitrary function in $K(\mathcal{F})$ and let f_2 be a primitive element for the extension $\mathbb{C}(z,f,f_1)$, i.e. $\mathbb{C}(z,f,f_1) = \mathbb{C}(z,f_2)$ (Section 7.5). By Theorem 11, $[\mathbb{C}(z,f_2) : \mathbb{C}(z)] \le n$, whence it follows that $f_1 \in \mathbb{C}(z,f)$. □

Exercises

11.1 Let $h_m(z)$ be a polynomial of mth degree without multiple zeros. Show that the Riemann surface $\mathcal{F}_g$ for the polynomial $w^2 - h_m(z)$ has genus $g = [(m-1)/2]$, and give a basis for the space of differentials of the first kind.

11.2 Give a basis for the space of differentials of the first kind in the case of the Fermat curve (Exercises 10.5 and 10.8).

11.3 Let $\mathcal{F}$ be a Riemann surface of genus g. Suppose that the function field of $\mathcal{F}$ contains a function which takes each value exactly twice.

a) Show that f represents the surface $\mathcal{F}$ as a two-sheeted covering of $\mathbb{C} \cup \{\infty\}$ with $2g+2$ branch points $t_1,\ldots,t_{2g+2}$.

b) Show that $\mathcal{F}$ is isomorphic to the $\mathcal{F}_g$ with $g = w^2 - (z-t_1)\ldots(z-t_{2g+2})$ (for $t_i = \infty$ one has to set $z-\infty := 1$).

c) Show that such a function f always exists for $g = 1$ or $g = 2$.

11.4 Let $\mathcal{F}$ be the Riemann surface of the polynomial $g(z,w) = w^2 = h(z)$, where $h(z)$ is a third degree polynomial without multiple zeros. By Chapter 10, Exercise 3, the points of $\mathcal{F}$ are in one-to-one correspondence with those of the complex plane curve $\{(z,w) \mid w^2=h(z)\}$, to which a point P_∞ at infinity has been added. We identify $\mathcal{F}$ with this curve.

a) Show that for any two points P_1,P_2 of $\mathcal{F}$ there is a unique point P_3 of $\mathcal{F}$ for which $P_1 + P_2 - P_3 - P_\infty$ is a principal divisor.

b) Show that $P_1,P_2 \mapsto P_1 \oplus P_2 := P_3$ defines a group operation with P_∞ as identity element.

c) Show that $\ominus P_1$ is the reflection of P_1 in the z-axis.

d) Show that $P_1,P_2,\ominus P_3$ lie on a line. (Hint: l is given by an equation $az + bw + c = 0$ with $a,b,c \in \mathbb{C}$; $az + bw + c$ is a meromorphic function on $\mathcal{F}$. Consider $(az + bw + c)$.)

12. The theorems of Abel and Jacobi

12.1 Abel's theorem

We now come back to the question raised in Section 10.1 about the integration of algebraic functions. We can now formulate this question precisely as the question of integrating meromorphic differentials; on a closed Riemann surface $\mathcal{F}$ of genus g.

For the mathematicians of the 18th century and the first half of the 19th century the integral of $\frac{dt}{\sqrt{1-t^2}}$, i.e. the function $y = \arcsin x$, often served as a model. The addition theorem for the inverse function $\sin y$ yields

$$\arcsin u + \arcsin v = \arcsin(u\sqrt{1-v^2}+v\sqrt{1-u^2}). \tag{1}$$

Euler found the analogue of (1) for the integral $\frac{dt}{\sqrt{f(t)}}$, where $f(t)$ is a fourth degree polynomial without multiple zeros (Exercise 12.3).

One of Abel's most important achievements was to find a generalisation of (1) to arbitrary integrals of meromorphic differentials, which for that reason are also known as *abelian integrals* (*Démonstration d'une propriété générale d'une certaine classe de fonctions transcendentes, J. reine angew. Math.* 4 (1829)).

Here we confine ourselves to differentials ω of the first kind. For them, Abel's result can be expressed as follows.

Let $(f) = \sum_{j=1}^{s} n_j P_j$ be a principal divisor and let O be an arbitrary point of $\mathcal{F}$. Then, with suitable choice of path of integration,

$$\sum_{j=1}^{s} n_j \int_{O}^{P_j} \omega = 0. \tag{2}$$

Equation (2) is an assertion about the values of the "function" $h(P) = \int_O^P \omega$. In the complex numbers one regards arcsin x as a many-valued function, i.e. as a mapping of $\mathbb{C}$ into $\mathbb{C}/2\pi\mathbb{Z}$. In the case $g = 1$, $h(P)$ proves to be similarly meaningful, since $\Lambda(\omega)$ in this case is a lattice in $\mathbb{C}$. (A subgroup Λ of the real n-dimensional vector space V is called a *lattice* in V when the rank of Λ (A1, Theorem 5) equals n). Thus $h(P)$ is a mapping of $\mathcal{F}$ into $\mathbb{C}/\Lambda(\omega)$. However, for $g > 1$, $\Lambda(\omega)$ is in general not a lattice in $\mathbb{C}$, and it is not useful to consider $h(P)$ as a many-valued function (cf. Section 10.1).

Jacobi found a way out of this situation by considering a basis $\omega_1,\dots,\omega_g$ of the complex vector space of differentials of the first kind (Chapter 11, Theorem 4). Then the closed curve C is associated with the period vector

$$\alpha(C) = \left(\int_C \omega_1,\dots,\int_C \omega_g\right).$$

Theorem 1. *Let* $A_1,B_1,\dots,A_g,B_g$ *be a canonical dissection of* $\mathcal{F}$. *Then the* 2g *period vectors* $\alpha(A_1),\alpha(B_1),\dots,\alpha(A_g),\alpha(B_g)$ *span a lattice* Γ *of the* 2g-*dimensional real vector space* $\mathbb{C}^g$.

Proof. Suppose that the abelian group Γ has a rank $< 2g$. Then there is an $\mathbb{R}$-vector space V of dimension $< 2g$, which contains Γ, and a non-trivial real linear form on $\mathbb{C}^g$ which vanishes on V. Let $z_1 = x_1 + iy_1,\dots,z_g = x_g + iy_g$ be the coordinates of $\mathbb{C}^g$. The linear form has the form

$$L(z_1,\dots,z_g) = \sum_{j=1}^{g} (c_jx_j+d_jy_j) = \operatorname{Re} \sum_{j=1}^{g} (c_j-id_j)z_j,$$

Where not all coefficients $c_1,d_1,\dots,c_g,d_g$ equal 0. It follows that

$$\operatorname{Re}\int_{A_k} \sum_{j=1}^{g} (c_j-id_j)\omega_j = \operatorname{Re}\int_{B_k} \sum_{j=1}^{g} (c_j-id_j)\omega_j = 0$$

for $k = 1,\dots,g$. By the uniqueness assertion of Chapter 11, Theorem 3, we therefore have $\sum_{j=1}^{g} (c_j-id_j)\omega_j = 0$, and hence $c_j = d_j = 0$ for $j = 1,\dots,g$, contrary to the choice of $c_1,d_1,\dots,c_g,d_g$. □

$\mathbb{C}^g/\Gamma$ is called the *Jacobian variety* of $\mathcal{F}$.

We define a mapping Φ of $\mathfrak{D}_0$ (Section 11.4) into $\mathbb{C}^g/\Gamma$ by

$$\Phi\left(\sum_{l=1}^{s} n_l P_l\right) = \left[\sum_{l=1}^{s} n_l \int_0^{P_l} \omega_1, \ldots, \sum_{l=1}^{s} n_l \int_0^{P_l} \omega_g\right] + \Gamma.$$

Here O is an arbitrary point of $\mathcal{F}$. The mapping Φ does not depend on the choice of O, since $D := \sum_{l=1}^{s} n_l P_l$ has degree 0, i.e. $\sum_{l=1}^{s} n_l = 0$. The integrals are to be taken along arbitrary curves from O to P_l. By definition of Γ, $\Phi(D)$ is independent of the choice of these curves, Φ is obviously a homomorphism of abelian groups. The following theorem is what we usually call *Abel's theorem* today,

Theorem 2. *The kernel of* Φ *is equal to the group* $\mathfrak{D}_H$ *of principal divisors.*

In fact, only the theorem that $\mathfrak{D}_H$ is contained in the kernel of Φ is due to Abel; see (2). The converse is due to Clebsch (*J. reine angew. Math.* 63 (1863)).

Proof of Theorem 2. Let f be a meromorphic function on $\mathcal{F}$ and let $(f) = n_1P_1 + \ldots + n_sP_s$. We want to show $\Phi((f)) = 0$.

We choose a canonical dissection of $\mathcal{F}$ such that no zeros or poles of f lie on $A_1, B_1, \ldots, A_g, B_g$. Let O be a point on the dissected surface $\mathcal{F}_0$. Then $f_k(P) = \int_O^P \omega_k$ is a regular function in $\mathcal{F}_0$ for $k = 1, \ldots, g$. We set $\gamma(P) = (f_1(P), \ldots, f_g(P))$. Then we have to show

$$n_1\gamma(P_1) + \ldots + n_s\gamma(P_s) \in \Gamma. \tag{3}$$

To do this we apply Chapter 11, (6), to f_k and df/f. Since $\operatorname{res}_{P_j}(f_k\, df/f) = n_j f_k(P_j)$ we get

$$2\pi i(n_1\gamma(P_1) + \ldots + n_s\gamma(P_s)) = \sum_{j=1}^{g} \left(\alpha(A_j)\int_{B_j} df/f - \alpha(B_j)\int_{A_j} df/f\right).$$

Hence to prove (3) it suffices to prove the following proposition.

Proposition. *Let* C *be a closed curve on* $\mathcal{F}$, *on which* f *has neither zeros nor poles. Then the integral of* df/f *over* C *equals an integer multiple of* $2\pi i$.

Proof. Let C have parametric representation C(t) for $t \in [0,1]$, and let log z be the principal value of the logarithm of z, i.e. $\log z = \log r + i\phi$ for $z = re^{i\phi}$ with $-\pi < \phi \le \pi$. Then

$$df/f = \log f(C(t)) - \log f(C(0)) + 2\pi i n(t), \tag{4}$$

where n(t) is the function with $n(0) = 0$ for which the right-hand side of (4) is continuous. The function n(t) takes integer values. It follows from (4) that

$$\int_C df/f = \int_0^1 df/f = 2\pi i n(1). \qquad \square$$

We now come to the proof of Clebsch's part of Theorem 2.

Let $D \in \mathfrak{D}_0$, $D = n_1P_1 + ... + n_sP_s$ and suppose $\Phi(D) = 0$. By Chapter 11, Theorem 3, there is a differential ω' with principal parts $n_jt_j^{-1}$ for $j = 1,...,s$, where t_j is the uniformising variable of P_j, and such that $\mathrm{Re}\int_{A_k} \omega' = \mathrm{Re}\int_{B_k} \omega' = 0$ for $k = 1,...,g$. By Chapter 11, (6) and the fact that $\Phi(D) = 0$ we get

$$\sum_{j=1}^{g} (\alpha(A_j)\int_{B_j} \omega' - \alpha(B_j)\int_{A_j} \omega') = 2\pi i(n_1\gamma(P_1) + ... + n_s\gamma(P_s)) \in 2\pi i\Gamma.$$

Hence by Theorem 1, $\int_{A_j} \omega'$ and $\int_{B_j} \omega'$ are in $2\pi i\mathbb{Z}$ for $j = 1,...,g$. It follows that $f(P) := \exp\int_O^P \omega'$ is a meromorphic function on $\mathcal{F}$ with $(f) = D$. $\square$

By Theorem 2, Φ induces a monomorphism Ψ of $\mathfrak{D}_0/\mathfrak{D}_H$ into $\mathbb{C}^g/\Gamma$. We now want to investigate the mathematical nature of the groups $\mathfrak{D}_0/\mathfrak{D}_H$, $\mathbb{C}^g/\Gamma$ and the mapping Φ in more detail.

12.2 Non-special divisors

We denote the set of divisors of the form $P_1 + ... + P_g$ by $S_g(\mathscr{F})$. A topology is defined on $S_g(\mathscr{F})$ in a natural way (A2) : if $P_1,...,P_g$ are points of $\mathscr{F}$ with neighbourhoods $U_1,...,U_g$ then

$$\{Q_1 + ... + Q_g | Q_j \in U_j \text{ for } j = 1,...,g\} \tag{5}$$

is a neighbourhood of $P_1 + ... + P_g$. We take the neighbourhoods of the form (5) as a basis for the system of neighbourhoods of $P_1 + ... + P_g$.

A divisor $D \in S_p(\mathscr{F})$ is called *special* when $\Omega_1(D) \neq \{0\}$ (Section 11.6).

Proposition 1. *There are distinct points* $Q_1,...,Q_g$ *such that* $Q_1 + ... + Q_g$ *is non-special.*

Proof. Let $\omega_1 \neq 0$ be a differential of the first kind. We choose a point Q_1 which is not a zero of ω_1. By Theorem 4, $\dim_{\mathbb{C}}(\Omega_1) = g\text{-}1$. When $g > 1$ we choose a differential $\omega_2 \neq 0$ in $\Omega_1(Q_1)$ and a point Q_2 which is not a zero of ω_2. Continuing in this way, we obtain a divisor $Q_1 + ... + Q_g$ with $\Omega_1(Q_1 + ... + Q_g) = \{0\}$. □

The proof shows, moreover, that the non-special divisors lie densely in $S_p(\mathscr{F})$.

In the case $g = 1$, $\Omega_1(P) = 0$ for all points $P \in \mathscr{F}$ (Chapter 11, Theorem 9 and (15)). Thus there are no special divisors.

For a general g the non-special divisors constitute an open subset in $S_g(\mathscr{F})$. This justifies calling them non-special.

We fix a non-special divisor $D_0 = Q_1 + ... + Q_g$ with distinct $Q_1,...,Q_g$ and define a mapping A of $\mathscr{F} \times ... \times \mathscr{F} = \mathscr{F}^g$ into $\mathfrak{D}_0/\mathfrak{D}_H$ by

$$A(P_1,...,P_g) = P_1 + ... + P_g - D_0 + \mathfrak{D}_H.$$

Proposition 2. A *is surjective and, in the case* $g = 1$, *also injective.*

Proof. Let $D \in \mathfrak{D}_0$. Then $D + D_0$ has degree g. By the Riemann inequality (Chapter 11, (21)), $l(D+D_0) \geq 1$. Hence there is a meromorphic function f for which $(f) + D + D_0$ is effective. The latter divisor is therefore of the form $P_1 + \ldots + P_g$.

For $g = 1$, $l(D+D_0) = 1$ (Chapter 11, Theorem 10), and this determines P_1 uniquely from D. □

For $g > 1$, A is never injective, since the same divisor class corresponds to the points $(P_{\pi(1)},\ldots,P_{\pi(g)})$ for each permutation π of the indices $1,\ldots,g$. We therefore consider, as well as $\mathcal{F}^g$, the *symmetric product* which one obtains from $\mathcal{F}^g$ by identifying all points which differ only by a permutation of the components. Obviously, the symmetric product can also be understood as the set $S_g(\mathcal{F})$ of effective divisors of degree g. We define a mapping B of $S_g(\mathcal{F})$ into $\mathfrak{D}_0/\mathfrak{D}_H$ by $B(D) = D - D_0 + \mathfrak{D}_H$ for $D \in S_g(\mathcal{F})$.

Theorem 3. B *is a mapping of* $S_g(\mathcal{F})$ *onto* $\mathfrak{D}_0/\mathfrak{D}_H$ *whose restriction to the set of non-special divisors is injective. More precisely, two divisors from* $S_g(\mathcal{F})$ *with the same image in* B, *one of which is non-special, are equal.*

Proof. Because of Proposition 2 it suffices to show that there is only one effective divisor in the class of a non-special divisor D. In fact, by the Riemann-Roch theorem (Chapter 11, Theorem 9 and (15)) $l(D) = 1$. □

12.3 The analytic nature of ΨA

The sets $\mathcal{F}^g$, $S_g(F)$ and $\mathbb{C}^g/\Gamma$ have a natural topological structure (A2), i.e. it is clear in each case what is meant by the neighbourhood of a point. A sufficiently small neighbourhood U of a point is homeomorphic in a natural way to a neighbourhood of the origin of $\mathbb{C}^g$ (one therefore speaks of g-dimensional analytic manifolds):

For $\mathbb{C}^g/\Gamma$ the projection $V \to \mathbb{C}^g/\Gamma$ is a homeomorphism for a sufficiently small neighbourhood V of O in $\mathbb{C}^g$, because Γ is a lattice in $\mathbb{C}^g$.

Let U_j be a neighbourhood of $P_j \in \mathcal{F}$ which is mapped onto a neighbourhood of O in $\mathbb{C}$ by a uniformising variable, for $j = 1,\ldots,g$. Then the neighbourhood $U_1 \times \ldots \times U_g$ of $(P_1,\ldots,P_g) \in \mathcal{F}^g$ is mapped onto a neighbourhood of the O in $\mathbb{C}^g$.

In order to prove the assertion for $S_g(\mathscr{F})$ it suffices, by what we have just said, to consider $S_g(\mathbb{C})$. We obtain a homeomorphism of $S_g(\mathbb{C})$ onto $\mathbb{C}^g$ by the mapping

$$(z_1,\dots,z_g) \mapsto (s_1(z_1,\dots,z_g),\dots,s_g(z_1,\dots,z_g)),$$

where $s_1(z_1,\dots,z_g) = z_1 + \dots + z_g,\dots,s_g(z_1,\dots,z_g)$ are the elementary symmetric functions (Section 7.3).

An analytic mapping of analytic manifolds is defined analogously to the case of Riemann surfaces (Section 10.4). It suffices to explain what an analytic mapping of a neighbourhood $U_\varepsilon(O) = \{(z_1,\dots,z_g) \,\big|\, |z_j| < \varepsilon \text{ for } j = 1,\dots,g\}$ in $\mathbb{C}^g$ is. This is a mapping

$$(z_1,\dots,z_g) \mapsto (f_1(z_1,\dots,z_g),\dots,f_g(z_1,\dots,z_g))$$

where $f_1,\dots,f_g$ are analytic functions on $U_\varepsilon(O)$, i.e. functions expandable in power series in $z_1,\dots,z_g$ (which converge in $U_\varepsilon(O)$).

Theorem 4. $\Psi A : \mathscr{F}^g \to \mathbb{C}^g/\Gamma$ *is an analytic mapping.*

Proof. Let $(P_1,\dots,P_g) \in \mathscr{F}^g$. Then the functions $f_{jk}(P_k') = \int_{P_k}^{P_k'} \omega_j$ are uniquely defined as regular functions for points P_k' in a sufficiently small neighbourhood of P_k, and

$$\Psi A(P_1',\dots,P_k') = \Psi A(P_1,\dots,P_g) + \left[\sum_{k=1}^{g} f_{1k}(P_k'),\dots,\sum_{k=1}^{g} f_{gk}(P_k')\right]. \tag{6}$$

Theorem 5. ΨA *is an analytic isomorphism in a neighbourhood of* $(Q_1,\dots,Q_g)$.

Proof. We have to show that the mapping $\Psi A|_U : U \to \Psi A(U)$ is invertible for a sufficiently small neighbourhood U of $(Q_1,\dots,Q_g)$. It will follow that the inverse mapping is also analytic. Let $t_1,\dots,t_g$ be uniformising variables for $Q_1,\dots,Q_g$. Then $\Psi A|_U$ is invertible for a sufficiently small U when the Jacobian of (6),

$$\det\begin{bmatrix} \frac{\omega_1}{dt_1}(Q_1) & \cdots & \frac{\omega_1}{dt_g}(Q_g) \\ \vdots & & \\ \frac{\omega_g}{dt_1}(Q_1) & \cdots & \frac{\omega_g}{dt_g}(Q_g) \end{bmatrix}$$

is non-zero. In order to convince ourselves that this is the case, we consider the homeomorphism

$$\omega \mapsto \left(\frac{\omega}{dt_1}(Q_1),\ldots,\frac{\omega}{dt_g}(Q_g)\right)$$

of the space of differentials of the first kind into $\mathbb{C}^g$. When ω is in the kernel of this homeomorphism, ω has the zeros $Q_1,\ldots,Q_g$ and is therefore 0. □

12.4 Jacobi's theorem and the Jacobi inversion problem

The following theorem is known as *Jacobi's theorem.*

Theorem 6. *The mapping* $\Phi : \mathfrak{D}_0 \to \mathbb{C}^g/\Gamma$ *is surjective.*

Proof. Let $z \in \mathbb{C}^g$. For a sufficiently large natural number n, z/n lies in an arbitrarily small neighbourhood of 0. Hence by Theorem 5 there is a $(P_1,\ldots,P_g) \in \mathcal{F}^g$ with $\Psi A(P_1,\ldots,P_g) = z/n + \Gamma$. This implies $\Phi(nP_1 + \ldots + nP_g) = z + \Gamma$. □

By Lemma 2, Theorem 6 implies

Theorem 7. *The mapping* $\Psi A : \mathcal{F}^g \to \mathbb{C}^g/\Gamma$ *is surjective.* □

To interpret our results further we first consider the case $g = 1$.

In this case the mapping $\Psi A : \mathcal{F} \to \mathbb{C}/\Gamma$ is an analytic isomorphism (Theorems 2 to 7). Let $\pi : \mathcal{F} \to \mathbb{C} \cup \{\infty\}$ be the projection associated with $\mathcal{F}$. Then

$$\mathbb{C} \to \mathbb{C}/\Gamma \xrightarrow[(\Psi A)^{-1}]{} \mathcal{F} \xrightarrow[\pi]{} \mathbb{C} \cup \{\infty\} \tag{7}$$

is a meromorphic function f on $\mathbb{C}$ with

$$f(z+w) = f(z) \quad \text{for} \quad z \in \mathbb{C},\ w \in \Gamma. \tag{8}$$

In particular, this proves the result of Abel, announced in Section 10.1, on the inversion of the integral $\int \frac{dx}{\sqrt{(1-x^2)(1-k^2x^2)}}$. Meromorphic functions f on $\mathbb{C}$ with property (8) for a lattice Γ of $\mathbb{C}$ are known as *elliptic functions*. The reason for the name is that the integral $\int \frac{dx}{\sqrt{(1-x^2)(1-k^2x^2)}}$ is connected with the arc length of the ellipse (Exercise 12.1). The lattice Γ is called the *lattice of periods* of f, as it consists of all the periods of f. Jacobi found a representation of the inverse function as the quotient of two functions regular in $\mathbb{C}$ (*Fundamenta nova functionum ellipticarum, Königsberg* 1829). We come back to this in the next chapter.

Jacobi also found the right way to generalise Abel's result to the case $g > 1$.

We have seen that the mapping ΨB is injective on the set of non-special divisors (Theorems 2, 3). Hence we can invert ΨB on the image of this set. Thus inversion is to be understood as a mapping of $\mathbb{C}^g/\Gamma$ which is defined only on an open, dense subset (Section 12.2). Such a phenomenon is natural in the function theory of several variables. It already occurs with the mapping $(z_1,z_2) \mapsto z_1/z_2$ of $\mathbb{C}^2$ into $\mathbb{C} \cup \{\infty\}$, where the point $(0,0)$ corresponds to no value in $\mathbb{C} \cup \{\infty\}$.

As a generalisation of (7) we have g inverse functions

$$f_j : \mathbb{C}^g \to \mathbb{C}^g/\Gamma \xrightarrow[(\Psi B)^{-1}]{} S_g(\mathcal{F}) \xrightarrow[\pi_j]{} \mathbb{C} \cup \{\infty\},\ j = 1,\dots,g, \tag{9}$$

where π_j gives the divisor $P_1 + \dots + P_g$ the value $s_j(\pi P_1,\dots,\pi P_g)$, if the latter is defined.

Following Riemann's function theory, Clebsch and Gordan (*Theorie der Abelschen Funktionen*, Leipzig 1866) generalised Jacobi's result for the case $g = 1$, that the functions f_j are representable as quotients of two functions regular on $\mathbb{C}^p$. This is the solution of the Jacobi inversion problem. Weierstrass worked on this problem throughout his life. His results found their final form in his *Vorlesungen über die Theorie der Abelschen Transzendenten*, 1875/76 (*Math. Werke* 4 (1902)).

Exercises

12.1 Let k be a real number with $|k| < 1$. Show that the arc length of the curve $y = \sqrt{1-(1-k^2)x^2}$ for $0 \le x \le h$ equals

$$\int_0^h \frac{(1-k^2x^2)dx}{\sqrt{(1-x^2)(1-k^2x^2)}}.$$

12.2 (Euler) Let u and v be continuously differentiable real functions of θ, with $u(\theta) = 0$, which satisfy the equation

$$u^2 + v^2 + 2buv - c^2 = 0,$$

where b and c are real numbers with $b^2 + c^2 = 1$.

a) Show that, with suitable signs for the square roots,

$$u\sqrt{1-v^2} + v\sqrt{1-u^2} = c \tag{10}$$

and

$$\frac{du/d\theta}{\sqrt{1-u^2}} + \frac{dv/d\theta}{\sqrt{1-v^2}} = 0. \tag{11}$$

b) Use (10) and (11) to derive the formula

$$\int_0^x \frac{dt}{\sqrt{1-t^2}} + \int_0^y \frac{dt}{\sqrt{1-t^2}} = \int_0^{x\sqrt{1-y^2}+y\sqrt{1-x^2}} \frac{dt}{\sqrt{1-t^2}}$$

for real x,y with $|x| \le 1$, $|y| \le 1$.

12.3 (Euler) Let u and v be continuously differentiable real functions of θ, with $u(\theta) = 0$, which satisfy the equation

$$u^2 + v^2 + c^2u^2v^2 + 2buv - c^2 = 0,$$

where b and c are real numbers with $b^2 + c^4 = 1$.

a) Show that, with suitable signs for the square roots,

$$u\sqrt{1-v^4} + v\sqrt{1-u^4} = c(1+u^2c^2) \tag{12}$$

and

$$\frac{du/d\theta}{\sqrt{1-u^4}} + \frac{dv/d\theta}{\sqrt{1-v^4}} = 0. \tag{13}$$

b) Use (12) and (13) to derive the formula

$$\int_0^x \frac{dt}{\sqrt{1-t^4}} + \int_0^y \frac{dt}{\sqrt{1-t^4}} = \int_0^{h(x,y)} \frac{dt}{\sqrt{1-t^4}}$$

for real x,y with $|x| \le 1$, $|y| \le 1$, where

$$h(x,y) = \frac{x\sqrt{1-y^4} + y\sqrt{1-x^4}}{1+x^2y^2}.$$

12.4 (Gauss) The lemniscate is the plane curve given in polar coordinates by $r^2 = \sin 2\phi$.

a) Show that the arc length of the lemniscate from the point $(0,0)$ to the point (ϕ,r), with $\phi \le \pi/4$, is

$$\int_0^r \frac{dt}{\sqrt{1-t^4}}. \tag{14}$$

b) Show that the inverse function to (14) may be continued to an elliptic function $sl(z)$ with period lattice $\Gamma = (w,iw)$, where $w = 4\int_0^1 \frac{dt}{\sqrt{1-t^4}}$.

c) Show

$$sl(-z) = -sl(z),\ sl(iz) = i\,sl(z),\ sl(z+\tfrac{w}{2}) = -sl(-z).$$

d) Show that $sl(z)$ takes each value exactly twice.

e) Show that the following addition theorem holds for $sl(z)$:

$$sl(z_1+z_2) = \frac{sl(z_1)\sqrt{1-sl^4(z_2)}+sl(z_2)\sqrt{1-sl^4(z_1)}}{1+sl^2(z_1)\,sl^2(z_2)}.$$

13. Elliptic functions

13.1 Elliptic functions in the framework of Riemann's function theory

In this chapter we consider the elliptic functions, introduced in the last section, without reference to the inversion problem. Liouville in 1844 first laid the foundations of the theory, in lectures not published until 1879. The modern theory of elliptic functions is in essentially the form it was given by Weierstrass in his lectures. We present an excerpt from them in Section 13.2.

In the present section we want to consider elliptic functions within the framework of Riemann's function theory.

In Section 12.4 we have seen that a Riemann surface $\mathcal{F}$ of genus 1 is analytically isomorphic to $\mathbb{C}/\Gamma$, where Γ is a lattice in $\mathbb{C}$. This suggests that $\mathbb{C}/\Gamma$ also be considered as a Riemann surface. Instead of the projection $\mathcal{F} \to \mathbb{C} \cup \{\infty\}$ we have the covering $\mathbb{C} \to \mathbb{C}/\Gamma$, which allows a natural uniformising variable to be introduced for each point P, i.e. a homeomorphism of a sufficiently small neighbourhood of P onto a neighbourhood of 0 in $\mathbb{C}$. The latter is sufficient to carry out the ideas of the preceding three chapters. This more general formulation of the concept of Riemann surface was first presented by Klein in his book *Über Riemanns Theorie der algebraischen Funktionen und Ihrer Integral, Leipzig* 1882. Since we shall prove the Riemann existence theorem for differentials (Chapter 11, Theorem 3) for Klein's conception of a Riemann surface (first put on a rigorous basis by Weyl in 1913) in Chapter 29, we can apply the results about functions on Riemann surfaces from Chapter 11 to $\mathbb{C}/\Gamma$. In this way one obtains the main theorems in the theory of elliptic functions. The elliptic functions for the lattice Γ are the meromorphic functions on $\mathbb{C}/\Gamma$.

A *canonical dissection* of $\mathbb{C}/\Gamma$ means a choice of fundamental region for Γ in $\mathbb{C}$. The latter means the following. Let F_0 be a region in $\mathbb{C}$ bounded by the closed curve $A_1B_1\tilde{A}_1^{-1}\tilde{B}_1^{-1}$. Let P_1 resp. $\tilde{P}_1$ be the initial point of A_1 resp. $\tilde{A}_1$, and let Q_1 resp. $\tilde{Q}_1$ be the final point of A_1 resp. $\tilde{A}_1$. Also let A_1 resp. B_1 be given parametrically by $A_1(t)$ resp. $B_1(t)$ for $t \in [0,1]$ and suppose

$$P_1 \equiv Q_1 \pmod{\Gamma},\ A_1(t) \equiv \tilde{A}_1(t) \pmod{\Gamma},\ B_1(t) \equiv \tilde{B}_1(t) \pmod{\Gamma}$$

for $t \in [0,1]$. Then

$$\mathcal{F} = (\mathcal{F}_0 \cup A_1 \cup B_1) - \{P_1\} - \{\tilde{Q}_1\}$$

is called a *fundamental region* of $\mathbb{C}/\Gamma$ when the projection $\mathcal{F} \to \mathbb{C}/\Gamma$ is a one-to-one mapping of $\mathcal{F}$ onto $\mathbb{C}/\Gamma$.

E.g., one can choose

$$A_1(t) = P_1 + w_1 t,\ B_1(t) = P_1 + w_1 + w_2 t \text{ for } t \in [0,1],$$

where w_1, w_2 is a basis for Γ and P_1 is a fixed complex number (Figure 12). In this case $\mathcal{F}$ is called a *period parallelogram*. For the

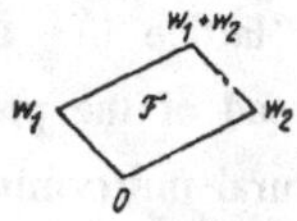

Fig. 12

complex variable z, dz in $\mathbb{C}/\Gamma$ is a differential of the first kind with periods w_1 resp. w_2 relative to the closed curves A_1 resp. B_1 (considered as curves in $\mathbb{C}/\Gamma$).

Theorem 1. *The sum of residues of an elliptic function in a fundamental region equals* 0 (Chapter 11, Theorem 2). □

Theorem 2. *A non-constant elliptic function takes each value (including* ∞*) only finitely often in a fundamental region, and each value is taken equally often* (Chapter 11, Theorems 5, 6). □

Theorem 3. *Let* $u_1,\dots,u_r$, $v_1,\dots,v_s$ *be arbitrary points in a fundamental region and let* $m_1,\dots,m_r$, $n_1,\dots,n_s$ *be natural numbers with* $m_1 + \dots + m_r = n_1 + \dots + n_s$. *Then there is an elliptic function* f *which has an* m_j*-tuple zero at* u_j *for*

$j = 1,\dots,r$ *and an* n_k*-tuple pole at* v_k *for* $k = 1,\dots,s,$ *and which is elsewhere regular and non-zero, if and only if*

$$m_1u_1 + \dots + m_ru_r \equiv n_1v_1 + \dots + n_sv_s \pmod{\Gamma} \tag{1}$$

The function f *is unique up to a constant factor* c (Chapter 12, Theorem 2). □

Let $K(\Gamma)$ be the field of all elliptic functions with period lattice Γ. By Theorem 3 there is an elliptic function $P(z)$ which has a double pole at $z = 0$ and which is regular elsewhere in a fundamental region. Because of Theorem 1, $P(z)$ has a Laurent expansion of the following form in the neighbourhood of 0 :

$$P(z) = \frac{c_{-2}}{z^2} + c_0 + c_1z + \dots \tag{2}$$

The normalisation $c_{-2} = 1$, $c_0 = 0$ determines $P(z)$ uniquely (Theorem 2). *This function is called the Weierstrass* P*-function.*

$P(z) - P(-z)$ is an entire function which vanishes for $z = 0$. Therefore $P(z) = P(-z)$, i.e. $P(z)$ is an even function. Let u be an arbitrary point of the fundamental region other than 0. Then $P(z) - P(u)$ has the zeros u and -u. By Theorem 3 these are all the zeros mod Γ. In particular $P(z) - P(u)$ has a double zero in the fundamental region if and only if $2u \in \Gamma$. There are three such points, congruent mod Γ to $w_1/2$, $w_2/2$ and $w_1/2 + w_2/2$.

More generally, a zero or a pole of an even elliptic function f always has even multiplicity at a point u such that $2u \in \Gamma$, because in this case f(z-u) is an even function. This yields

Theorem 4. *Let* f(z) *be an even elliptic function with zeros* $u_1,u_1',\dots,u_s,u_s'$ *and poles* $v_1,v_1',\dots,v_s,v_s'$ *in a fundamental region, listed according to multiplicity, where* $u_j' \equiv -u_j \pmod{\Gamma}$ *and* $v_j' \equiv -v_j \pmod{\Gamma}$ *for* $j = 1,\dots,s.$ *Then* f(z) *has the representation*

$$f(z) = c\,\frac{(P(z)-P(u_1))\dots(P(z)-P(u_s))}{(P(z)-P(v_1))\dots(P(z)-P(v_s))}, \tag{3}$$

where c *is a constant and* $P(z) - P(0)$ *is interpreted as the constant* 1. □

The derivative $P'(z)$ of $P(z)$ is an odd function. Hence by Theorem 4 an arbitrary elliptic function may be expressed in the form $f_1(z) + f_2(z)P'(z)$, where $f_1(z)$ and $f_2(z)$ are rational functions of $P(z)$. In particular, $K(\Gamma)$ is generated over $\mathbb{C}$ by $P(z)$ and $P'(z)$. Now $P'(z)^2$ is an even elliptic function with a sextuple pole at $z = 0$ and no other pole in the fundamental region. Since $P(z) - P(u)$ has a double zero for $u = w_1/2, w_2/2$ and $w_1/2 + w_2/2$, $P'(z)$ has simple zeros at these points. Looking back at (2), one then finds

$$P'(z)^2 = 4(P(z) - P(w_1/2))(P(z) - P(w_2/2))(P(z) - P(w_1/2+w_2/2)) \tag{4}$$

and

$$P'(z)^2 = 4P(z)^3 - 20c_2P(z) - 28c_4, \tag{5}$$

with coefficients c_2, c_4 from the expansion (2) of $P(z)$.

13.2 Construction of elliptic functions

First we want to construct an entire function which has simple zeros for all $w \in \Gamma$. To do this we need the following

Proposition 1. *Let $\sigma \in \mathbb{R}$. The series $\Sigma' \frac{1}{|w|^\sigma}$, where the sum is taken over all $w \in \Gamma - \{0\}$, converges for $\sigma > 2$.*

For a later application we need the following Proposition 2, from which Proposition 1 follows immediately.

Proposition 2. *Let d_1,d_2 be positive constants and let $\sigma > 2$. Also let w_1,w_2 be generators of Γ with $\mathrm{Im}(w_1/w_2) > 1$. We set $\tau := w_1/w_2$. Then the series $\Sigma' \frac{1}{|m+n\tau|^\sigma}$, where the summation is over all pairs $(m,n) \in \mathbb{Z} \times \mathbb{Z} - \{(0,0)\}$, is uniformly convergent for all τ with $\mathrm{Im}\,\tau \geq d_1$, $|\tau| \leq d_2$.*

Proof. First we consider the parallelogram $\mathfrak{P}$ with vertices $w_1 + w_2$, $-w_1 + w_2$, $-w_1 - w_2$, $w_1 - w_2$. Without loss of generality assume $d_2 \geq 1$. Then each boundary point of $\mathfrak{P}$ has absolute value greater than or equal to $|w_2|d_1/d_2$. In order to see this, one has to determine the minima of the quadratic polynomials $|x+\tau|^2$ and $|1+x\tau|^2$ which equal $(\text{Im }\tau)^2$ and $(\text{Im }\tau)^2/|\tau|^2$. It follows that the boundary points of $n\mathfrak{P}$ have absolute value greater than or equal to $n|w_2|d_1/d_2$. Each lattice point of $\Gamma - \{0\}$ is a boundary point of exactly one parallelogram $n\mathfrak{P}$, $n = 1, \ldots$. On the boundary of $n\mathfrak{P}$ there are $8n$ points of Γ. This yields the estimate

$$\Sigma'\frac{1}{|m+n\tau|^{\sigma}} \le \sum_{n=1}^{\infty}\frac{8nd_2^{\sigma}}{d_1^{\sigma}n^{\sigma}} = \frac{8d_2^{\sigma}}{d_1^{\sigma}}\sum_{n=1}^{\infty}\frac{1}{n^{\sigma-1}} .$$

The series $\Sigma\frac{1}{n^{\sigma-1}}$ is convergent for $\sigma > 2$ (Chapter 6, (6)). □

By the Weierstrass product theorem (Chapter 9, Theorem 2) and Proposition 1,

$$\sigma(z) := z\Pi'\left(\left(1-\frac{z}{w}\right)\exp\left(\frac{z}{w}+\frac{1}{2}\left(\frac{z}{w}\right)^2\right)\right)$$

is an entire function. The product on the right converges uniformly in each bounded region of the complex plane. We can therefore form the logarithmic derivative and obtain

$$\zeta(z) := \frac{1}{z} + \Sigma'\left(\frac{1}{z-w} + \frac{1}{w} + \frac{z}{w^2}\right). \qquad (6)$$

A second differentiation yields

$$-\zeta'(z) := \frac{1}{z^2} + \Sigma'\left(\frac{1}{(z-w)^2} - \frac{1}{w^2}\right).$$

For $z_0 = 0$ the function $-\zeta'(z)$ has a Laurent expansion

$$-\zeta'(z) = \frac{1}{z^2} + c_2z^2 + \ldots .$$

Thus if we are able to show that $\zeta'(z+w) = \zeta'(z)$ for $w \in \Gamma$ then it will follow that $-\zeta'(z) = P(z)$.

First,

$$\zeta''(z) = \frac{2}{z^3} + \Sigma'\frac{2}{(z-w)^3}$$

and therefore $\zeta''(z+w) = \zeta''(z)$ for $w \in \Gamma$. It follows that $\zeta'(z+w) = \zeta'(z) + c(w)$. But since $\zeta'(-z) = \zeta'(z)$ and $\zeta'(w/2) = \zeta'(-w/2) + c(w)$, we have $c(w) = 0$.

Thus we have found the following expression for $P(z)$:

$$P(z) = \frac{1}{z^2} + \Sigma'\left(\frac{1}{(z-w)^2} - \frac{1}{w^2}\right). \tag{7}$$

One easily calculates that the coefficients of the Laurent expansion $P(z) = 1/z^2 + c_2 z^2 + \ldots$ satisfy

$$c_{2n} = (2n+1)G_{2n+2}, \text{ where } G_{2n+2} := \Sigma'\frac{1}{w^{2n+2}} \text{ for } n = 1,2,\ldots . \tag{8}$$

Because of (5) we have

$$P'(z)^2 = 4P(z)^3 - g_2 P(z) - g_3, \tag{9}$$

with $g_2 := 60G_4$, $g_3 := 140G_6$. Equation (6) gives a recurrence formula for G_n. All the G_n are expressible as polynomials in G_4 and G_6 with rational coefficients.

We shall now express an arbitrary elliptic function in terms of the σ-function. First we have

$$\zeta(z+w) = \zeta(z) + \eta(w) \quad \text{for} \quad w \in \Gamma, \tag{10}$$

with constant of integration $\eta(w)$ independent of w. Also

$$\sigma(z+w) = \sigma(z)c(w)e^{\eta(w)z}.$$

Here $c(w) = -e^{\eta(w)w/2}$ because $\sigma(z)$ is an odd function. The following theorem now follows easily from Theorem 3.

Theorem 5. *Let* f *be an elliptic function which has an* m_j*-tuple zero for* $u_j + \Gamma$, $j = 1,\ldots,r$, *and an* n_k*-tuple pole for* $v_k + \Gamma$, $k = 1,\ldots,s$, *and which is elsewhere regular and non-zero. Then*

$$f(z) = c\,\frac{\sigma(z-u_1)^{m_1}\cdots\sigma(z-u_r)^{m_r}}{\sigma(z-v_1)^{n_1}\cdots\sigma(z-v_s)^{n_s}} \tag{11}$$

for some constant c, *where* $u_1,\ldots,u_r$, $v_1,\ldots,v_s$ *are chosen in their classes* mod Γ *in such a way that*

$$m_1u_1 + \ldots + m_ru_r = n_1v_1 + \ldots + n_sv_s$$

(cf. (1)). □

In his construction of the theory of elliptic functions, Jacobi used a periodic function $\theta(z)$, known as a *Jacobi theta function*, which plays an important rôle in various aspects of what follows. We want to derive this function from the function $\sigma(z)$. To do this we use the Legendre relation

$$w_1\eta(w_2) - w_2\eta(w_1) = -2\pi i \tag{12}$$

for a pair w_1,w_2 of generators of Γ. One obtains (12) by applying Chapter 11, (6) to the differentials $w := dz$ and $w' := -P(z)dz$ (Figure 12).

We assume that the order of w_1,w_2 is chosen so that $\operatorname{Im} w_1/w_2 > 0$. We set $\tau = w_1/w_2$ and let

$$\theta(z) = c\cdot\exp(i\pi z - w_2\eta(w_2)z^2/2)\cdot\sigma(w_2z), \tag{13}$$

where the constant c will be determined below. The exponential factor in (13) is chosen so that $\theta(z)$ has period 1 :

$$\theta(z+1) = \theta(z). \tag{14}$$

With the help of (12) one also obtains

$$\theta(z+\tau) = -(\exp(-2\pi iz))\theta(z). \tag{15}$$

By (14), $\theta(\frac{1}{2\pi i}\log q)$ is a function of the complex variable q which is regular in $\mathbb{C} - \{0\}$, and hence is expandable in a Laurent series (Section 8.8). This means

$$\theta(z) = \sum_{n=-\infty}^{\infty} c_n e^{2\pi inz}. \tag{16}$$

The coefficients c_n satisfy the recurrence relation

$$c_{n+1} = -c_n e^{2\pi in\tau} \quad \text{for } n \in \mathbb{Z} \tag{17}$$

because of (15). This implies

$$c_n = (-1)^n c_0 \exp(\pi i(n^2-n)\tau) \quad \text{for } n \in \mathbb{Z}.$$

We set $c_0 = \exp \pi i\tau/4$ and obtain

$$\theta(z) = \sum_{n=-\infty}^{\infty} (-1)^n \exp(\pi i(n-\tfrac{1}{2})^2\tau + 2\pi inz). \tag{18}$$

To determine c we note that $\theta(0) = 0$. Hence by (13), $\theta'(0) = \lim_{z\to 0} \theta(z)/z = cw_2$, hence $c = \theta'(0)/w_2$.

Since

$$|e^{2\pi i\tau}| = e^{-2\pi \operatorname{im}\tau} < 1,$$

$\theta(z)$ is a rapidly converging series in $q = e^{2\pi i\tau}$.

To conclude this section we shall derive the addition theorem for $P(z)$.

Theorem 6. $$P(z_1+z_2) = -P(z_1) - P(z_2) + \frac{1}{4}\left[\frac{P'(z_1)-P'(z_2)}{P(z_1)-P(z_2)}\right]^2, \tag{19}$$

where z_1 *and* z_2 *are complex numbers for which the left and right sides of* (19) *are defined.*

Proof. We consider the elliptic function

$$f(z,z_2) = P(z+z_2) + P(z) + P(z_2) - \frac{1}{4}\left[\frac{P'(z_1)-P'(z_2)}{P(z_1)-P(z_2)}\right]^2 \tag{20}$$

as a function of z. It has poles at most for $z \equiv 0 \pmod{\Gamma}$ or for $z \equiv \pm z_2 \pmod{\Gamma}$. Computation of the principal part at these places shows that $f(z,z_2)$ is regular and hence constant. Since (20) is symmetric with respect to z and z_2, $f(z,z_2) = f(z_2,z)$ also cannot depend on z_2. We now set $z = w_1/2$ and $z_2 = w_2/2$. Then by (4) and (5) we get

$$f(w_1/2,w_2/2) = P(w_1/2+w_2/2) + P(w_1/2) + P(w_2/2) = 0. \quad \square$$

13.3 Classification of Riemann surfaces of genus 1

In this section we consider the isomorphism classes of Riemann surfaces of genus 1, each associated with a complex number called its *modulus*.

In Section 12.4 we have seen that every Riemann surface of genus 1 is isomorphic to $\mathbb{C}/\Gamma$ for a certain lattice Γ. It therefore suffices to consider the surfaces $\mathbb{C}/\Gamma$, which will be called *tori*.

Theorem 7. *Let* Γ_1 *and* Γ_2 *be lattices in* $\mathbb{C}$. *Each analytic mapping* α *of* $\mathbb{C}/\Gamma_1$ *into* $\mathbb{C}/\Gamma_2$ *has the form*

$$\alpha(z+\Gamma_1) = c_0 + c_1 z + \Gamma_2 \tag{21}$$

with $c_0,c_1 \in \mathbb{C}$ *and* $c_1\Gamma_1 \subset \Gamma_2$.

Proof. In a sufficiently small neighbourhood U of 0, α defines a mapping of U into a neighbourhood of a representative c_0 of $\alpha(0)$. The regular function $\tilde{\alpha}$ given in this way can be continued along arbitrary curves in $\mathbb{C}$. By the monodromy theorem (Section 8.11), $\tilde{\alpha}$ is therefore a regular function on $\mathbb{C}$ with $\tilde{\alpha}(z+\Gamma_1) = \tilde{\alpha}(z) + \Gamma_2$. It follows that the derivative of $\tilde{\alpha}(z)$ is a regular elliptic function, and hence constant. $\square$

Now α is injective if and only if $c_1\Gamma_1 = \Gamma_2$, hence

Theorem 8. *Two tori* $\mathbb{C}/\Gamma_1$ *and* $\mathbb{C}/\Gamma_2$ *are isomorphic if and only if* $\Gamma_2 = c\Gamma_1$ *for some* $c \in \mathbb{C} - \{0\}$. $\square$

In the last section we found a lattice Γ to be associated with the numbers $G_{2n}(\Gamma) = \Sigma' w^{-2n}$ for $n = 2,3,\ldots$ (cf. (8)). We have seen that they are polynomials in

$g_2 = 60G_4$ and $g_3 = 140G_6$. The polynomial $4x^3 - g_2x - g_3$ has three different zeros (cf. (4)). Hence the discriminant $g_2^3 - 27g_3^2 \neq 0$. Conversely, we have

Theorem 9. *Let* h_2, h_3 *be arbitrary complex numbers with* $h_2^3 - 27h_3^2 \neq 0$ *and let* Γ *be the period lattice associated with the differential* $\omega = dz/\sqrt{4z^3-h_2z-h_3}$. *Then* $h_2 = g_2(\Gamma)$ *and* $h_3 = g_3(\Gamma)$.

Proof. The Riemann surface $\mathcal{F}$ associated with $w^2 - 4z^3 + h_2z + h_3$ has genus 1 (Section 10.5). The differential ω is of first kind on $\mathcal{F}$. By Section 12.4, the inverse function $z(u)$ of the associated integral is an elliptic function with lattice Γ. We choose the initial point for integration to be the branch point P_∞ of $\mathcal{F}$ over ∞. Then $z(u)$ has a pole at all points of Γ, and nowhere else. We determine the initial terms of the Laurent expansion of $z(u)$ at the point 0 : as uniformising variable at the point P_∞ we have to take the function $t = 1/\sqrt{z}$. From it one obtains $\omega/dt = -2/\sqrt{4-h_2t^4-h_3t^6}$ and hence $u = -t + c_5t^5 + \ldots$, where c_5, as well as the d_5, e_6 and f_2 about to appear, is a certain complex number. It follows that for the inverse function $t(u)$ we have $t = -u + d_5u^5 + \ldots$ and $t^2 = u^2 + e_6u^6 + \ldots$. This yields the desired expansion $z(u) = 1/u^2 + f_2u^2 + \ldots$, i.e. $z(u)$ is the Weierstrass function $\wp(u)$ for the lattice Γ. Since $(dz/du)^2 = 4z^3 - h_2z - h_3$ it follows that $h_2 = g_2(\Gamma)$, $h_3 = g_3(\Gamma)$. □

We now seek a number which depends only on the lattice similarity class $\{c\Gamma | c \in \mathbb{C} - \{0\}\}$. We have

$$G_{2n}(c\Gamma) = c^{-2n}G_{2n}(\Gamma),$$

i.e. G_{2n} is homogeneous of degree $-2n$. We seek a non-trivial rational function of g_2 and g_3 which is homogeneous of degree 0. The simplest such function is the quotient of two polynomials which are homogeneous of degree -12. We set

$$j(\Gamma) := \frac{g_2^3}{g_2^3-27g_3^2}.$$

(The simpler function g_2^3/g_3^2, which was first studied by Hermite, has the disadvantage of not being finite for all lattices.)

Theorem 10. $j(\Gamma)$ *establishes a one-to-one correspondence between the lattice classes* $\{c\Gamma | c \in \mathbb{C} - \{0\}\}$ *and the set of complex numbers.*

Proof. It follows from Theorem 9 that $j(\Gamma)$ takes all complex numbers as values. Now suppose g_2,g_3 and g_2',g_3' are number pairs with

$$g_2^3/(g_2^3-27g_3^2) = g_2'^3/(g_2'^3-27g_3'^2).$$

Then $(g_2^3:g_3^2) = (g_2'^3:g_3'^2)$, i.e. there is a complex number $c \neq 0$ with $g_2' = c^{-2}g_2$, $g_3' = c^{-3}g_3$. The associated lattices Γ and Γ' are therefore proportional. □

13.4 The elliptic modular function

Let w_1,w_2 be a basis for the lattice Γ with $\mathrm{Im}(w_1/w_2) > 0$. We set $\tau := w_1/w_2$ and $j(\tau) = j(\Gamma)$, $G_{2n}(\tau) = G_{2n}(\Gamma)w^{2n}$. By Proposition 2 these functions are regular in the complex upper half plane H (Chapter 8, Theorem 15). The function $j(\tau)$ is called the *elliptic modular function* and $G_{2n}(\tau)$ the $2n^{th}$ *Eisenstein series.* These series, as well as the Weierstrass *P*-function, were introduced by Eisenstein in 1847 in his work "*Beiträge zur Theorie der elliptischen Funktionen*" (*J. reine angew. Math.* 35 (1847)). With Weierstrass, $P(z)$ first appears in his lectures of 1862.[1)]

Let

$$w_1' = aw_1 + bw_2,\ w_2' = cw_1 + dw_2 \quad \text{with} \quad a,b,c,d \in \mathbb{Z} \tag{22}$$

be another basis of Γ, which is the case precisely when $\det\begin{pmatrix} a & b \\ c & d\end{pmatrix} = ad - bc = \pm 1$. The additional condition $\mathrm{Im}\, w_1'/w_2' > 0$ means that $ad - bc = 1$. The group of all integer matrices $\begin{pmatrix} a & b \\ c & d\end{pmatrix}$ with $ad - bc = 1$ is denoted by $SL_2(\mathbb{Z})$. In general, $SL_n(\Lambda)$ is defined for any commutative ring Λ with unit as the group of all $n \times n$ matrices with determinant 1. $SL_n(\Lambda)$ is called the nth *special linear group of* Λ.

Let $A = \begin{pmatrix} a & b \\ c & d\end{pmatrix} \in SL_2(\mathbb{Z})$. Then by definition of $j(\tau)$ and $G_{2n}(\tau)$

---

1) See Weil, A., *Elliptic functions according to Eisenstein and Kronecker*, *Springer-Verlag* 1976.

$$j\left(\frac{a\tau+b}{c\tau+d}\right) = j(\tau)$$

and

$$G_{2n}\left(\frac{a\tau+b}{c\tau+d}\right) = (c\tau+d)^{2n}G_{2n}(\tau),\ n = 2,3,\ldots,\ \text{for}\ \ \tau \in H.$$

The linear fractional transformation

$$f_A(\tau) = \frac{a\tau+b}{c\tau+d} \tag{23}$$

is an analytic isomorphism of the upper half plane H. As one easily sees, $f_A = f_B$ for two matrices $A,B \in SL_2(\mathbb{Z})$ if and only if $A = B$ or $A = -B$. The group G of all linear fractional transformations (23) is therefore isomorphic to $SL_2(\mathbb{Z})/\{E,-E\}$. The group G is called the *modular group*. When A is a representative of $g \in G$ in $SL_2(\mathbb{Z})$ we set $g\tau := f_A(\tau)$. For a fixed $t \in H$, $G\tau = \{g\tau \mid g \in G\}$ is called the *orbit* of τ. Obviously, two orbits are either equal or disjoint. The set of orbits in H is denoted by $G\backslash H$ and is called the *quotient* or *orbit space* of H by G. The set $G\backslash H$ is a topological space in a natural way. As neighbourhoods of a point $G\tau$ one takes the sets of the form $\{Gz \mid z \in U\}$, where U is a neighbourhood of τ. By Theorem 10 we have

Theorem 11. $j(\tau)$ *establishes a one-to-one continuous mapping of* $G\backslash H$ *onto* $\mathbb{C}$. □

We now want to investigate when $j(\tau)$ is locally invertible, i.e. when a sufficiently small neighbourhood of τ is mapped one-to-one onto a neighbourhood of $j(\tau)$. The inverse function is then analytic in this neighbourhood of $j(\tau)$ (Chapter 8, Theorem 14).

Theorem 12. $j(\tau)$ *is locally invertible if and only if* τ *is not in the orbit of* i *or* $\rho = -\frac{1}{2} + \frac{1}{2}\sqrt{3}$.

To prove Theorem 12 we construct a *fundamental region* of $G\backslash H$, i.e. a connected set F in H such that the projection F $G\backslash H$ is a one-to-one mapping (cf. Section 13.1). We set $F = F_0 \cup \partial F$, where

$$F_0 = \{\tau \in H \mid\ |\tau| > 1,\ |\mathrm{Re}\ \tau| < \tfrac{1}{2}\}$$

and

$$\partial F = \{\tau \in H|\ |\tau| > 1, \operatorname{Re}\tau = -\tfrac{1}{2}\} \cup \{\tau \in H|\ |\tau| = 1, -\tfrac{1}{2} \le \operatorname{Re}\tau \le 0\}.$$

Proposition 3. *Each orbit* $G\tau$ *contains a point in the closure* $\overline{F}$ *of* F *in* H.

Proof For all matrices $A = \begin{pmatrix} a & b \\ c & d \end{pmatrix} \in SL_2(\mathbb{Z})$,

$$\operatorname{Im} f_A(\tau)\operatorname{Im}\frac{a\tau+b}{c\tau+d} = \operatorname{Im}\frac{(a\tau+b)(c\bar{\tau}+d)}{|c\tau+d|^2} = \frac{\operatorname{Im}\tau}{|c\tau+d|^2}. \tag{24}$$

Let G' be the subgroup of G generated by $f_1(\tau) := \tau+1$ and $f_2(\tau) := -\frac{1}{\tau}$. There are only finitely many pairs c,d with

$$|c\tau+d|^2 = (c\operatorname{Re}\tau+d)^2 + c^2(\operatorname{Im}\tau)^2 \le 1$$

for a given $\tau \in H$. Hence there is an $f_A \in G'$ with

$$\operatorname{Im} f_A(\tau) \ge \operatorname{Im} g\tau$$

for all $g \in G'$. For suitable $n \in \mathbb{Z}$ we get

$$|\operatorname{Re} f_1^n f_A(\tau)| \le \tfrac{1}{2}.$$

Then $\tau_1 := f_1^n f_A(\tau) \in \overline{F}$. For if $|\tau_1| < 1$ we should have

$$\operatorname{Im} f_2(\tau_1) = \frac{\operatorname{Im}\tau_1}{|\tau_1|^2} > \operatorname{Im}\tau_1,$$

contrary to the maximality of $\operatorname{Im}\tau_1$. □

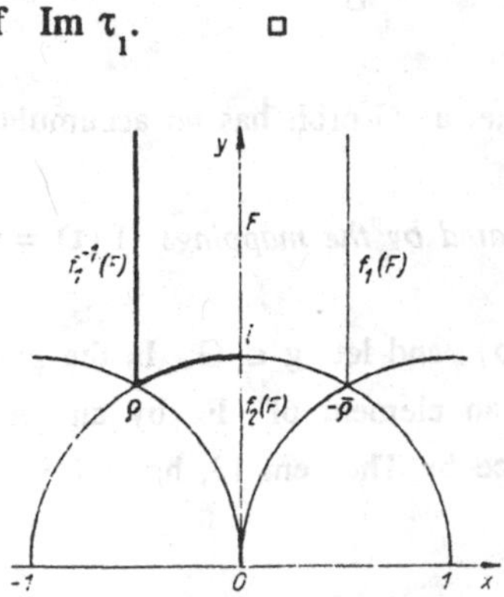

Figure 13.

Proposition 4. *Each orbit* $G\tau$ *contains exactly one point of* F.

Proof. It follows easily from Proposition 3 that each orbit contains a point of F. Now let

$$\tau \in F,\ g\tau = \frac{a\tau+b}{c\tau+d} \in F \text{ and } \tau \neq g\tau. \tag{25}$$

Then there is a pair $\tau \in F$, $g \in G$ which satisfies not only (25) but also $\mathrm{Im}(g\tau) \geq \mathrm{Im}\,\tau$: if at first $\mathrm{Im}(g\tau) < \mathrm{Im}\,\tau$, then one goes to the pair $g\tau, g^{-1}$.

By (24) we then have $|c\tau+d| \leq 1$. This implies c is 0, 1 or -1. If $c = 0$ then $g\tau = \tau+b$, contrary to (25). If $c = \pm 1$ then

$$1 \geq |\tau|^2 + 2cd\,\mathrm{Re}\,\tau + d^2 \geq |\tau|^2 - 2|d|\,|\mathrm{Re}\,\tau| + s^2 \geq 1 - |d| + d^2.$$

Hence d is 0, 1 or -1. In each case $1 = |\tau|^2 - 2|d|\,|\mathrm{Re}\,\tau| + d^2 = |\tau|^2$, which easily leads to a contradiction of (25). □

The proof of Proposition 4 shows that the mapping $H \to G\backslash H$ is locally invertible for all $\tau \in F - \{i,\rho\}$, i.e. a sufficiently small neighbourhood of τ is mapped one-to-one onto a neighbourhood of $G\tau$. When combined with Theorem 11, this gives Theorem 12.□

One calls the group $\{g \in G \mid g\tau = \tau\}$ the *isotropy* or *fixed point group* of τ in G.

As a corollary of the above considerations we obtain

Theorem 13. *The isotropy group of* $\tau \in F - \{i,\rho\}$ *consists of the identity element alone. The isotropy group of* i *resp.* ρ *is cyclic of order* 2 *resp.* 3, *with generator* $-1/\tau$ *resp.* $-1/(\tau+1)$. □

G acts discretely on H, i.e. a G-orbit has no accumulation point.

Theorem 14. G *is generated by the mappings* $f_1(\tau) = \tau+1$ *and* $f_2(\tau) = -1/\tau$.

Proof. Let $\tau \in F - \{i,\rho\}$ and let $g \in G$. In the proof of Proposition 1 we have seen that $g\tau$ can be sent to an element of F by an h from the subgroup of G generated by f_1 and f_2. Hence by Theorem 13, $hg = 1$. □

Theorem 15. $G_6(i) = 0$, $G_4(\rho) = 0$, $j(i) = 1$, $j(\rho) = 0$.

Proof.

$$G_6(i) = \sum_{a,b\in\mathbb{Z}}{}' (ai+b)^{-6} = \sum_{a,b\in\mathbb{Z}}{}' (i(ai+b))^{-6} = -G_6(i) = 0,$$

$$G_4(\rho) = \sum_{a,b\in\mathbb{Z}}{}' (a\rho+b)^{-4} = \sum_{a,b\in\mathbb{Z}}{}' (\rho(a\rho+b))^{-4} = \rho^{-1}G_4(\rho) = 0. \qquad \square$$

13.5 Picard's theorem in the theory of entire functions

In 1879 Picard found a surprising application of the function $j(\tau)$ in the theory of entire functions (Chapter 9) (*Sur une propriété des fonctions entières, C.R. Acad. Sci. Paris* 88 (1879)):

Theorem 16 *(Little Picard theorem). An entire function* $f(z)$ *either takes every complex number as value, with at most one exception, or it is a constant.*

Proof. Suppose that there are two values not taken by $f(z)$. Without loss of generality let these be the numbers 0 and 1. By Theorem 12, $j(\tau)$ is locally invertible except at the orbits of i and ρ, where it takes the values 1 and 0 by Theorem 15. By Theorem 11, $j(\tau)$ takes all complex numbers as values. It then follows from the monodromy theorem (Chapter 8, Theorem 16) that there is an entire function $g(z)$ which takes values in the upper half plane H and which satisfies the equation

$$j(g(z)) = f(z). \tag{26}$$

By the Casorati-Weierstrass theorem (Section 8.8), $g(z)$ is a constant. Hence by (26) $f(z)$ is also constant. $\square$

Still in 1879, Picard was able to sharpen Theorem 16 to the following result, which is called the *great Picard theorem.*

Let $f(z)$ *be a function which is defined and meromorphic in the neighbourhood of the point* a *and essentially singular at* a. *Then, in each neighbourhood of* a, $f(z)$ *takes each value in* $\mathbb{C} \cup \{\infty\}$, *with at most two exceptions, infinitely often.*

This is a sharpening of the Casorati-Weierstrass theorem (Section 8.8). For a proof, see Exercise 13.7.

Exercises

13.1 Derive Theorems 1 to 3 by the methods of Chapter 8, without using Chapters 11, 12.

13.2 Show that

$$P(z_1) - P(z_2) = -\frac{\sigma(z_1+z_2)\sigma(z_1-z_2)}{\sigma^2(z_1)\sigma^2(z_2)}.$$

13.3 Let $f_A(z) = \frac{az+b}{cz+d}$ be a linear fractional transformation, different from the identity, with $A = \begin{pmatrix} a & b \\ c & d \end{pmatrix} \in SL_2(\mathbb{R})$. One calls f_A *hyperbolic, elliptic* or *parabolic* according as $(a+d) > 4$, < 4 or $= 4$, respectively. Show

a) $f_A(z)$ is hyperbolic if and only if f_A has two real fixed points.

b) f_A is elliptic if and only if f_A has two distinct conjugate fixed points.

c) f_A is parabolic if and only if f_A has a real fixed point, where ∞ is considered to be a real point.

13.4 Now let $A \in SL_2(\mathbb{Z})$.

a) Let f_A be hyperbolic with fixed points α_1, α_2. Without loss of generality assume $\alpha_1 \neq \infty$. We set $f_B(z) = \frac{z-\alpha_1}{z-\alpha_2}$ if $\alpha_2 \neq \infty$, $f_B = (z) = z-\alpha_1$ if $\alpha_2 = \infty$. Show $f_B f_A(z) = \mu f_B(z)$ with a positive real number $\mu \neq 1$.

b) Let f_A be elliptic. By Theorem 13, f_A is conjugate in G to a transformation in G with fixed point i or ρ. Suppose this is already the case for f_A. Show that, in that case

$$\frac{f_A(z)-i}{f_A(z)+i} = -\frac{z-i}{z+i} \text{ or } \frac{f_A(z)-\rho}{f_A(z)-\rho^{-1}} = \rho^{-1}\frac{z-\rho}{z-\rho^{-1}}.$$

c) Let f_A be parabolic with fixed point α. We set $f_B(z) = \frac{1}{z-\alpha}$ if $\alpha \neq 0$, $f_B(z) = z$ if $\alpha = \infty$. Show that $f_B f_A(z) = f_B(z) + \nu$ for a real number ν different from 0.

13.5 Let $g(z)$ be a function defined in a neighbourhood U of the origin in the complex plane. Let $g(z)$ be regular in U except on the negative real semiaxis, suppose it has an essential singular point at 0, and also let it be continued over the negative real semiaxis to the whole of $U - \{0\}$. If one makes a circuit once around the origin, starting from $z \in U - \{0\}$, then $g(z)$ becomes $f_A(g(z))$, where f_A is in G (see Exercise 13.3).

a) Let f_A be hyperbolic. Show that there is a function $\phi(z)$, meromorphic in $U - \{0\}$, with (notation as in Exercise 13.4)

$$f_B(g(z)) = z^k\phi(z),\ k = \log \mu/2\pi i.$$

b) Let f_A have the fixed point i resp. ρ. Show that there is a function $\phi_i(z)$ resp. $\phi_\rho(z)$, meromorphic in $U - \{0\}$, with

$$\frac{g(z)-i}{g(z)+i} = \sqrt{z}\phi_i(z) \text{ and } \frac{g(z)-\rho}{g(z)-\rho^{-1}} = \frac{1}{\sqrt[3]{z}}\phi_\rho(z).$$

c) Let f_A be parabolic. Show that there is a function $\phi(z)$, meromorphic in $U - \{0\}$, with

$$f_B(g(z)) = \frac{\nu}{2\pi i}\log z + \phi(z).$$

13.6 We now assume also that $\operatorname{Im} g(z) > 0$ for $U - \{0\}$. Use the Casorati-Weierstrass theorem (Section 8.8) to derive a contradiction. (Hint: in case a) first show that $f_B(g(z))$ can be written in the form

$$f_B(g(z)) = \exp((k+m)\log z + h(z))$$

with $m \in \mathbb{Z}$ and $h(z)$ a regular function in $U - \{0\}$. In case b) show first that $\phi_i(z)$ resp. $\phi_\rho(z)$ is regular at 0.)

13.7 Prove the great Picard theorem (Section 13.5) using the idea of Theorem 16 and Exercises 13.3 to 13.6.

14. Riemannian geometry

14.1 Riemann's inaugural lecture

On 10 June 1854 Riemann, in the presence of Gauss, delivered his inaugural lecture to the Göttingen philosophical faculty, with the title "*On the hypotheses which lie at the foundations of geometry.* Starting from Gauss's intrinsic geometry of surfaces (Section 4.7), Riemann developed fundamentally new ideas on the structure of space, which can be regarded as precursors of Einstein's general theory of relativity. Due to the nature of the audience, the lecture aimed to get across Riemann's ideas without the use of mathematical formulae. It was first published in 1866, though some of the results appeared in mathematical form in a prize entry on thermodynamics submitted by Riemann to the Paris Academy in 1861. He did not win the prize, since the jury did not consider his work to be sufficiently detailed. After 1866, Riemann's ideas were worked out and developed further by Christoffel, Lipschitz and especially Ricci and Levi-Civita.

We begin by repeating the introduction to Riemann's lecture, the "*Plan of the investigation*".

"*It is well known that geometry assumes as given not only the concept of space, but also the basic principles of construction in space. It gives only nominal definitions of these things; their determination being in the form of axioms. As a result, the relationships between these assumptions are left in the dark; one does not see whether, or to what extent, connections between them are necessary, or even whether they are* **a priori** *possible.*

This darkness has persisted from Euclid to Legendre, to name the most famous modern writer on geometry, not being lifted by either the mathematicians or philosophers who have investigated it. The reason is undoubtedly that the general concept of multiply extended quantities, which includes the concept of spatial magnitude, remains quite unexplored. I have therefore begun by setting myself the task of constructing the concept of a multiply extended quantity from the general concept of magnitude. It turns out that a multiply extended quantity can admit various metric relations, so that space is only one special case of a triply extended quantity. This has the necessary consequence that the theorems of geometry cannot be derived from general concepts of quantity, and those properties which distinguish space from other conceivable triply extended quantities can only be gathered from experience. This raises the problem of finding the simplest facts which determine the metric relations of space – a problem which from the nature of things is not completely definite, for there may be many systems of simple facts which suffice to determine the metric relations of space. The most important for present purposes is that laid down by Euclid. These facts, like all facts, are not necessary, but only of empirical certainty. They are hypotheses. Thus one can investigate their likelihood, which is certainly very great within the bounds of observation, and afterwards judge whether they can be extended beyond the bounds of observation, not only in the direction of the immeasurably large, but also in the direction of the immeasurably small".

14.2 n-dimensional Riemannian manifolds

Gauss's intrinsic geometry of a surface (Section 4.7) contains everything which is determined by the line element

$$ds^2 = E\,dp^2 + F\,dpdq + G\,dq^2$$

alone. Riemann similarly considers the case of *"n-tuply extended quantities"*. His natural generalisation of Gaussian intrinsic geometry therefore consists of the following. The points of a *Riemannian manifold* M are determined by coordinate vectors $x = (x^1,\ldots,x^n)$ which vary over a region U, i.e. an open and connected subset of $\mathbb{R}^n$. A curve C is given by a smooth mapping of an interval [0,a] into U, i.e. by n continuously differentiable functions $x^1(t),\ldots,x^n(t)$, with values in $\mathbb{R}$, which are defined for $t \in [0,a]$. Each such curve has a length

$$l(C) := \int_0^a \sqrt{\sum_{i,j=1} g_{ij}\frac{dx^i}{dt}\frac{dx^j}{dt}}\,dt. \qquad (1)$$

Here the g_{ij} are (infinitely) differentiable functions of $x^1,\ldots,x^n$, with $g_{ij} = g_{ji}$. It is also assumed that the quadratic form

$$\sum_{i,j=1}^{n} g_{ij}(x)c^ic^j$$

in the variables $c^1,\ldots,c^n$ is positive definite for $x = (x^1,\ldots,x^n) \in U$.

$\{U, g_{ij}\}$ is the realisation of our Riemannian manifold M in the coordinates $x^1,\ldots,x^n$. If we go over to other coordinates $x'^1,\ldots,x'^n$, which have to be differentiable functions of $x^1,\ldots,x^n$ for which the Jacobian

$$\det\left(\frac{\partial x'^i}{\partial x^j}\right)_{ij}$$

is non-zero and for which the mapping $U \to U'$ given by $(x^1,\ldots,x^n) \mapsto (x'^1,\ldots,x'^n)$ is one-to-one, then we obtain a realisation of our manifold by the region U' of $\mathbb{R}^n$. (The condition $\det\left(\frac{\partial x'^i}{\partial x^j}\right)_{ij} \neq 0$ for a point $(x_0^1,\ldots,x_0^n) = x_0$ guarantees the invertibility of the mapping in a sufficiently small neighbourhood of x_0.) The g'_{ij} associated with U' are obtained from the demand that the length $l(C)$ remain invariant for all curves C. This implies

$$\sum_{i,j} g_{ij} \frac{dx^i}{dt}\frac{dx^j}{dt} = \sum_{i,j} g_{ij} \sum_{i',j'} \tilde{c}^i_{i'} \tilde{c}^j_{j'} \frac{dx^{i'}}{dt}\frac{dx^{j'}}{dt} = \sum_{i',j'} g'_{i'j'} \frac{dx^{i'}}{dt}\frac{dx^{j'}}{dt}, \tag{2}$$

where all sums extend from 1 to n and $(\tilde{c}^i_{i'})_{ii'} := \left(\frac{dx^i}{\partial x'^{i'}}\right)_{ii'}$ denotes the inverse of the matrix $(c^{i'}_i)_{i'i} := \left(\frac{\partial x'^{i'}}{dx^i}\right)_{i'i}$. It follows from (2) that

$$g'_{i'j'} = \sum_{i,j} g_{ij} \tilde{c}^i_{i'} \tilde{c}^j_{j'}. \tag{3}$$

Equation (3) describes the transformation behaviour of the quantities g_{ij} on passing to another coordinate system. The differential form

$$\sum_{i,j} g_{ij}\, dx^i dx^j$$

which is invariant in the sense of (3) is called the *fundamental form* or *metric* of the Riemannian manifold.

To simplify the notation we use the Einstein summation convention, which says that summation is to be performed over identical upper and lower indices in a formula. For example, instead of $\sum_{i,j} g_{ij} \frac{dx^i}{dt}\frac{dx^j}{dt}$ we write $g_{ij} \frac{dx^i}{dt}\frac{dx^j}{dt}$. We assume that all the functions appearing are infinitely differentiable. The derivative of a function $f(x)$ with respect to x is denoted by f_x.

14.3 Tangent vectors

In Section 4.5 we considered the tangent vectors to curves on a surface at a point P, and established that they lie in a plane, i.e. that they form a vector space of dimension 2. Since our Riemannian manifold M is not embedded in a euclidean space, we have *a priori* no tangent vectors at a point P. We have to get hold of them in an abstract way.

Since it is the case for surfaces in E^3 that the tangent vector to a curve through P is determined by a direction at the point P and the "speed" of traversal of the curve, which is given by the length of the tangent vector, we shall conversely use these properties as a basis for the definition of tangent vector:

A *tangent vector* v at a point P of M is given by a curve C through the point P. Two curves C and C′ through P define the same tangent vector when $\left.\frac{dx^i}{dt}\right|_P = \left.\frac{dx^i}{dt'}\right|_P$ for the coordinates $x^i(t)$ resp. $x^i(t')$ of C resp. C′. The derivatives

$$\left.\frac{dx^1}{dt}\right|_P \cdots \left.\frac{dx^n}{dt}\right|_P$$

are the components of v relative to the coordinate system $x^1,\dots,x^n$.

This definition is obviously independent of the choice of coordinate system. One sees easily that the tangent vectors at a point P constitute a vector space T_P of dimension n. One obtains a basis for T_P with the help of the coordinate lines, i.e. the curves x^i, for $i = 1,\dots,n$, with the coordinate functions $x^j(t) = x_0^i + \delta_{ij}t$, for $j = 1,\dots,n$, where $x_0^1,\dots,x_0^n$ are the coordinates of the point P. The associated tangent vector is denoted by $\left.\frac{\partial}{\partial x_i}\right|_P$. For an arbitrary tangent vector v, given by a curve with coordinate functions $x^i(t_1)$, for $i = 1,\dots,n$, we then have

$$v = \frac{dx^i}{dt_1}\left.\frac{\partial}{\partial x^i}\right|_P$$

(the index i in $\frac{\partial}{\partial x^i}$ is taken to be a lower index).

A *vector field* on M is a differentiable mapping which associates with each $P \in M$ a vector v_P on T_P. Here, differentiable means that in

$$v_P = v^i(P)\frac{\partial}{\partial x^i}\bigg|_P$$

the functions $v^i(P) = v^i(x^1,\dots,x^n)$ are differentiable. In particular, $\partial/\partial x^i$ is the vector field with the vector $\frac{\partial}{\partial x^i}\big|_P$ at P.

The fundamental form defines a scalar product of two tangent vectors v_1 and v_2 : when v_1 resp. v_2 is given by the curve with coordinates $x^i(t_1)$ resp. $x^i(t_2)$, for $i = 1,\dots,n$, it is

$$<v_1,v_2> := g_{ij}\frac{dx^i}{dt_1}\cdot\frac{dx^j}{dt_2}\bigg|_P .$$

14.4 Geodesics

We introduce the concept of a *geodesic* first as the shortest path between points P and Q.

Let $C : [0,1] \to U$ be a curve from P to Q with coordinates $y^1(t),\dots,y^n(t)$, so $P = C(0)$, $Q = C(1)$. For C to be a geodesic, each curve C_h with coordinates $x^i(t) = y^i(t) + hz^i(t)$, for $i = 1,\dots,n$, where the $z^i(t)$ are functions with $z^i(0) = z^i(1) = 0$ and h is a real number, must have a length $\int_0^1 \sqrt{g_{ij}\frac{dx^i}{dt}\frac{dx^j}{dt}}dt$ which, as a function of h, takes an extreme value for $h = 0$, i.e. the h derivative must vanish. We make the abbreviation

$$F(x,\dot{x}) = \sqrt{g_{ij}\frac{dx^i}{dt}\frac{dx^j}{dt}},$$

where $\dot{x}$ denotes the vector $\left(\frac{dx^1}{dt},\dots,\frac{dx^n}{dt}\right)$. Then one obtains the condition

$$\int_0^1\left(F_{x^k}z^k + F_{\dot{x}^k}\frac{dz^k}{dt}\right)dt = 0.$$

Partial integration gives

$$\int_0^1 \left(F_{x^k} - \frac{dF_{\dot{x}^k}}{dt} \right) z^k dt = 0.$$

Since the functions z^i can be chosen arbitrarily, it follows that

$$F_{x^k} = \frac{dF_{\dot{x}^k}}{dt} \quad \text{for } k = 1,\ldots,n.$$

For $F(x,\dot{x}) = \sqrt{g_{ij} \frac{dx^i}{dt} \frac{dx^j}{dt}}$ this yields

$$\frac{1}{2F} \frac{\partial g_{ij}}{\partial x^k} \frac{dx^i}{dt} \frac{dx^j}{dt} = \frac{d}{dt} \left(\frac{1}{F} g_{ik} \frac{dx^i}{dt} \right)$$

$$= -\frac{1}{F^2} \frac{dF}{dt} g_{ik} \frac{dx^i}{dt} + \frac{1}{F} \left(\frac{\partial g_{ik}}{\partial x^j} \frac{dx^i}{dx} \frac{dx^j}{dt} + g_{ik} \frac{d^2x^i}{dt^2} \right). \tag{4}$$

This necessary condition for a geodesic simplifies when we choose the parameter t to be the arc length of the curve. Then $dF/dt = 0$ and (4) is equivalent to

$$\frac{1}{2} \left(-\frac{\partial g_{ij}}{\partial x^k} + \frac{\partial g_{ik}}{\partial x^j} + \frac{\partial g_{jk}}{\partial x^i} \right) \frac{dx^i}{dx} \frac{dx^j}{dt} + g_{ik} \frac{d^2x^i}{dt^2} = 0. \tag{5}$$

We make the abbreviations

$$\Gamma_{kij} := \frac{1}{2} \left(-\frac{\partial g_{ij}}{\partial x^k} + \frac{\partial g_{ik}}{\partial x^j} + \frac{\partial g_{jk}}{\partial x^i} \right) \tag{6}$$

and

$$\Gamma^k_{ij} = g^{kl} \Gamma_{lij}, \tag{7}$$

where $(g^{kl})_{kl}$ is the matrix inverse to $(g_{kl})_{kl}$. The quantities Γ_{kij} resp. Γ^k_{ij} are called *Christoffel symbols* of the *first* resp. *second kind.* They were introduced in 1869 by Christoffel. In Riemann's prize submission mentioned above one finds the expression $p_{kij} := 2\Gamma_{kij}$.

With (7), (5) is equivalent to

$$\Gamma^k_{ij} \frac{dx^i}{dx} \frac{dx^j}{dt} + \frac{d^2x^k}{dt^2} = 0. \tag{8}$$

One computes immediately that

$$\Gamma^l_{ki} g_{jl} + \Gamma^l_{kj} g_{il} = \Gamma_{jki} + \Gamma_{ikj} = \frac{\partial g_{ij}}{dx^k}. \tag{9}$$

This identity is called *Ricci's lemma.*

For a sufficiently small parameter $t \geq 0$ the system of differential equations (8) has a unique solution $x(t,v)$ with the initial conditions $x(0,v) = x_0$, $\dot{x}(0,v) = v$, where x_0 is the coordinate vector of a point P and $v = (v^1,...,v^n)$ is a tangent vector from a neighbourhood V_P of the O of T_P. The function $x^i(t,v)$ is a differentiable function of $v^1,...,v^n$ for $i = 1,...,n$.

We can assume that V_P is starlike, i.e. that it includes sv along with v for $0 \leq s \leq 1$. The uniqueness of the solution implies

$$x(t,sv) = x(st,v) \quad \text{for} \quad 0 \leq s \leq 1. \tag{10}$$

For a positive number $s > 1$ such that $sv \in V_P$, $x(ts^{-1},sv)$ is a solution of (8) which continues $x(t,v)$. We can therefore assume, without loss of generality, that $x(t,v)$ is defined for all $t \in [0,1]$. A mapping ϕ of V_P into M is then defined by $v \mapsto x(1,v)$. This mapping, which does not depend on the choice of coordinate system, is called the *exponential mapping.*

We compute the Jacobian of ϕ for $v = o$. Differentiating (10) by s one obtains

$$v^i \frac{\partial x^i}{\partial v^j}(t,sv) = t \frac{\partial x^i}{dt}(st,v),$$

$$v^j v^k \frac{\partial^2 x^i}{\partial v^j \partial v^k}(t,sv) = t^2 \frac{\partial^2 x^i}{\partial t^2}(st,v).$$

The Taylor formula gives

$$x^i(1,v) = x_0^i + v^i + \frac{1}{2}\frac{\partial^2 x^i}{\partial t^2}(t_i,v)$$

for some t_i with $0 < t_i \le 1$. Thus

$$x^i(1,v) = x_0^i + v^i + \frac{1}{2}v^j v^k \frac{\partial^2 x^i}{\partial v^j \partial v^k}(1,t_i v)\frac{1}{t_i^2}.$$

This shows that the Jacobian of ϕ equals 1 for $v = o$; thus ϕ is invertible in a sufficiently small neighbourhood of o. Hence for each point Q in the image of this neighbourhood U_P under ϕ there is a unique vector v such that x(t,v) is a solution of (8), where x(1,v) is the coordinate vector of Q. This is the geodesic connecting P to Q. In general we understand a geodesic to be any solution of the system of equations (8).

14.5 Riemannian normal coordinates

With the help of the exponential mapping we shall define a special coordinate system in the neighbourhood of a point O of our Riemannian manifold. The point O itself is given the coordinates (0,...,0). Also, let $v_1,\dots,v_n$ be an orthonormal basis for the tangent space T_O at the point O, with respect to the scalar product on T_O introduced in Section 14.3. The exponential mapping sends $x^i v_i$, for a small $(x^1,\dots,x^n) \in \mathbb{R}^n$, to a point of U_O. In this way we obtain coordinates $x^1,\dots,x^n$, called *Riemannian normal coordinates*, for the points of U_O. The geodesics are thereby parametrised by the lines in T_O. Relative to these coordinates, the g_{ik} have an especially simple form:

Theorem 1. *In Riemannian normal coordinates* $x^1,\dots,x^n$,

$$g_{ij}(O) = \delta_{ij} \quad \textit{and} \quad \frac{\partial g_{ij}}{\partial x^k}(O) = 0. \quad \textit{If} \quad c_{ij,kl} := \frac{1}{2}\frac{\partial^2 g_{ij}}{\partial x^k \partial x^l}(O) \quad \textit{then}$$

$$c_{ij,kl} + c_{ik,lj} + c_{il,jk} = 0 \quad \textit{and} \quad c_{ij,kl} = c_{kl,ij}.$$

Proof. By definition we have

$$g_{ij}(O) = \langle v_i, v_j \rangle = \delta_{ij}.$$

The geodesics originating from O are given in a unique way by $x^i(t) = \xi^i(t)$ for $i = 1,\ldots,n$ and $t \geq 0$, with $(\xi^1,\ldots,\xi^n) \in \mathbb{R}^n$ and $(\xi^1)^2 + \ldots + (\xi^n)^2 = 1$. From (5), (6) we get

$$\Gamma_{kij}\xi^i\xi^j = 0.$$

Multiplying this by t^2 yields

$$\Gamma_{kij}x^ix^j = \left(\frac{\partial g_{kj}}{\partial x^i} - \frac{1}{2}\frac{\partial g_{ij}}{\partial x^k}\right)x^ix^j = 0 \tag{11}$$

for all x in a neighbourhood of O.

We can split the latter equation into two equations as follows: since we have chosen the parameter t to be arc length,

$$g_{ij}\frac{dx^i}{dt}\frac{dx^j}{dt} = g_{ij}\xi^i\xi^j = 1 = (\xi^1)^2 + \ldots + (\xi^n)^2,$$

hence

$$g_{ij}x^ix^j = (x^1)^2 + \ldots + (x^n)^2 \tag{12}$$

for all x in a neighbourhood of O. We set $g_{ij}x^j := y_i$ for short. Then $\frac{\partial g_{ij}}{\partial x^k}x^j = \frac{\partial y_i}{\partial y^k} - g_{ik}$, hence by (11) and (12)

$$\frac{\partial y_k}{\partial x^i}x^i - x^k = \frac{\partial(y_k - x^k)}{\partial x^i}x^i = 0.$$

If we now confine ourselves again to a geodesic $x^i(t) = \xi^i(t)$, then

$$\frac{\partial(y_k - x^k)}{dt} = \frac{\partial(y_k - x^k)}{\partial x^i}\xi^i = 0.$$

Since $y_k(0) = x^k(0) = 0$ it follows that

$$y_k = g_{ik}x^i = x^k \tag{13}$$

and hence $g_{ik}\xi^i = \xi^k$. Differentiation with respect to t gives

$$\frac{\partial g_{ik}}{\partial x^j}\xi^i\xi^j = 0. \tag{14}$$

From this and (11) it follows that

$$\frac{\partial g_{ij}}{\partial x^k}\xi^i\xi^j = 0. \tag{15}$$

For $t = 0$, one first obtains from (15), since $g_{ij} = g_{ji}$, that $\frac{\partial g_{ij}}{\partial x^k}(0) = 0$. If one differentiates (14) with respect to t and afterwards sets $t = 0$, then the result is

$$c_{ik,ja}\xi^i\xi^j\xi^a = 0.$$

By partial differentiation of this cubic form with respect to ξ^l one finds

$$c_{lk,ja}\xi^j\xi^a + c_{ik,la}\xi^i\xi^a + c_{ik,jl}\xi^i\xi^l = 0,$$

i.e.

$$(c_{lk,ji} + c_{jk,li} + c_{ik,jl})\xi^i\xi^j = 0.$$

In view of the symmetry of $c_{ij,kl}$ in i, j and k, l, it follows that

$$c_{kl,ij} + c_{kj,li} + c_{ki,jl} = 0. \tag{16}$$

Similarly, one obtains from (15) that

$$c_{kl,ij} + c_{li,kj} + c_{ki,jl} = 0. \tag{17}$$

Comparison of (16) and (17) finally yields

$$c_{kj,li} = c_{li,kj}. \quad \square \tag{18}$$

For a euclidean space the Riemannian normal coordinates are the ordinary coordinates with respect to an orthonormal basis. The $c_{ij,kl}$ equal 0 in this case, so in general they can be regarded as a measure of the deviation of M from euclidean space at a point O.

Let a,b be vectors from T_O, with components a^i, b^i relative to Riemannian normal coordinates. By Theorem 1 the biquadratic form

$$Q(a,b) := c_{ij,kl}a^i a^j b^k b^l$$

can also be written as follows:

$$Q(a,b) = \frac{1}{3}c_{ij,kl}(a^i b^k - a^k b^i)(a^j b^l - a^l b^j). \tag{19}$$

Let a and b be linearly independent and suppose a_0, b_0 are vectors which span the same vector space V as a, b. Also let A be the 2×2 matrix which transforms a_0, b_0 into a, b. Then

$$Q(a,b) = (\det A)^2 Q(a_0,b_0).$$

When a_0, b_0 is an orthonormal basis of V, det A equals the area F(a,b) of the parallelogram spanned by a and b. Therefore

$$Q(a,b)/F(a,b)^2$$

is independent of the choice of a and b in V.

A sufficiently small neighbourhood U of O in V is mapped by the exponential map onto a two-dimensional submanifold exp U of M. By the *Theorema egregium* (Chapter 4, Theorem 2), U has Gaussian curvature K at O. Riemann gives (without proof) the following theorem.

Theorem 2. $K = -\frac{1}{3}Q(a,b)/F(a,b)^2$. ◼

We prove Theorem 2 in Section 14.10.

14.6 The Riemann curvature tensor

We consider a region M of the euclidean space E^n as a Riemannian space with arbitrary coordinates $x^1,\ldots,x^n$ and associated fundamental form $g_{ij}dx^i dx^j$. We would like to know how one can derive the fact that M is a euclidean space from the g_{ij}.

Let $y^1,\ldots,y^n$ be the euclidean coordinates of M relative to an orthonormal basis $e_1,\ldots,e_n$ of the associated vector space V. We then have

$$g_{ij}dx^i dx^j = \sum_{k=1}^{n} (dy^k)^2 = \sum_{k=1}^{n} \left[\frac{\partial x^k}{\partial x^i} dx^i \right]^2 = \sum_{k=1}^{n} \frac{\partial x^k}{\partial x^i} \frac{\partial x^k}{\partial x^j} dx^i dx^j$$

and hence

$$g_{ij} = \sum_{k=1}^{n} \frac{\partial x^k}{\partial x^i} \cdot \frac{\partial x^k}{\partial x^j}, \tag{20}$$

$$\frac{\partial g_{ij}}{\partial x^l} = \sum_{k=1}^{n} \frac{\partial^2 y^k}{\partial x^i \partial x^l} \cdot \frac{\partial y^k}{\partial x^j} + \sum_{k=1}^{n} \frac{\partial^2 y^k}{\partial x^j \partial x^l} \cdot \frac{\partial y^k}{\partial x^i}. \tag{21}$$

The tangent space of a point P is identified, by parallel displacement, with the vector space V for E^n. The vectors

$$f_j(P) = \frac{\partial y^i}{\partial x^j}(P) e_i, \quad j = 1,\ldots,n, \tag{22}$$

constitute a basis of V. In particular, $\frac{\partial f_j}{\partial x^i}$, the change in these vectors along the coordinate lines, must be expressible as a linear combination of $f_1,\ldots,f_n$.

Theorem 3. $\frac{\partial f_j}{\partial x^i} = \Gamma^k_{ij} f_k$, *where the* Γ^k_{ij} *are the Christoffel symbols introduced above.*

Proof. By the remark just before this theorem there are functions $H^k_{ij}(P)$ with

$$\frac{\partial f_j}{\partial x^i} = H^k_{ij} f_k.$$

If one substitutes the expressions (22) for f_j and f_k in this expression, then one obtains

$$\frac{\partial^2 y^l}{\partial x^i \partial x^j} = H^k_{ij} \frac{\partial y^l}{\partial x^k}.$$

We multiply these equations by $\frac{\partial y^l}{\partial x^k}$ and sum over l. It follows easily from the resulting equations and (20), (21) that

$$\frac{\partial g_{ij}}{\partial x^k} = H^l_{ik} g_{lj} + H^l_{kj} g_{li}. \tag{23}$$

We now recall that, by definition of f_j,

$$H^k_{ij} = H^k_{ji}.$$

Therefore

$$-\frac{\partial g_{ij}}{\partial x^k} + \frac{\partial g_{ik}}{\partial x^j} + \frac{\partial g_{jk}}{\partial x^i} = 2H^l_{ij} g_{lk}.$$

The assertion now follows by comparison of (6) and (7). □

It follows from the equation

$$\frac{\partial}{\partial x^l}\left(\frac{\partial f_j}{\partial x^i}\right) = \frac{\partial}{\partial x^i}\left(\frac{\partial f_j}{\partial x^l}\right)$$

and Theorem 3 that

$$\frac{\partial \Gamma^k_{ij}}{\partial x^l} f_k + \Gamma^k_{ij} \Gamma^h_{lk} f_h = \frac{\partial \Gamma^k_{lj}}{\partial x^i} f_k + \Gamma^k_{lj} \Gamma^h_{ik} f_h$$

and hence

$$\frac{\partial \Gamma^i_{lj}}{\partial x^k} - \frac{\partial \Gamma^i_{kj}}{\partial x^l} + \Gamma^h_{lj} \Gamma^i_{kh} - \Gamma^h_{kj} \Gamma^i_{lk} = 0. \tag{24}$$

Thus we have

Theorem 4. *The* n^4 *functions*

$$K^i_{jkl} := \frac{\partial \Gamma^i_{lj}}{\partial x^k} - \frac{\partial \Gamma^i_{kj}}{\partial x^l} + \Gamma^h_{lj}\Gamma^i_{kh} - \Gamma^h_{kj}\Gamma^i_{lh}$$

vanish for a region of the euclidean space E^n. □

Theorem 4 has the following converse:

Let M *be a Riemannian manifold and suppose* $K^i_{jkl} = 0$, *for all* i,j,k,l, *in a neighbourhood of a point* P. *Then there is a, possibly smaller, neighbourhood* U *of* P *which has coordinates such that* $g_{ij} = \delta_{ij}$ *in* U, *i.e.* U *is a region of euclidean space.*

We do not go into the proof of this theorem here.

We have associated various systems of numbers with the points P of a Riemannian manifold M, such as g_{ij}, Γ^k_{ij}, K^i_{jkl}, which depend on the choice of coordinates $x^1,...,x^n$. The simplest such non-trivial system consists of the components $v^i = \left(\frac{dx^i}{dt}\right)_{t=0}$ of the tangent vector v associated with a curve C(t) with C(0) = P. On passing to another coordinate system $x'^1,...,x'^n$ one finds the corresponding components v'^i of v to be, in the notation of 14.2,

$$v'^{i'} = c^{i'}_i v^i.$$

This transformation behaviour is called *contravariant*.

A differential form of the first degree, $\sigma := a_i dx^i$, where the a_i are functions on M, is a mapping which associates with each point $P \in M$ a linear form on the tangent space T_P. The tangent vector with components $\left(\frac{dx^i}{dt}\right)_{t=0}$ is associated, via σ, with

$$\left(\frac{\sigma}{dt}\right)_{t=0} = a_i\left(\frac{dx^i}{dt}\right)_{t=0}.$$

On passing to the coordinates $x'^1,\dots,x'^n$ the a_i exhibit the transformation behaviour

$$a'_{i'} = \tilde{c}^i_{i'} a_i \,, \tag{25}$$

which is called *covariant.*

By (3), the differential form $g_{ij}dx^i dx^j$ satisfies

$$g'_{ij} = \tilde{c}^i_{i'}\tilde{c}^j_{j'}G_{ij}.$$

One says that the g_{ij} are the components of a *tensor of the second kind* which is covariant with respect to i and j. If v^i, a_i, g_{ij} are viewed as functions on M, then one speaks of a *tensor field on* M.

After this it should be clear what one means by a tensor and a tensor field of arbitrary kind. On the components of a tensor, upper indices indicate contravariant transformation behaviour, and lower indices indicate covariant transformation behaviour.

The Christoffel symbols Γ^k_{ij} have the transformation behaviour

$$\Gamma'^{k'}_{i'j'} = \tilde{c}^i_{i'}\tilde{c}^j_{j'}c^{k'}_k\Gamma^k_{ij} + c^{k'}_k\tilde{c}^i_{i'}\frac{\partial \tilde{c}^k_{j'}}{\partial x^i} \tag{26}$$

and are therefore not the components of a tensor.

To derive (26) it is convenient to show first that

$$\Gamma'_{k'i'j'} = \tilde{c}^i_{i'}\tilde{c}^j_{j'}\tilde{c}^k_{k'}\Gamma_{kij} + c^l_k\frac{\partial \tilde{c}^k_{i'}}{\partial x'^j}g'_{lk'} \,,$$

bearing in mind the relation

$$\frac{\partial \tilde{c}^i_j}{\partial x'^k} = \frac{\partial \tilde{c}^i_k}{\partial x'^j}.$$

Also,

$$K'^{i'}_{j'k'l'} = c^{i'}_i\tilde{c}^j_{j'}\tilde{c}^k_{k'}\tilde{c}^l_{l'}K^i_{jkl}. \tag{27}$$

Hence the K^i_{jkl} are the components of a tensor which is contravariant with respect to i and covariant with respect to j, k, *l*. The K^i_{jkl} were investigated by Riemann in his above-mentioned prize entry to the Paris Academy. The corresponding tensor is called the *Riemann curvature tensor.*

If a^j_i is a tensor then so is the *trace* a^i_i, and hence the latter is a number independent of the coordinate system. The same holds for more complicated tensors. In particular,

$$K^i_{jki}g^{jk}(P) \tag{28}$$

is an invariant of M at the point P, known as the *scalar curvature.*

14.7 Tensors as multilinear forms

The concept of tensor presented in 14.6 – as something which transforms in a certain way – was developed by Ricci in the years 1887 to 1896. It is far more elegant to define tensors as multilinear forms.

Instead of the tangent space T_P we consider an arbitrary vector space V of dimension n. The choice of a coordinate system in M corresponds to a basis of T_P, so we consider an arbitrary basis $e_1,\ldots,e_n$ of V. Let A^k_{ij} be the components of a tensor relative to this basis, and let a^i, b^j be components of contravariant vectors while c_k are components of a covariant vector. Then a^i resp. b^j correspond to the vector a^ie_i resp. b^je_j. On the other hand, c_k is associated with the linear form on V which associates the number a^kc_k with the vector with components a^i. Thus c_k corresponds to a vector in the vector space V* dual to V. The tensor A^k_{ij} corresponds to the multilinear mapping which associates the number $a^ib^jc_kA^k_{ij}$ with the a^i,b^j and c_k. We can identify the tensor with this multilinear mapping, which is independent of the choice of basis. In general, we have the following definition of tensor:

A *tensor of type* (p,q) is a multilinear mapping of p times the vector space V and q times the vector space V* into $\mathbb{R}$.

Here we also have to view the vectors $v \in V$ as linear forms on V* : v corresponds to the linear form which associates the number f(v) with each $f \in V^*$.

14.8 The relation between the curvature tensor and the form Q(a,b)

We now return to the Riemann curvature tensor. For Riemannian normal coordinates, Theorem 1 gives us, in the notation of 14.5,

$$K_{jkl}(O) = \frac{\partial \Gamma^i_{lj}}{\partial x^k}(O) - \frac{\partial \Gamma^i_{kj}}{\partial x^l}(O) = 2(c_{kj,il} - c_{jl,ik}). \tag{29}$$

Since O can be chosen arbitrarily, it follows from Theorem 1 that

Theorem 5. *The Riemann curvature tensor satisfies the following symmetry conditions:*

$$K^j_{ikl} = -K^i_{jkl}, \quad K^i_{jlk} = -K^i_{jkl}, \quad K^k_{lij} = K^i_{jkl},$$

$$K^i_{jkl} + K^i_{klj} + K^i_{ljk} = 0. \quad \square$$

Now we want to establish the relation between the Riemann curvature tensor and the form $Q(a,b)$. We set

$$K_{ijkl} := K^{i'}_{jkl} g_{i'i}.$$

Theorem 6. $Q(a,b) = -\frac{1}{12} K_{ijkl}(O)(a^i b^j - a^j b^i)(a^k b^l - a^l b^k)$

$$= -\frac{1}{3} K_{ijkl}(O) a^i b^j a^k b^l.$$

Proof. Because of (19) we have

$$Q(a,b) = \frac{1}{3} c_{ik,jl}(a^i b^j - a^j b^i)(a^k b^l - a^l b^k)$$

$$= \frac{1}{6}(c_{ik,jl} - c_{il,jk})(a^i b^j - a^j b^i)(a^k b^l - a^l b^k).$$

The assertion now follows from (29). $\square$

14.9 Orthonormal bases

Previously we only admitted bases of the tangent space T_P which correspond to coordinate systems. However, since we have at our disposal a positive-definite bilinear form on T_P, it is often preferable to work with orthonormal bases. We shall give an example of this in Section 14.10.

Starting from the basis $\frac{\partial}{\partial x^1}(P),\ldots,\frac{\partial}{\partial x^n}(P)$, we obtain an orthonormal basis $g_1(P),\ldots,g_n(P)$ by the *Gram-Schmidt orthonormalisation process*. In

$$g_i = \tilde{c}_i^j \frac{\partial}{\partial x^j}$$

the $\tilde{c}_i^j$ are differentiable functions of $x^1,\ldots,x^n$.

We now return to the special case of a region M in the euclidean space E^n. In this case we can set $\partial/\partial x^i = f_i$, where the f_i, for $i = 1,\ldots,n$, are the tangential vector fields introduced in 14.6, which we also view as vector functions on M with values in V.

In Theorem 3 we have described the variation of the f_i along the coordinate lines. We now consider the variation of the g_i along a curve C_i through $P \in M$ with tangent vector $g_i(P)$ at the point P. Let t be the parameter of C_i and let $C_i(0) = P$. We set $a(t) := a(C_i(t))$ for functions a on M. Then we get

$$g_i(0) = \frac{dx^h}{dt}(0)f_h$$

and therefore

$$\tilde{c}_i^h(0) = \frac{dx^h}{dt}(0).$$

It follows from this that

$$\lim_{t\to 0} \frac{g_j(t)-g_j(0)}{t} = \lim_{t\to 0} \frac{\tilde{c}_j^k(t)f_k(t)-\tilde{c}_j^k(0)f_k(0)}{t}$$

$$= \tilde{c}_j^k(0)\tilde{c}_i^h(0)\frac{\partial f_k}{\partial x^h}(0) + \frac{\partial \tilde{c}_j^k}{\partial x^h}(0)\tilde{c}_i^h(0)f_k(0)$$

$$= \tilde{c}_j^k(0)\tilde{c}_i^h(0)c_{l'}^{l}(0)\Gamma_{hk}^{l'}(0)g_l(0) + \frac{\partial \tilde{c}_j^k}{\partial x^h}(0)\tilde{c}_i^h(0)\tilde{c}_k^l(0)g_l(0). \tag{30}$$

We see that $\lim_{t\to 0} \frac{g_j(t)-g_j(0)}{t}$ depends only on the vector $g_i(0)$ associated with C_i, which we write as $(\nabla_{g_i} g_j)(P)$. The latter vector is a linear combination of the basis $g_1(P),\ldots,g_n(P)$:

$$(\nabla_{g_i} g_j)(P) = \Gamma_{ij}^{\cdot k}(P) = \Gamma_{ij}^{\cdot k}(P) g_k(P). \tag{31}$$

Because of (30) we have

$$\Gamma_{i'j'}^{\cdot k'} = \tilde{c}_{i'}^{i} \tilde{c}_{j'}^{j} c_{k}^{k'} \Gamma_{ij}^{\cdot k} + c_{k}^{k'} \tilde{c}_{i'}^{i} \frac{\partial \tilde{c}_{j'}^{k}}{\partial x^{i}}. \tag{32}$$

Thus the Christoffel symbols transform as in (26).

Ricci's lemma (9) takes the form

$$\Gamma_{ij}^{\cdot k} + \Gamma_{ik}^{\cdot j} = 0. \tag{33}$$

To derive (33) one needs only to remark that in cartesian coordinates the fundamental tensor g_{ij} is constant. Hence by differentiation of $\langle g_j, g_k\rangle = \delta_{jk}$ along a curve in the direction g_i one obtains

$$\langle \nabla_{g_i} g_j, g_k\rangle + \langle g_j, \nabla_{g_i} g_k\rangle = 0. \qquad \square$$

14.10 The Gaussian curvature

In the case of a surface the symmetry conditions of Theorem 5 reduce the Riemann curvature tensor to the scalar curvature S (28): from $K_{kl}^{ij} = K_{j'kl}^{i} g^{j'j}$ we get

$$S = K_{ji}^{ij} = 2K_{21}^{12}. \tag{34}$$

In what follows we show that, in the case of a surface in three-dimensional euclidean space, K_{12}^{12} equals the Gaussian curvature. In this way we obtain not only the desired invariant meaning of the Gaussian curvature (*Theorema egregium*), but also the proof of Theorem 2.

We consider a surface F in E^3 in the neighbourhood of a fixed point O with coordinates $0, 0, 0$. In a sufficiently small neighbourhood U of O in E^3 we choose space coordinates in the following way: for $P \in U$ there is exactly one normal to F through P (Chapter 4, (10)). Let the intersection of this normal with F be P_1. Then we choose, as coordinates of P, the surface coordinates x^1, x^2 of P_1 and the oriented distance of P from P_1 as the third coordinate x^3. For points on the surface we have $g_{3i} = \delta_{3i}$, for $i = 1,2,3$. Hence the Christoffel symbols of the surface equal the corresponding symbols of the space. Since the curvature tensor of the space vanishes (Theorem 4), one finds the curvature tensor K^i_{jkl} of the surface to be

$$K^i_{jkl} = \Gamma^3_{kj}\Gamma^i_{l3} - \Gamma^3_{lj}\Gamma^i_{k3}. \tag{35}$$

We now pass to an orthonormal basis $g_1, g_2, g_3 = \partial/\partial x^3$. For points of the surface c^l_k is independent of x^3 and $c^i_3 = c^3_i = \delta_{i3}$ for $i = 1,2,3$. It follows that

$$\Gamma'^3_{i'j'} = \tilde{c}^i_{i'}\tilde{c}^j_{j'}\Gamma^3_{ij}$$

and

$$\Gamma'^{k'}_{i'3'} = \tilde{c}^i_{i'} c^{k'}_k \Gamma^k_{i3},$$

i.e. these quantities transform tensorially. Hence the components K'^i_{jkl} of the curvature tensor relative to the basis g_1, g_2 also satisfy the relation

$$K'^i_{jkl} = \Gamma'^3_{kj}\Gamma'^i_{l3} - \Gamma'^3_{lj}\Gamma'^i_{k3}. \tag{36}$$

We now exploit the remaining freedom in the choice of basis g_1, g_2. Γ'^3_{ij} is symmetric in i and j. By a principal axis transformation we can arrange that $\Gamma'^3_{12} = 0$. Consideration of (33) then yields $K'^1_{212} = K$, with $K = \Gamma'^3_{11}\Gamma'^3_{22}$.

We shall now show that g_1, g_2 are the directions of principal curvature and that Γ'^3_{11}, Γ'^3_{22} are the corresponding principal curvatures (Section 4.6). Let the surface be given by the equation $x^3 = f(x^1, x^2)$ as in Section 4.4. Then one finds (with the notation of Section 14.6) that

$$f_i = e_i + \frac{\partial f}{\partial x^i} e_3, \quad \Gamma^3_{ij}(0,0) = \frac{\partial^2 f}{\partial x^i \partial x^j}(0,0) \quad \text{for } i,j = 1,2.$$

Since the basis for the tangent space at the origin given by the coordinates x^1, x^2 is already orthonormal, the assertion follows, because the principal curvatures result from the principal axis transformation of the matrix $\left(\frac{\partial^2 f}{\partial x^i \partial x^j}(0,0)\right)_{ij}$. By Chapter 4, Theorem 1, K is the Gaussian curvature. Hence by the remark at the beginning of this section we have

Theorem 7. *The scalar curvature equals the Gaussian curvature multiplied by* -2.

14.11 Spaces of constant curvature

In two sections of his inaugural lecture Riemann considered manifolds M with constant curvature in the sense of Theorem 2 : at each point and for each two-dimensional submanifold of the form exp U the Gaussian curvature has the same value K. The metric properties of such a manifold are locally determined by K.

This can be formulated more precisely as follows: let M and N be n-dimensional Riemannian manifolds with coordinates $x^1,\ldots,x^n$ and $y^1,\ldots,y^n$ and fundamental forms $g_{ij}dx^i dx^j$ and $h_{ij}dy^i dy^j$. A differentiable mapping ϕ of M into N is a mapping given by differentiable functions ϕ^i:

$$y^i = \phi^i(x^1,\ldots,x^k),\quad i = 1,\ldots,n.$$

By means of ϕ, the fundamental form $h_{ij}dy^i dy^j$ is transported to M:

$$h^*_{kl}(x)dx^k dx^l = h_{ij}(\phi(x))\frac{\partial\phi^i}{\partial x^k}(x)\frac{\partial\phi^i}{\partial x^l}dx^k dx^l.$$

The mapping ϕ is called *isometric* when $h^*_{ij} = g_{ij}$. One easily sees that this definition is independent of the choice of coordinate system.

An *isometry* of M onto N is a one-to-one isometric mapping whose inverse is also isometric. Finally, a *local isometry* at the point P of M is an isometry of a neighbourhood of P onto a neighbourhood of the image of P in N.

Riemann's assertion then is that two manifolds of equal constant curvature are locally isometric at each point.

It is easy to see the surface of a sphere of radius r has constant curvature $1/r^2$. Here we consider an example of a surface with constant negative curvature.

Let H denote the upper half plane, i.e. the set of all points $z = (x,y)$ with $x,y \in \mathbb{R}$, $y > 0$. In H we consider the differential form

$$\frac{1}{y^2}(dx^2+dy^2). \tag{37}$$

We want to determine the Gaussian curvature and the geodesics of the Riemannian surface thus defined.

We have

$$(g_{ij})_{ij} = \begin{pmatrix} y^{-2} & 0 \\ 0 & y^{-2} \end{pmatrix}, \quad (g^{ij})_{ij} = \begin{pmatrix} y^2 & 0 \\ 0 & y^2 \end{pmatrix},$$

$$(\Gamma^1_{ij})_{ij} = \begin{pmatrix} 0 & -y^{-1} \\ -y^{-1} & 0 \end{pmatrix}, \quad (\Gamma^2_{ij})_{ij} = \begin{pmatrix} y^{-1} & 0 \\ 0 & -y^{-1} \end{pmatrix},$$

from which the Gaussian curvature K^1_{212} may be calculated to be -1.

By (8), a geodesic $(x(t),y(t))$ on H satisfies the system of equations

$$\frac{d^2x}{dt^2} = \frac{2}{y}\frac{dx}{dt}\frac{dy}{dt}, \quad \frac{d^2x}{dt^2} = \frac{1}{y}\left(\frac{dy}{dt}\right)^2 - \frac{1}{y}\left(\frac{dx}{dt}\right)^2. \tag{38}$$

A particular solution of this system is $x = 0$, $y = e^t$, i.e. the positive y-axis. The distance between two points $(0,y_1)$ and $(0,y_2)$ on this geodesic comes to

$$l(y_1,y_2) = \log y_2 - \log y_1.$$

It tends to ∞ as y_1 0, i.e. the x-axis is infinitely far from each point of H.

The mapping

$$\phi(z) = \frac{az+b}{cz+d} \quad \text{with} \quad a,b,c,d \in \mathbb{R},\ \det\begin{bmatrix} a b \\ c d \end{bmatrix} = 1,\ z = x + iy, \tag{39}$$

is an isometry of H onto itself. To show this one has to verify the equation

$$\frac{d\phi(z)\overline{d\phi(z)}}{(\operatorname{Im}\phi(z))^2} = \frac{dz\,d\bar{z}}{(\operatorname{Im} z)^2}, \text{ i.e. } \frac{|\phi'(z)|^2}{(\operatorname{Im}\phi(z))^2} = \frac{1}{(\operatorname{Im} z)^2},$$

which causes no difficulties.

It follows that ϕ sends geodesics to geodesics. The positive y-axis is mapped by ϕ onto either a semicircle with midpoint on the x-axis or else a half line parallel to the y-axis. All such semicircles and half lines appear as images under certain isometries ϕ. We denote the set of these geodesics on H by G. For each point P of H and each tangent vector v at the point P there is exactly one curve from G through P which has tangent vector v at P. Therefore G is the set of all geodesics in H (Section 14.4).

In addition to the isometries of the form (39) there are also the following: let

$$\psi(z) = \frac{a\bar{z}+b}{c\bar{z}+d} \quad \text{with} \quad a,b,c,d \in \mathbb{R},\ \det\begin{bmatrix} a b \\ c d \end{bmatrix} = -1,\ z = x + iy. \tag{40}$$

Then ψ maps H onto itself and is an isometry, by the same verification as above.

Theorem 8. *Any isometry of* H *has the form* (39) *or* (40).

Proof. Let (ϕ^1,ϕ^2) be an isometry of H. Then

$$\frac{1}{y^2}(dx^2+dy^2) = \frac{1}{(\phi^2)^2}\left[\left(\frac{\partial\phi^1}{\partial x}dx + \frac{\partial\phi^1}{\partial y}dy\right)^2 + \left(\frac{\partial\phi^2}{\partial x}dx + \frac{\partial\phi^2}{\partial y}dy\right)^2\right]. \tag{41}$$

We make the abbreviations

$$\frac{\partial\phi^1}{\partial x} = a^1_1,\quad \frac{\partial\phi^1}{\partial y} = a^1_2,\quad \frac{\partial\phi^2}{\partial x} = a^2_1,\quad \frac{\partial\phi^2}{\partial y} = a^2_2.$$

It follows from (41) that

$$(a^1_1)^2 + (a^2_1)^2 = (a^1_2)^2 + (a^2_2)^2, \qquad a^1_1a^1_2 + a^2_1a^2_2 = 0$$

and hence

$$(a^1_1+ia^1_2)^2 = (a^2_2-ia^2_1)^2.$$

Because of (41) the two sides of this equation cannot vanish. It then follows on continuity grounds that

$$a^1_1 + ia^1_2 = a^2_2 - ia^2_1 \quad \text{for all } (x,y) \in H \tag{42}$$

or

$$a^1_1 + ia^1_2 = -(a^2_2-ia^2_1) \quad \text{for all } (x,y) \in H. \tag{43}$$

Equation (42) yields the Cauchy-Riemann differential equations for the function $\phi = \phi^1 + i\phi^2$, which is therefore regular in H as a function of $z = x+iy$ (Chapter 8, Theorem 1). By hypothesis, ϕ is invertible in H. The inverse function is also regular in H. Hence when (42) is satisfied it suffices, for the proof of Theorem 8, to prove the following lemma of Schwarz.

Lemma 1. *Let* $\phi(z)$ *be a regular function which maps* H *one-to-one onto itself. Then* $\phi(z)$ *is of the form* (39).

Proof. With the help of an isometry of the form (39) one can send any point of H to a given point of H. We can therefore assume without loss of generality that $\phi(i) = i$. The function $w(z) = \frac{z-i}{z+i}$ maps H one-to-one into the interior of the unit circle. Hence $f(z) = w(\phi(w^{-1}z))$ is a regular function which maps the interior of the unit circle one-to-one onto itself. Also, $f(0) = 0$, and hence $f(z)/z$ is a regular function.

Since a regular function takes its supremum on the boundary (Chapter 8, Theorem 9), $|f(z)/z| \leq 1$. The same inequality for the inverse function yields $|z/f(z)| \leq 1$. Therefore $|f(z)/z| = 1$, whence it follows that $f(z) = cz$, where c is a constant with $|c| = 1$. This gives $\phi(z) = \frac{az+b}{-bz+a}$ with $a,b \in \mathbb{R}$, $a^2 + b^2 = 1$. □

Now let (ϕ^1,ϕ^2) be an isometry satisfying (43). Then $(-\phi^1,\phi^2)$ is an isometry satisfying (42), and hence (ϕ^1,ϕ^2) is of the form (40). □

The isometries (39) and (40) are distinguished by the fact that the former preserves the orientation of the tangent space of H, while the latter reverses it.

14.12 Conformal mapping

Let $z_0 = x_0 + iy_0$ be an arbitrary point of the complex plane, and let C_1, C_2 be two curves through z_0. The *angle* α between C_1 and C_2 at the point z_0 is defined to be the angle between the corresponding tangent vectors v_1, v_2:

$$\cos \alpha = \frac{\langle v_1, v_2 \rangle}{\| v_1 \| \; \| v_2 \|}, \; 0 \leq \alpha \leq \pi.$$

This angle is obviously the same whether one takes the underlying metric to be the euclidean metric or (37).

Let $w(z) = w(x,y)$ be a differentiable mapping of a neighbourhood of z_0 onto a neighbourhood of $w(z_0)$. Using the ideas introduced in 14.11 one can easily prove the following.

The function $w(z)$ is regular at z_0 with $dw/dz(z_0) \neq 0$ if and only if $w(z)$ is angle- and orientation-preserving at z_0.

Such a mapping $w(z)$ is called *conformal* at the point z_0.

14.13 Non-euclidean geometry

In the 4th century BC Euclid wrote the *Elements,* in which geometry is built up from definitions, axioms and general principles. Of the five axioms, the last is significantly more complicated than the others. It can be formulated as saying that, for line g and a point P outside it, there is exactly one line h, in the same plane, which contains P and has no point in common with G. This axiom is called the *parallel axiom.*

Long after the emergence of the *Elements*, many mathematicians attempted to prove the parallel axiom from the remaining axioms and principles. They even included Legendre, who from 1794 onwards gave proofs of the parallel axiom in eight editions of his geometry textbook. All turned out to be false, and in the ninth edition he returned to presenting the theory of parallels in essentially the same way as Euclid. It was this to which Riemann referred in the introduction to his inaugural lecture, cited in 14.1.

Gauss also concerned himself with the parallel axiom and, after futile attempts to prove it, came to the conclusion that it is not provable, and moreover that this axiom can be replaced by the requirement that there be infinitely many parallels to a line through a given point, and that in this way one comes to a distinctly new kind of geometry, known today as *non-euclidean geometry*. Gauss published none of his results, unfortunately, though we can think of various reasons for this. One was that the new geometry was only hypothetical. The possibility was not excluded that a contradiction could be found in it, thus leading to a proof of the parallel axiom. Another reason was that euclidean geometry had been declared by Kant to be an *a priori* assumption of human thought. Gauss shied away from opposing such a generally accepted philosophical position. E.g. on 25.8.1818 he wrote to Gerling, in answer to a question concerning the parallel axiom: "*I am glad that you have the courage to speak up when you recognise the possibility that our theory of parallels, and hence our whole geometry, may be false. However, the wasps whose nest you stir will fly at your head.*"

Non-euclidean geometry was rediscovered independently by J. Bolyai in 1823 and Lobachevsky in 1826. Both published their results:

Bolyai, J., *Scientiam spatii absolute veram exhibens: a veritate aut falsitate Axiomatis XI Euclidei (a priori haud unquam decidenda) independentem: adjecta ad casum falsitatis, quadratura circuli geometrica* (Appendix to a geometry textbook of his father, W. Bolyai, 1832).

Lobachevsky, N.I., *O natschalach geometrii, Kasanskii Vestnik* 1829-1830.

However, they received no support from mathematicians apart from Gauss, who expressed his enthusiasm in letters to them both, but contributed nothing to their public recognition. It is noteworthy that Riemann, as the introduction to his inaugural lecture shows, knew nothing of the non-euclidean geometry of Gauss, Bolyai and Lobachevsky.

Non-euclidean geometry first became a more secure part of mathematics with the construction of a model, defined with the help of the real numbers (and thus assuming that the real numbers themselves were secure), established its freedom from contradiction. The first such model was given by Beltrami (*Saggio di interpretazione della geometria non-euclidea, G. mat. Napoli* 1868). The concept of model appeared with full clarity in Klein (*Über die sogenannte Nicht-Euklidische Geometrie, Math. Ann.* 4 (1871)), based on a metric for projective space introduced by Cayley (*A sixth memoir upon quantics, Philos. Trans.* 1859).

The euclidean axiom system is incomplete. It implicitly uses the concept of motion of a geometric figure without incorporating it in the axioms. In addition, the axiomatisation overlooks the ordering of points on a line. A complete axiom system for euclidean geometry was given in 1899 by Hilbert in his book *Grundlagen der Geometrie*.

We shall now give a model of the non-euclidean geometry of the plane. As a Riemannian manifold this model is the same as the upper half plane H, considered in Section 14.12, with the metric $(1/y^2)(dx^2+dy^2)$. We define the lines of our geometry to be the geodesics of H, i.e. the semicircles with midpoints on the x-axis and lines perpendicular to the x-axis. One easily sees that for two distinct points of H there is exactly one line of our geometry which contains them both. On the other hand, for each line g there are infinitely many parallels which go through a point outside g. The distance between two points is given as in Section 14.12. The angle between two lines through a point is equal to the euclidean angle between the corresponding semicircles. The admissible orientation-preserving motions of the geometry (cf. Section 4.2) are the linear fractional transformations (39) (Theorem 8).

This model was pointed out by Poincaré. Hence H is often called the *Poincaré upper half plane.*

Exercises

14.1 Let M be a region of the euclidean space E^n with coordinates $x^1,\dots,x^n$. Also, let $f_1,\dots,f_n$ be the vector field on M introduced in Section 14.6. For a curve $C(t)$ we consider $f_1,\dots,f_n$ as functions of the parameter t.

a) Show that

$$\frac{df_j}{dt} = \Gamma^k_{ij}\frac{dx^i}{dt}f_k.$$

b) Let $a = a^i f_i$ be the vector field on M given by parallel displacement of a vector from the point $O = C(0)$. We consider a along the curve C(t). Show that the components a^i, as functions of the parameter t, satisfy the system of differential equations

$$\frac{da^k}{dt} + a^j \Gamma^k_{ij} \frac{dx^i}{dt} = 0, \quad k = 1,\dots,n. \tag{44}$$

14.2 (Levi-Civita). The result of exercise 14.1(b) suggests that we define parallel displacement along any curve C(t) in a Riemannian manifold by the condition (44): thus we now let the Γ^k_{ij} be the Christoffel symbols of a Riemannian manifold. In a neighbourhood of $O = C(0)$, (44) has a unique solution $a^j(t)$. Show that $a^j(t)\partial/\partial x^j$ is a vector at the point C(t), i.e. that on passing to other coordinates the $a^j(t)$ have the transformation behaviour of a contravariant vector.

14.3 Show that parallel displaced vectors along a geodesic are tangent vectors to the geodesic.

14.4 Let ϕ_1, ϕ_2, r, a be real numbers with $0 < \phi_1 \le \phi_2 < \pi$ and $r > 0$, so that $z_1 := re^{i\phi_1} + a$ and $z_2 := re^{i\phi_2} + a$ are points of the Poincaré upper half plane. Show that the non-euclidean distance between z_1 and z_2 equals $\log \tan(\phi_2/2) - \log \tan(\phi_1/2)$.

14.5 Let z_1 and z_2 be arbitrary points of the Poincaré upper half plane, and let r_1 and r_2 be the intersections with the real axis of the geodesic L through z_1 and z_2. Let the order of these points on L be r_1, z_1, z_2, r_2. Show that the non-euclidean distance between z_1 and z_2 equals

$$\log\left(\frac{z_1 - r_2}{z_1 - r_1} : \frac{z_2 - r_2}{z_2 - r_1} \right).$$

15. On the number of primes less than a given magnitude

15.1 Formulation of the problem

Let $\pi(x)$ be the number of primes less than or equal to the positive real number x. The asymptotic behaviour of this function had already interested the young Gauss. As a result of computing its value up to the argument $x = 3.10^6$, he arrived at the conjecture that

$$\lim_{x\to\infty} \pi(x)\Big/\frac{x}{\log x} = 1, \tag{1}$$

and that $\pi(x)$ is still better approximated by the function $\displaystyle\int_2^x \frac{dt}{\log t}$.

Chebyshev was the first able to prove a result in the direction of this conjecture. He showed, e.g., that sufficiently large x satisfy the inequality

$$0.92129\,\frac{x}{\log x} < \pi(x) < 1.10555\,\frac{x}{\log x}$$

(*Sur la totalité des nombres premiers inférieurs à une limite donnée, J. math. pures appl.* 17 (1852)).

Gauss's conjecture was proved in 1896 by Hadamard (*Sur la distribution des zéros de la fonction* $\zeta(s)$ *et ses conséquences arithmétiques, Bull. soc. math. France* 24 (1896)) in the form (1), and independently by de la Vallée Poussin (*Récherches analytiques sur la théorie des nombres premiers, Ann. soc. sci. Bruxelles* 20, 2 (1896)) in a sharper form (Chapter 27). Both proofs are based on the work of Riemann (*Über die Anzahl der Primzahlen unter einer gegebenen Grösse* (*Monatsber. der Preuss. Akad. Wiss.*, November 1859), to which we now turn.

15.2 The functional equation of the ζ-function

Riemann's starting point is the investigation of the ζ-function, already considered by Euler for real arguments (Chapter 6), in the realm of complex function theory:

$$\zeta(s) = \sum_{n=1}^{\infty} \frac{1}{n^s} = \prod_p \left(1 - \frac{1}{p^s}\right). \tag{2}$$

We have shown in Section 6.3 that the sum and product in (2) converge uniformly to the same function for $s > s_0 > 1$, and hence represent a continuous function for $s > 1$. The same considerations show immediately that this is also true for complex $s = \sigma + it$ with $\sigma > 1$ (in what follows σ resp. t always denotes the real resp. imaginary part of s). With the help of the Weierstrass double series theorem (Section 8.10), which is also used several times below, one shows that $\zeta(s)$ is a regular function for $\sigma > 1$. Because of the product representation (2), $\zeta(s)$ is non-zero for $\sigma > 1$. Riemann proved first that $\zeta(s)$ can be extended to a function meromorphic over the whole plane. To do this he used the Euler definition of the Γ-function:

$$\Gamma(s) = \int_0^\infty x^{s-1}e^{-x}dx \quad \text{(Chapter 9, Theorem 3).}$$

For $\sigma > 1$ we have

$$\int_0^\infty e^{-nx}x^{s-1}dx = n^{-s}\int_0^\infty e^{-x}x^{s-1}dx = n^{-s}\Gamma(s)$$

and hence

$$\Gamma(s)\zeta(s) = \sum_{n=1}^{\infty}\int_0^\infty e^{-nx}x^{s-1}dx = \int_0^\infty x^{s-1}(e^{x}-1)^{-1}dx.$$

We consider the function $\phi(z,s) := e^{(s-1)\log(-z)}(e^{z}-1)^{-1}$, where the xy-plane ($z = x + iy$) is cut along the positive x-axis and $\log(-z)$ takes real values for negative real z, so that for positive real z one has $\log(-z) = \log z - i\pi$ as one approaches the x-axis from above (upper sheet) and $\log(-z) = \log z + i\pi$ as one approaches from below (lower sheet). We then integrate along the cut positive x-axis, starting at a number r on the upper sheet, describing a circle C of radius ε around the origin, for a small $\varepsilon > 0$, then returning to r along the real axis on the lower sheet (see Figure 14).

Fig. 14

Then the integral over the whole curve $S(r)$ is

$$\int_{S(r)} \phi(z,s)dz = \int_r^{\varepsilon} e^{(s-1)(\log x - i\pi)}(e^x-1)^{-1}dx + \int_C \phi(z,s)dz$$

$$+ \int_{\varepsilon}^{r} e^{(s-1)(\log x + i\pi)}(e^x-1)^{-1}dx.$$

By the Cauchy integral theorem, the value of the integral is independent of ε, and as $\varepsilon \to 0$ it tends to

$$(e^{-\pi si} - e^{\pi si})\int_0^r x^{s-1}(e^x-1)^{-1}dx.$$

Hence as $r \to \infty$ we get

$$i \lim_{r\to\infty} \int_{S(r)} \phi(z,s)dz = 2 \sin \pi s \cdot \Gamma(s) \cdot \zeta(s). \tag{3}$$

This holds initially for $\sigma > 1$. However, the left-hand side represents a regular function for arbitrary s, and hence defines $\zeta(s)$ as a single-valued meromorphic function over the whole plane. Since we know the zeros and poles of $\sin \pi s$ and $\Gamma(s)$, we can deduce from (3) that $\zeta(s)$ is regular except at $s = 1$, and that for $s = 1$ it has at most a simple pole. Since $\lim_{\sigma\to\infty} \zeta(\sigma) = \infty$, there is in fact a pole at $s = 1$.

In the case $\sigma < 0$ we also consider the integral of $\phi(z,s)$ over $S(r)$ together with a circle of radius r about the origin in the clockwise sense, where we choose $r = \pi + 2n\pi$ in order to avoid having poles on the circle. As one easily sees, the integral over this circle tends to 0 as $n \to \infty$. Hence by the residue theorem

$$i \lim_{r\to\infty} \int_{S(r)} \phi(z,s)dz = 2\pi \sum_{m=-\infty}^{\infty} \text{Res}\, \phi(z,s),$$

where $m = 0$ is omitted from the summation. For $m > 0$ we have

$$\text{Res}_{2\pi mi}\phi(z,s) = \lim_{z\to 2\pi mi} \phi(z,s)(z-2\pi mi)$$

$$= e^{(s-1)\log(-2\pi mi)}$$

$$= (2\pi m)^{s-1} e^{-(\pi/2)i(s-1)}$$

and for $m < 0$

$$\operatorname{Res}_{2\pi mi}\phi(z,s) = (2\pi|m|)^{s-1}e^{(\pi/2)i(s-1)},$$

hence

$$2\sin\pi s\cdot\Gamma(s)\cdot\zeta(s) = (2\pi)^s\zeta(1-s)2\cos\frac{\pi}{2}(s-1)$$

and

$$2\cos\frac{\pi}{2}s\cdot\Gamma(s)\cdot\zeta(s) = (2\pi)^s\zeta(1-s). \tag{4}$$

This equation holds initially for $\sigma < 0$. But since only regular functions appear, it holds for arbitrary s.

By Chapter 9, (14), we can also write (4) in the symmetrical form

$$\Gamma(\frac{s}{2})\pi^{-s/2}\zeta(s) = \Gamma(\frac{1-s}{2})\pi^{-(1-s)/2}\zeta(1-s). \tag{5}$$

For $\sigma > 1$ the function on the left is regular and non-vanishing. Hence $\zeta(s)$ has simple zeros for $s = -2, -4, \ldots$ and it is non-zero for all other s with $\sigma < 0$.

Remark. In Section 5.6 we have computed $\zeta(s)$ for even positive integers s :

$$\zeta(2n) = (-1)^{n+1}\frac{(2\pi)^{2n}}{2(2n)!}B_{2n}.$$

The functional equation (4) gives

$$\zeta(1-2n) = 2(-1)^n(2n-1)!(2\pi)^{-2n}\zeta(2n) = -\frac{B_{2n}}{2n}, \tag{6}$$

a rational number!

15.3 On the zeros of the ζ-function

Riemann realised the connection between the distribution of primes in the sequence of natural numbers and the distribution of zeros of $\zeta(s)$ in the complex plane. We have already convinced ourselves that $\zeta(s)$ has only the zeros $s = -2, -4, \ldots$ outside the "critical strip" $0 \le \sigma \le 1$. The latter are called the *trivial* zeros.

Riemann asserted that the number of zeros of $\zeta(\frac{1}{2} + iz)$ whose real part lies between 0 and T is "about" equal to $\frac{T}{2}\log\frac{T}{2} - \frac{T}{2}$. He actually wrote:

"In fact one now finds about as many real values between these bounds, and it is very probable that all roots are real. Naturally, a rigorous proof of this is desirable; however, I have put this question aside after a few brief unsuccessful attempts, since it does not seem to be necessary for the immediate objective of my investigation."

The assertion that the zeros of the ζ-function in the critical strip have the form $\frac{1}{2} + iy$ with real y is the *Riemann hypothesis*, which remains unproved to the present day. Riemann also asserted that the entire function

$$\xi_1(z) := \frac{s}{2}(s-1)\Gamma(\tfrac{s}{2})\pi^{-s/2}\zeta(s) \quad \text{with} \quad s = \frac{1}{2} + iz \tag{7}$$

has the product representation

$$\xi_1(z) = \xi_1(0) \prod_{\alpha} \left(1 - \frac{z^2}{\alpha^2}\right)$$

where the product is taken over all zeros α with $\mathrm{Re}\,\alpha > 0$. This assertion was first proved in 1893 by Hadamard (Chapter 26, Theorem 5).

15.4 Riemann's exact formula

The further content of Riemann's work consists of an exact formula for a function closely related to the prime number function.

First we go from $\pi(x)$ to a function $\pi_0(x)$ which equals $\pi(x)$ when x is not a prime (i.e., when $\pi(x)$ has no jump), and which equals $\pi(x) - \frac{1}{2}$ when x is a prime (cf. Section 5.2 and Section 27.3). We also set

$$f(x) := \sum_{n=1}^{\infty} \frac{1}{n}\, \pi_0(x^{1/n}).$$

One gets back to $\pi_0(x)$ from $f(x)$ with the help of the *Möbius function* $\mu(m)$, which is defined for natural numbers m as follows:

$\mu(m) = 0$ when m is divisible by the square of a prime.
$\mu(m) = (-1)^r$ when m is the product of r distinct primes.

This function satisfies

$$\sum_{d|m} \mu(d) = \begin{cases} 1 & \text{for } m = 1 \\ 0 & \text{for } m > 1 \end{cases}, \tag{8}$$

where the sum is taken over all divisors d of m. From this it easily follows that

$$\pi_0(x) = \sum_{m=1}^{\infty} \mu(m)\frac{1}{m} f(x^{1/m}). \tag{9}$$

To state Riemann's exact formula we first define the logarithmic integral Li(x), which differs from the function $\int_2^x \frac{dt}{\log t}$ only by a constant. We begin by proving a simple proposition.

Theorem 1. *Let* $x > 1$ *be a positive real number. Then the limit*

$$\lim_{\varepsilon\to 0}\left(\int_0^{1-\varepsilon} \frac{dt}{\log t} + \int_{1+\varepsilon}^{x} \frac{dt}{\log t}\right) \tag{10}$$

exists and is equal to

$$\lim_{\varepsilon\to 0}\left(\int_{-\infty}^{-\varepsilon} \frac{e^s ds}{s} + \int_{\varepsilon}^{\log x} \frac{e^s ds}{s}\right) \tag{11}$$

Proof. Under the substitution $t = e^s$, (10) becomes

$$\lim_{\varepsilon\to 0}\left(\int_{-\infty}^{\log(1-\varepsilon)} \frac{e^s ds}{s} + \int_{\log(1+\varepsilon)}^{\log x} \frac{e^s ds}{s}\right) \tag{12}$$

We write (12) in the form

$$\int_{-\infty}^{1} \frac{e^s ds}{s} + \lim_{\varepsilon\to 0}\left(\int_{-1}^{\log(1-\varepsilon)} \frac{(e^s-1)ds}{s} + \int_{\log(1+\varepsilon)}^{\log x} \frac{(e^s-1)ds}{s}\right) + \lim_{\varepsilon\to 0}\int_{-1}^{\log(1-\varepsilon)} \frac{ds}{s} + \int_{\log(1+\varepsilon)}^{\log x} \frac{ds}{s}. \tag{13}$$

The limit $\lim_{\varepsilon\to 0}\left(\int_{-1}^{\log(1-\varepsilon)} \frac{ds}{s} + \int_{\log(1-\varepsilon)}^{\log x} \frac{ds}{s}\right)$ is easily computed to be log log x, while the integrand $\frac{e^s-1}{s}$ is regular in the whole plane. One also sees easily that

$$\lim_{\varepsilon\to 0}\left(\int_{-1}^{-\varepsilon} \frac{ds}{s} + \int_{\varepsilon}^{\log x} \frac{ds}{s}\right) = \log\log x. \qquad \square$$

By (13), the function $\mathrm{Li}(e^z)$ has the representation

$$\mathrm{Li}(e^z) = \int_{-\infty}^{-1} \frac{e^s ds}{s} + \int_{-1}^{0} \frac{(e^s-1)}{s} ds + \log z + \int_0^z \frac{(e^s-1)ds}{s} \tag{14}$$

for $z > 0$, and hence as an analytic function it has the same multi-valued behaviour as $\log z$. We cut the complex plane along the negative real axis and can then define $\mathrm{Li}(e^z)$ by (14) as a single-valued function for all z not on the negative real axis. We define $\mathrm{Li}(x^\rho)$ for $x > 1$ as $\mathrm{Li}(e^{\rho \log x})$.

Finally we convince ourselves that $c := \int_{-\infty}^{-1} \frac{e^s ds}{s} + \int_{-1}^{0} \frac{(e^s-1)}{s} ds$ is equal to Euler's constant γ (Chapter 9.3). To do this we use integration by parts:

$$-c = \int_1^\infty \frac{e^{-s} ds}{s} + \int_0^1 \frac{(e^{-s}-1)}{s} ds$$

$$= \int_1^\infty e^{-s} \log s \, ds + \int_0^1 e^{-s} \log s \, ds = \Gamma'(1) = -\gamma.$$

Equation (14) shows that one has the highly convergent series

$$\mathrm{Li}(e^z) = \gamma + \log z + \sum_{n=1}^{\infty} \frac{z^n}{n!n} \tag{15}$$

Riemann's *exact formula* for $f(x)$ reads

$$f(x) = \mathrm{Li}(x) - \sum_\alpha (\mathrm{Li}(x^{1/2+i\alpha}) + \mathrm{Li}(x^{1/2-i\alpha})) + \int_x^\infty \frac{dt}{(t^2-1)t \log t} - \log 2. \tag{16}$$

Here the sum is over all zeros α of $\xi_1(z)$ with $\mathrm{Re}\,\alpha > 0$, ordered by increasing real parts.

The reason for using, instead of $\int_2^x \frac{dt}{\log t}$, a more complicated definition of $\mathrm{Li}(x)$, is that only with the latter definition of the constant of integration does the sum (16) converge.

The formula (16) was rigorously proved by Mangoldt in 1895 (Exercise 27.9). The formulae (9) and (16) suggest that we consider, in place of Li(x), the function

$$R(x) := \sum_{m=1}^{\infty} \mu(m)\frac{1}{m}\,\mathrm{Li}(x^{1/m})$$

$$= 1 + \sum_{n=1}^{\infty} \frac{1}{n\zeta(n+1)}\,\frac{(\log x)^n}{n!} \tag{17}$$

as an approximation to $\pi(x)$. The following table shows the good agreement between $\pi(x)$ and $R(x)$:

x	10^8	2.10^8	3.10^8	4.10^8	5.10^8
$\pi(x)$	5761455	11078937	16252325	21336326	26355867
$R(x)$	5761552	11079090	16252355	21336185	26355517

x	6.10^8	7.10^8	8.10^8	9.10^8	10^9
$\pi(x)$	31324703	36252931	41146179	46009215	50847534
$R(x)$	31324622	36252719	41146248	46009949	50847455

Exercises

15.1 Let n be a natural number. Show that the product of all primes p with $n < p < 2n$ is smaller than 2^{2n}. (Hint: use the binomial coefficient $\binom{2n}{n}$.)

15.2 Show that $\pi(2n) - \pi(n) < \dfrac{2n \log 2}{\log n}$.

15.3 Show that $\pi(x) < 1.7 \dfrac{x}{\log x}$ for $x \geq 2$ under the assumption that this is already verified for all $x \leq 1200$.

15.4 Show that $\binom{n}{k} \leq n^{\pi(n)}$ for all natural numbers k with $1 \leq k \leq n$.

15.5 Show that $2^n \leq (n+1)n^{\pi(n)}$.

15.6 Show that $\pi(x) > \dfrac{2}{3}\dfrac{x}{\log x}$ for $x \geq 200$.

Remark: The method of proof for Exercises 15.1 to 15.6 goes back to Chebyshev (cf. Section 15.1). We have taken them from the inaugural lecture of D. Zagier at the University of Bonn, in which many interesting results on the distribution of primes are cited (Zagier, D., *The first 50 million prime numbers,* The Mathematical Intelligencer 0, August 1977. *Lebendige Zahlen, fünf Exkursionen,* Mathematischen Miniaturen 1, Birkhäuser 1981, pp. 39-73).

15.7 Show that

$$\frac{1}{\zeta(s)} = \sum_{n=1}^{\infty} \frac{\mu(n)}{n^s} \quad \text{for} \quad \operatorname{Re} s \geq 1.$$

15.8 Show that $\sum_{n=1}^{\infty} \mu(n)\frac{\log n}{n} = -1.$

15.9 Prove (17).

16. The origins of algebraic number theory

16.1 The Gaussian integers

In his effort to construct a theory of biquadratic residues analogous to the theory of quadratic residues (Chapter 1), Gauss realised that it would be necessary to pass from the domain $\mathbb{Z}$ of integers to the domain $\mathbb{Z}[\sqrt{-1}]$ of numbers of the form $x + y\sqrt{-1}$ with $x,y \in \mathbb{Z}$. In his 1832 work *Theoria residuorum biquadraticum* II, (*Comm. Soc. Reg. Sci. Gottingensis* 7) he showed that the theorem of unique prime decomposition (Section A1.1) also held for the ring $\mathbb{Z}[\sqrt{-1}]$. We shall follow Dedekind below, and prove this theorem with the help of the euclidean algorithm.

We set $i := \sqrt{-1}$ as usual. The complex numbers of the form $x + yi$ with $x,y \in \mathbb{Q}$ constitute a field $\mathbb{Q}(i)$, which therefore has degree 2 as a vector space over $\mathbb{Q}$. Let α' denote the *complex conjugate* of $\alpha \in \mathbb{Q}(i)$. $N(\alpha) = \alpha\alpha'$ is called the *norm* of α. If $\alpha = x + yi$ then $\alpha' = x - yi$ and $N(\alpha) = x^2 + y^2$. Hence the norm is always ≥ 0, and equals 0 only when $\alpha = 0$. As one easily sees, $N(\alpha\beta) = N(\alpha)N(\beta)$.

The number $\alpha = x + yi$ is called a *Gaussian integer* when x and y are from $\mathbb{Z}$. These numbers obviously constitute a ring $\mathbb{Z}[i]$, the ring of *Gaussian integers*. They make up the lattice of points with integer coordinates in the plane of complex numbers. It is obvious from this that each number α in $\mathbb{Q}(i)$ may be represented as a sum $\beta + \gamma$ where β is from $\mathbb{Z}[i]$ and $N(\gamma) \leq \frac{1}{2}$.

Let $\varepsilon = x + yi$ be a unit in $\mathbb{Z}[i]$. Then $N(\varepsilon) = x^2 + y^2 = 1$. Hence $\mathbb{Z}[i]$ has only four units, ± 1, $\pm i$.

Arithmetic in $\mathbb{Z}[i]$ is based on the following theorem about division with remainder.

Theorem 1. *Let* α *be arbitrary and let* β *be a non-zero number from* $\mathbb{Z}[i]$. *Then there are numbers* γ *and* ν *in* $\mathbb{Z}[i]$ *such that*

$$\alpha = \nu\beta + \gamma \text{ with } N(\gamma) < N(\beta).$$

Proof. We write α/β in the form $\nu + \mu$ with $\nu \in \mathbb{Z}[i]$ and $N(\mu) \leq \frac{1}{2}$. Then $\gamma = \beta\mu \in \mathbb{Z}[i]$ and $N(\gamma) = N(\beta)N(\mu) < N(\beta)$. □

By Theorem 1, $\mathbb{Z}[i]$ is a euclidean ring. Hence the Theorem of unique prime decomposition (A1, Theorem 3) holds in $\mathbb{Z}[i]$.

We now want to gain an overview of all prime elements π in $\mathbb{Z}[i]$. Since $\pi\pi' \in \mathbb{Z}$, the multiples of π include exactly one prime number p in $\mathbb{Z}$. It therefore suffices to determine the prime divisors, in $\mathbb{Z}[i]$, of each prime number p.

Theorem 2. (i) *If* $p \equiv 1 \pmod 4$ *then* $p = \pi\pi'$, *where* π *and* π' *are non-associated conjugate prime elements.*

(ii) *If* $p \equiv 3 \pmod 4$ *then* p *is a prime element in* $\mathbb{Z}[i]$.

(iii) $2 = (1-i)^2 i$ *with the prime element* $1-i$.

Proof. It follows from $\pi | p$ that $N(\pi) | N(p) = p^2$. Therefore $N(\pi) = p$ or $N(\pi) = p^2$. In the second case, unique prime factorisation implies that π is an associate of p, i.e. p is a prime element in $\mathbb{Z}[i]$.

For $p \equiv 3 \pmod 4$ the second case must occur, since with $\pi = x + yi$ and $x,y \in \mathbb{Z}$ the congruence $N(\pi) = x^2 + y^2 \equiv 3 \pmod 4$ is unsolvable.

For $p \equiv 1 \pmod 4$, Chapter 1, Theorem 7 shows that -1 is a quadratic residue mod p, i.e. there is an $x \in \mathbb{Z}$ with $p | (x^2+1) = (x+i)(x-i)$. Since p divides neither of the factors $x + i$ and $x - i$, p is not a prime element of $\mathbb{Z}[i]$. Thus we have $N(\pi) = p$. The prime elements π and π' are not associates, since $\pi - \pi' = 2yi$ is relatively prime to p. □

Theorem 2 contains Theorem 10 from Chapter 2, hence we have just obtained a new proof of the latter.

16.2 Introduction to Chapters 17 to 21

Basing himself on Gauss's theory of circle division (Chapter 3) and results of Jacobi, Kummer began in 1844 to study arithmetic in the ring $\mathbb{Z}[\zeta_p]$, where p is a prime number and ζ_p is a pth root of unity different from 1 (*De numeris complexis, qui radicibus unitatis et numeris integris realibus constant*, in: *Gratulationsschrift der Breslauer Universität zur Jubelfeier der Königsberger Universität, Breslau* 1844). After it had become clear to him that unique prime factorisation did not generally hold in these rings, he introduced "ideal numbers" to reinstate it. In a later work he applied the results so obtained to prove the Fermat conjecture in a number of cases.[1)]

Dirichlet studied more general rings of the form $\mathbb{Z}[\alpha]$, where α is an algebraic integer, i.e. a number satisfying an equation $\alpha^n + a_1\alpha^{n-1} + \ldots + a_n = 0$ with integer coefficients $a_1,\ldots,a_n$. He was able to elucidate the structure of the group of units (Section A1.1) in this ring (*Zur Theorie der complexen Einheiten, Monatsber. Preuss. Akad. Wiss., March* 1846). We present this result of Dirichlet in Chapter 19.

The next important work in the theory of algebraic numbers appeared in 1871 as Dedekind's *Supplement* X to his edition of Dirichlet's *Vorlesungen über Zahlentheorie.* Dedekind's goal was to develop a general theory of algebraic numbers. Kronecker (*Grundzüge einer arithmetischen Theorie der algebraischen Grössen, J. reine angew. Math.* 92 (1882)) and Zolotarev (*Sur la théorie des nombres complexes, J. math. pures appl.* 6 (1880)) arrived at such a theory in other ways. Kronecker's method was initially the most fruitful and was presented, e.g. by Weber in his *Lehrbuch der Algebra*, volume II, in 1896, while Zolotarev's results can be regarded as precursors of Hensel's valuation-theoretic method, whose influence did not begin to be felt until the 20's of the present century, above all through the work of Hasse. We take up these methods in the second volume of this book. Beginning in the 1890's, Dedekind's theory asserted itself, and became an essential part of mathematics in the twentieth century.

In the fourth edition of Dirichlet's *Vorlesungen* (1894), the earlier *Supplement* X appeared in revised form as *Supplement* XI, containing for the first time a presentation of Galois theory (Chapter 7) as a theory of finite extensions of fields. Striking gains in simplicity and clarity were achieved thereby. Dedekind's ideas here go back to his Göttingen lectures from 1857 to 1858.

We shall now allow ourselves to be guided by *Supplement* XI, and divide the material into a number of chapters. In Chapter 17 we treat the algebraic foundations of field theory. Dedekind confined himself to number fields, i.e. fields contained in the field $\mathbb{C}$ of complex numbers. However, since we want here to close gaps remaining from our earlier treatment of Galois theory (Chapter 7) we shall, as we did there, allow the fields to be arbitrary of characteristic 0. Chapter 18 contains Dedekind's general theory of

---

[1)] See Neumann, O., *Über die Anstösse zu Kummers Schöpfung der "idealen complexen Zahlen", in: Mathematical Perspectives (Biermann-Festschrift), Academic Press* 1981.

algebraic integers, which is called *ideal theory*. This theory may be carried over to function fields, i.e. to fields which are finite extensions of the field $F_0(x)$ of rational functions over a field F_0. Here the polynomial ring $F_0[x]$ plays a rôle analogous to that of the ring $\mathbb{Z}$ of integers in the theory of algebraic numbers. One property both rings have in common is that of being euclidean (Section A1.2). This property suffices for the development of Dedekind's ideal theory. We are thereby able simultaneously to lay the foundations for the theory of algebraic functions, which we treat in Chapter 23.

Chapters 19 and 20 contain those parts of the theory which are specific to number fields. Finally, in Chapter 21 we treat the connection between quadratic forms and quadratic fields, including a lucid theory of composition of classes of forms (Section 2.9).

Exercises

16.1 Determine the prime factorisation of $17 + 19i$ in $\mathbb{Z}[i]$.

16.2 Let π be a prime in $\mathbb{Z}[i]$. Prove that $\alpha^{N(\pi)} \equiv \alpha \pmod{\pi}$ for all $\alpha \in \mathbb{Z}[i]$.

16.3 Find a solution of the equation

$$\xi(3+5i) + \eta(4+5i) = 1 \quad \text{with } \xi,\eta \in \mathbb{Z}[i].$$

16.4 Show that, for each $\alpha \in \mathbb{Z}[i]$ which is relatively prime to 3, there is a natural number x with $(1+i)^x \equiv \alpha \pmod 3$.

16.5 Let p be a prime in $\mathbb{Z}$ with $p \equiv 3 \pmod 4$. Show that the group $(\mathbb{Z}[i]/p\mathbb{Z}[i])^\times$ is cyclic.

16.6 Let p be a prime in $\mathbb{Z}$ with $p \equiv 3 \pmod 4$ and let $a \in \mathbb{Z}$ with $a \not\equiv 0 \pmod p$. Show that the congruence $x^2 + y^2 \equiv a \pmod p$ has exactly $p+1$ solutions $x,y \in \mathbb{Z}$.

16.7 Show that $\mathbb{Z}[\sqrt{-3}]$ is not a ring with unique prime factorisation.

16.8 Show that $\mathbb{Z}[(1+\sqrt{-3})/2]$, $\mathbb{Z}[(1+\sqrt{-7})/2]$, $\mathbb{Z}[\sqrt{-2}]$ and $\mathbb{Z}[(1+\sqrt{-11})/2]$ are rings with unique prime factorisation.

16.9 By analogy with the procedure for $\mathbb{Z}[i]$, give a survey of the primes in $\mathbb{Z}[(1+\sqrt{-3})/2]$, $\mathbb{Z}[(1+\sqrt{-7})/2]$, $\mathbb{Z}[\sqrt{-2}]$ and $\mathbb{Z}[(1+\sqrt{-11})/2]$.

17. Field theory

17.1 Field isomorphisms

Paragraphs 160 to 167 of *Supplement* XI presented the "*most important foundations of today's algebra*". In §164 Dedekind declared the proper objective of contemporary algebra to be "*the precise investigation of the relationship between different fields*".

Let K_1 and K_2 be two fields. A *field homomorphism* ϕ of K_1 into K_2 is a mapping of K_1 into K_2 with

$$\phi(a+b) = \phi(a) + \phi(b),\ \phi(ab) = \phi(a)\phi(b) \quad \text{for all } a,b \in K_1. \tag{1}$$

$$\phi(1) = 1. \tag{2}$$

Obviously the image $\phi(K_1)$ is a subfield of K_2.

Theorem 1. *A homomorphism* ϕ *of* K_1 *into* K_2 *is injective.*

Proof. Suppose there were $a,b \in K_1$ with $a \neq b$ and $\phi(a) = \phi(b)$. Then $\phi(1) = \phi(1/(a-b))\phi(a-b) = 0$, contrary to (2). □

Each homomorphism ϕ of K_1 into K_2 is therefore an isomorphism of K_1 onto $\phi(K_1)$. "*Investigation of the relationships between different fields*" means investigation of the isomorphisms between fields.

An algebraic number field L of degree n is a subfield of $\mathbb{C}$ which has degree n over $\mathbb{Q}$.

Theorem 2. *There are exactly* n *different isomorphisms of* L *into* $\mathbb{C}$.

Proof. By the theorem of the primitive element (Chapter 7, Theorem 4) there is a $\theta \in L$ with $L = \mathbb{Q}(\theta)$. Let f(x) be the irreducible polynomial with coefficients in $\mathbb{Q}$ associated with θ and let θ_i be one of the n zeros of f(x) in $\mathbb{C}$. Then by Section A1.5 the correspondence ϕ_i :

$$\sum_{k=0}^{n-1} a_k\theta^k \mapsto \sum_{k=0}^{n-1} a_k\theta_i^k \quad \text{for } a_0,\dots,a_{n-1} \in \mathbb{Z}$$

is an isomorphism of L into $\mathbb{C}$.

In this way we have found n isomorphisms. On the other hand, if ϕ is an isomorphism of L into $\mathbb{C}$, then $\phi(\theta)$ is a zero of $f(x)$, and ϕ is uniquely determined by $\phi(\theta)$. □

We want to generalise Theorem 2 to an arbitrary field of characteristic 0.

Let L be a finite extension of a field K of characteristic 0, $L = K(\theta)$. Also let Z be a splitting field of the minimal polynomial of θ over K (A.1, Theorem 10). The desired generalisation of Theorem 2 reads:

Theorem 3. *Let* M *be a subfield of* L *which includes* K, *and let* ϕ *be an isomorphism of* M *into* Z *which fixes the elements of* K. *Then there are exactly* $[L:M]$ *extensions of* ϕ *to an isomorphism of* L *into* Z.

Proof. Let $[L:M] = n$ and let $\theta_1,\ldots,\theta_n$ be the zeros of the minimal polynomial of θ over M. Then the desired n extensions ϕ_i of ϕ $(i = 1,\ldots,n)$ are given by

$$\phi_i\left(\sum_{k=0}^{n-1} \alpha_k\theta^k\right) = \sum_{k=0}^{n-1} \phi(\alpha_k)\theta_i^k \quad \text{with} \quad \alpha_0,\ldots,\alpha_{n-1} \in M. \qquad \square$$

Remark. The proof of Theorem 3 rests on the theorem that an irreducible polynomial of degree n has n distinct zeros in a splitting field. This is not generally true over a field of characteristic $p \neq 0$.

In the following sections of this chapter we always assume that all fields have characteristic 0.

17.2 Normal extensions and Galois groups

A finite extension N/K is called *normal* when each irreducible polynomial $f(x) \in K[x]$ which has a zero in N splits into linear factors in N.

Theorem 4. *Let* $f(x) \in K[x]$ *be a polynomial which splits into linear factors* $x-\alpha_i$, $i = 1,\ldots,n$, *in an extension field of* K. *Then* $K(\alpha_1,\ldots,\alpha_n)/K$ *is a normal extension.*

Proof. Let $\alpha \in K(\alpha_1,...,\alpha_n)$, let f_α be the minimal polynomial of α over K, and let Z be a splitting field of f_α which contains $K(\alpha_1,...,\alpha_n)$. By Theorem 3 there are $[K(\alpha) : K]$ isomorphisms ϕ of $K(\alpha)$ into Z which fix the elements of K. There is an extension of ϕ to Z, which will also be denoted by ϕ. Since $\alpha \in K(\alpha_1,...,\alpha_n)$, $\phi(\alpha) \in K(\phi(\alpha_1),...,\phi(\alpha_n))$. On the other hand, $\phi(\alpha_i)$ is a zero of f_α for all $i = 1,...,n$, and hence $K(\phi(\alpha_1),...,\phi(\alpha_n)) \subset K(\alpha_1,...,\alpha_n)$. It follows that f_α already splits into linear factors in $K(\alpha_1,...,\alpha_n)$. □

Let $L = K(\theta)$. By Theorem 4, a splitting field Z for the minimal polynomial of θ over K is also a splitting field for each irreducible polynomial which has a zero in L. Because of this we also call Z a splitting field of L/K when L is contained in Z.

For splitting fields we have the following uniqueness theorem.

Theorem 5. *Let* $f(x) \in K[x]$ *and let* Z *and* Z' *be two splitting fields of* f. *Suppose the zeros of* f *in* Z *resp.* Z' *are* $\alpha_1,...,\alpha_n$ *resp.* $\alpha'_1,...,\alpha'_n$. *Then* $K(\alpha_1,...,\alpha_n)$ *is isomorphic to* $K(\alpha'_1,...,\alpha'_n)$.

Proof. By Theorem 3 and Theorem 4 there is an isomorphism of $K(\alpha_1,...,\alpha_n)$ into Z'; its image is $K(\alpha'_1,...,\alpha'_n)$. □

Let N/K be a normal extension of degree n. By Theorem 3 there are exactly n automorphisms of N (i.e. isomorphisms of N onto itself) which fix the elements of K. They obviously constitute a group under composition. This group is called the *Galois group of* N/K and is denoted by $G(N/K)$.

Let H be a subgroup of $G(N/K)$. The field N^H of all elements of N which remain fixed under the automorphisms in H is called the *fixed field of* H.

17.3 The fundamental theorem of Galois theory

In the framework of field theory, the fundamental theorem of Galois theory is the following.

Theorem 6. *Let* N/K *be a finite normal extension. The correspondence* $L \mapsto G(N/L)$ *defines a one-to-one mapping* Φ *of the set of intermediate fields* L *of* N/K *onto the set of subgroups of* $G(N/K)$. *The inverse mapping is given by* $H \mapsto N^H$.

The following rules hold:

(i) *If* L_1, L_2 *are intermediate fields of* N/K *then* $L_1 \subset L_2$ *if and only if* $G(N/L_1) \supset G(N/L_2)$.

(ii) *If* $g \in G(N/K)$ *then* $G(N/gL) = gG(N/L)g^{-1}$.

Proof. Rules (i) and (ii) are immediately clear. It therefore suffices to show that $L = N^{G(N/L)}$ and $H = G(N/N^H)$ for all intermediate fields L of N/K and all subgroups H of G(N/K). This follows easily, provided one can show that

$$[N:N^H] = |H|$$

for all subgroups H of G(N/K).

By definition of the Galois group, $G(N/N^H) \supset H$, and hence $[N:N^H] \geq |H|$ (Theorem 3). As well as this, we also consider the polynomial

$$f_H(x) = \prod_{h \in H} (x - h\theta),$$

where θ is a primitive element of N/K. The coefficients of $f_H(x)$ are in N^H, and $N = N^H(\theta)$. Therefore $[N:N^H] \leq |H|$. □

When L/K is a normal extension, $gL = L$ for all $g \in G(N/K)$. By (iii), G(N/L) is a normal subgroup of G(N/K) in this case. Restriction of the automorphisms of G(N/K) to L defines a homomorphism π of G(N/K) into G(L/K). The latter is surjective by Theorem 3 and has kernel G(N/L). Hence π induces an isomorphism of G(N/K)/G(N/L) onto G(L/K). We call π a *projection*.

Conversely, if $H \subset G(N/K)$ is a normal subgroup, then N^H/K is a normal extension.

17.4 The group of an equation

Let f(x) be a polynomial in K[x] which splits into distinct linear factors $x - \alpha_i$, for $i = 1,\dots,n$, in an extension field L. In Chapter 7 we have defined the group G(f) of f. On the other hand, $K(\alpha_1,\dots,\alpha_n)/K$ is a normal extension by Theorem

4. The connection between $G(f)$ and $G := G(K(\alpha_1,...,\alpha_n)/K)$ is given by the following theorem:

Theorem 7. *The mapping* ϕ *which associates each* $g \in G$ *with the permutation of the zeros* $\alpha_1,...,\alpha_n$ *induced by* g *is an isomorphism of* G *onto* $G(f)$.

Proof. Obviously ϕ is a homomorphism of G into $G(f)$. Let $K(\alpha_1,...,\alpha_n) = K(\theta)$, let f_θ be the minimal polynomial of θ over K, let $\psi(x_1,...,x_n)$ be a polynomial from $K[x_1,...,x_n]$ with $\theta = \psi(\alpha_1,...,\alpha_n)$ and let $\pi \in G(f)$. Then $f_\theta(\psi(\pi\alpha_1,...,\pi\alpha_n)) = 0$ by definition of π, i.e. $\theta_\pi = \psi(\pi\alpha_1,...,\pi\alpha_n)$ is a zero of f_θ. The correspondence $\theta \mapsto \theta_\pi$ defines an automorphism $\phi'\pi$ of $K(\theta)/K$. One easily sees that the mapping ϕ' of $G(f)$ into G defined in this way is inverse to ϕ. □

One now proves the theorems in Section 7.6 as immediate consequences of the preceding results of this chapter. We only remark that to prove Theorem 9, which was incompletely proved by Galois, one has to call on the Galois group of the splitting field of ef.

17.5 The composite of two fields

In this section we prove a theorem which will be needed in a later chapter.

Let M and N be finite extensions of a field K which are contained in a field L. The smallest subfield of L which contains M and N is called the *composite* of M and N and is denoted by MN.

Theorem 8. *Let* N/K *normal and let* $N \cap M = K$. *Then*

$$[MN:M] = [N:K]. \tag{3}$$

If M/K *is also normal, then the projection*

$$G(MN/K) \to G(M/K) \times G(N/K) \tag{4}$$

defines an isomorphism.

Proof. Let $N = K(\theta)$. When the minimal polynomial f_θ of θ over M, with coefficients in K, splits into the factors f_1 and f_2, then the coefficients of these polynomials lie in N, and hence in K because $N \cap M = K$. It follows that f_θ is irreducible over M, i.e. $[NM:M] = [N:K]$.

The mapping (4) is obviously injective. By (3), G(NM/K) and G(N/K) × G(M/K) have equally many elements. Hence (4) is also surjective. □

17.6 Trace, norm, different and discriminant

Let L/K be a finite extension of degree n. Important tools in algebraic number theory are the *trace*, the *norm* and the *different* of an element α of L with respect to L/K, as well as the *discriminant*, with respect to L/K, of a sequence of elements from L.

Let Z be a splitting field of L/K. For $\alpha \in L$ the images $\alpha_1 = \alpha, \dots, \alpha_n$ of α under the n isomorphisms of L into Z which fix K are called the *conjugates* of α. We define the trace, norm and different of α with respect to L/K by

$$Tr_{L/K}(\alpha) = \sum_{i=1}^{n} \alpha_i, \quad N_{L/K}(\alpha) = \prod_{i=1}^{n} \alpha_i, \quad D_{L/K}(\alpha) = \prod_{i=2}^{n} (\alpha - \alpha_i). \tag{5}$$

Since $f(x) = \prod_{i=1}^{n} (x - \alpha_i)$ is a power of the minimal polynomial of α over K, $S_{L/K}(\alpha)$ and $N_{L/K}(\alpha)$ are in K and $D_{L/K}(\alpha) = f'(\alpha)$ is in L. By Theorem 5 the trace, norm and different are independent of the choice of splitting field Z. When no confusion is likely to result, we omit the index L/K in what follows.

The following basic properties of trace, norm and different are immediate from the definitions

(i) $N(\alpha) = 0$ *if and only if* $\alpha = 0$.

(ii) $Tr(\alpha+\beta) = Tr(\alpha) + Tr(\beta)$, $N(\alpha\beta) = N(\alpha)N(\beta)$ *for* $\alpha, \beta \in L$.

(iii) $Tr(a) = na$, $N(a) = a^n$, $Tr(a\alpha) = aTr(\alpha)$ *for* $a \in K$, $\alpha \in L$.

(iv) $D(\alpha) \neq 0$ *if and only if* $L = K(\alpha)$.

Let $\omega_1, \dots, \omega_n$ be arbitrary elements from L. We define the discriminant $\Delta_{L/K}(\omega_1, \dots, \omega_n) = \Delta(\omega_1, \dots, \omega_n)$ by

$$\Delta(\omega_1, \dots, \omega_n) := \det(Tr(\omega_i \omega_j))_{ij}.$$

Let $g_1,...,g_n$ be the n isomorphisms of L/K into Z. Then

$$(Tr(\omega_i\omega_j))_{ij} = WW^T$$

where $W = (g_j\omega_i)_{ij}$, and hence

$$\Delta(\omega_1,...,\omega_n) = (\det W)^2. \tag{6}$$

Let $\omega_i' = \sum_{j=1}^{n} a_{ij}\omega_j$ for $i = 1,...,n$, where $a_{ij} \in K$ and $A = (a_{ij})_{ij}$. Then $(g_j\omega_i') = AW$ and therefore

$$\Delta(\omega_1',...,\omega_n') = (\det A)^2\Delta(\omega_1,...,\omega_n). \tag{7}$$

Theorem 9. $\omega_1,...,\omega_n$ *is a basis for the vector space* L *over* K *if and only if* $\Delta(\omega_1,...,\omega_n) \neq 0$.

Proof. Let $L = K(\theta)$. Then $1,\theta,...,\theta^{n-1}$ is a basis of L/K, and $\det(g_j\theta^{i-1})_{ij}$ is a *Vandermonde determinant*, i.e., as one easily sees,

$$\det(g_j\theta^{i-1})_{ij} = \prod_{i<j}(g_i\theta - g_j\theta) \neq 0 \text{ (Theorem 3)}$$

and hence also

$$\Delta(\theta) := \Delta(1,\theta,...,\theta^{n-1}) = (\det(g_j\theta^{i-1})_{ij})^2 \neq 0. \tag{8}$$

Now let $\omega_1,...,\omega_n$ be arbitrary elements from L and let $\omega_i = \sum_{j=1}^{n} a_{ij}\theta^{j-1}$ for $i = 1,...,n$. Then $\omega_1,...,\omega_n$ is a basis of L/K if and only if $\det(a_{ij})_{ij} \neq 0$. The assertion therefore follows from (7) and (8). □

Now let $\omega_1,...,\omega_n$ be a basis of L/K and let $\omega_i' = \alpha\omega_i = \sum_{j=1}^{n} a_{ij}\omega_j$ with $\alpha \in L$ and $a_{ij} \in K$. Then

$$\det(g_j\omega_i') = \det(g_j\alpha g_j\omega_i) = N(\alpha)\det(g_j\omega_i) = \det(a_{ij})\det(g_j\omega_i)$$

and therefore

$$N(\alpha) = \det(a_{ij}). \tag{9}$$

For $t \in K$ we get

$$N(t-\alpha) = \prod_{i=1}^{n} (t-g_i\alpha) = \det(tE-(a_{ij})),$$

where E denotes the n-rowed identity matrix. We obtain a polynomial in t of degree n. If one compares the coefficients of t, then one finds

$$Tr(\alpha) = Tr(a_{ij}) = \sum_{i=1}^{n} a_{ii}. \tag{10}$$

The polynomial

$$\chi_\alpha(x) = \det(xE-(a_{ij})) = (\prod_{i=1}^{n} -xg_i\alpha)$$

is called the *characteristic polynomial* of α. As one easily sees, $D(\alpha) = \chi'_\alpha(\alpha)$ and $\Delta(\alpha) = (-1)^{n(n-1)/2}N(D(\alpha))$.

$Tr(\alpha\beta)$, for $\alpha,\beta \in L$, is a non-degenerate bilinear form on the vector space L over K. Let $\omega_1,\ldots,\omega_n$ be a basis of L/K. Then by Theorem 9 there are unique elements $\kappa_1,\ldots,\kappa_n$ with

$$Tr(\omega_i\kappa_j) = \delta_{ij} \quad \text{for } i,j = 1,\ldots,n. \tag{11}$$

The elements $\kappa_1,\ldots,\kappa_n$ form a basis of L/K called the *complementary basis* of $\omega_1,\ldots,\omega_n$. For $\alpha = a_1\omega_1 + \ldots + a_n\omega_n$ with $a_1,\ldots,a_n \in K$ we have $a_i = Tr(\alpha\kappa_i)$, $i = 1,\ldots,n$. For a generating element θ of L/K one can give the complementary basis of $1,\theta,\ldots,\theta^{n-1}$ explicitly:

Theorem 10. *Let* $L = K(\theta)$. *Then the complementary basis of* $1,\theta,\ldots,\theta^{n-1}$ *equals*

$$\beta_0/D(\theta),\ \beta_1/D(\theta),\ldots,\beta_{n-1}/D(\theta),$$

where the elements $\beta_0,\ldots,\beta_{n-1}$ *are given by the equation*

$$\chi_\theta(x)/(x-\theta) = \sum_{i=1}^{n-1} \beta_i x^i.$$

Proof. In our case, condition (11) may be written in the form

$$\mathrm{Tr}(\chi_\theta(x)\theta^m/((x-\theta)D(\theta))) = x^m \quad \text{for} \quad m = 0,\dots,n-1. \tag{12}$$

Since

$$\mathrm{Tr}(\chi_\theta(x)\theta^m/((x-\theta)D(\theta))) = \sum_{i=1}^{n-1} \chi_\theta(x)g_i\theta^m/((x-g_i\theta)D(g_i\theta)),$$

(12) is satisfied the the n distinct values $g_j\theta$, $j = 1,\dots,n$. Since both sides of (12) are polynomials of degree $\leq n-1$, the assertion follows. □

Exercises

17.1 Show that a finite extension of a field of characteristic 0 has only finitely many intermediate fields.

17.2 Let N/K be a normal extension with Galois group A_4 (Section 7.3). Give a survey of all the intermediate fields of N/K.

17.3 Determine the Galois group of the normal extension belonging to

$$\mathbb{Q}(\sqrt{2+\sqrt{2}})/\mathbb{Q} \quad \text{resp.} \quad \mathbb{Q}(\sqrt{3+\sqrt{2}})/\mathbb{Q}.$$

17.4 Prove Theorem 8 with the help of Galois theory (one can assume, without loss of generality, that M and N are contained in a finite normal extension of K).

17.5 Show that each finite group is isomorphic to the Galois group of a certain normal extension.

17.6 Let L/K be a finite extension of a field of characteristic 0. Show that $\Delta_{L/K}(\alpha+a) = \Delta_{L/K}(\alpha)$ holds for all $\alpha \in L$ and $a \in K$.

17.7 (Brill) Let L be a finite extension of $\mathbb{Q}$ of degree n with $2r_2$ complex conjugates ($2r_2$ isomorphisms into $\mathbb{C}$, which do not map into $\mathbb{R}$). Also let $\alpha_1,\dots,\alpha_n$ be linearly independent elements of L. Show that the sign of $\Delta_{L/\mathbb{Q}}(\alpha_1,\dots,\alpha_n)$ is $(-1)^{r_n}$.

18. Dedekind's theory of ideals

18.1 Integral elements

A complex number α is called an *algebraic number* when it satisfies an equation

$$\alpha^m + a_1\alpha^{m-1} + \ldots + a_m = 0 \tag{1}$$

with rational coefficients $a_1,\ldots,a_m$. We call α an *algebraic integer* when there is such an equation whose coefficients are integers.

Likewise, α is called an *algebraic function (integral algebraic function) of the indeterminate* x when the coefficients of (1) are rational functions (polynomials) in x over a field F_0.

In order to be able to handle both cases together, we proceed from a euclidean ring Γ (Section A1.2) whose field of fractions P has characteristic 0. Let K be an extension field of P. The cases of interest to us are $\Gamma = \mathbb{Z}$ and $\Gamma = F_0[x]$.

An element α of K is called *integral* (over Γ) when α satisfies an equation (1) with coefficients $a_1,\ldots,a_m$ from Γ. We denote the collection of integral elements of K by O_K.

We first remark that the above definition of algebraic integer is compatible with the ordinary concept of integer, i.e. $O_{\mathbb{C}} \cap \mathbb{Q} = \mathbb{Z}$. More generally, we have

Theorem 1. $O_K \cap P = \Gamma$.

Proof. It is clear that each element a of Γ lies in O_K. Conversely, suppose that α is in $O_K \cap P$. Then α may be represented in the form $\alpha = b/c$, where $b,c \in \Gamma$ are relatively prime, and there is an equation $\alpha^m + a_1\alpha^{m-1} + \ldots + a_m = 0$ with $a_1,\ldots,a_m \in \Gamma$. Therefore $b^m + a_1b^{m-1}c + \ldots + a_mc^m = 0$, whence it follows that $c\,|\,b^m$. It then follows, by A1, Theorem 3, that c is a unit of Γ. □

An important rôle in Dedekind's ideal theory is played by the finitely generated Γ-modules contained in K. A basic idea of Dedekind is to operate with such modules as with algebraic quantities.

Let $\mathfrak{M}_K$ be the collection of finitely generated Γ-modules in K. If such a module $\mathfrak{a}$ is generated by the elements $\alpha_1,\dots,\alpha_s$, then we write $\mathfrak{a} = (\alpha_1,\dots,\alpha_s)$.

Two modules $\mathfrak{a}$ and $\mathfrak{b}$ from $\mathfrak{M}_K$ have a product $\mathfrak{ab}$, defined as the module generated by all products $\alpha\beta$ where $\alpha \in \mathfrak{a}$ and $\beta \in \mathfrak{b}$. When $\mathfrak{a} = (\alpha_1,\dots,\alpha_s)$ and $\mathfrak{b} = (\beta_1,\dots,\beta_t)$ then obviously

$$\mathfrak{ab} = (\alpha_i\beta_j \mid i = 1,\dots,s\ ;\ j = 1,\dots,t).$$

Also, $\mathfrak{a} + \mathfrak{b}$ denotes, as usual, the Γ-module generated by $\mathfrak{a}$ and $\mathfrak{b}$. The multiplication just defined is associative and distributive. We set $\alpha\mathfrak{a} := (\alpha)\mathfrak{a}$ for $\alpha \in O_K$.

We now give a useful characterisation of integral elements in K with the help of modules in K.

Theorem 2. *An element* α *of* K *belongs to* O_K *if and only if there is a module* $\mathfrak{m} \neq \{0\}$ *in* $\mathfrak{M}_K$ *with* $\alpha\mathfrak{m} \subset \mathfrak{m}$.

Proof. For $\alpha \in O_K$ satisfying (1), one such module is $(1,\alpha,\dots,\alpha^{m-1})$. Conversely, if $\alpha \in K$ and $\alpha\mathfrak{m} \subset \mathfrak{m}$ for a module $\mathfrak{m} \neq \{0\}$ in $\mathfrak{M}_K$, then $\alpha \in O_K$. Because if $\mathfrak{m} = (\alpha_1,\dots,\alpha_m)$ and

$$\alpha\alpha_i = \sum_{j=1}^{m} a_{ij}\alpha_j \quad \text{with} \quad a_{ij} \in \Gamma$$

then

$$\det(\alpha E - (a_{ij})_{ij}) = 0.$$

This is an equation for α with coefficients from Γ and highest coefficient 1. □

The collection of elements α from K with $\alpha\mathfrak{m} \subset \mathfrak{m}$ for a fixed module $\mathfrak{m} \neq \{0\}$ is called the *order* $O(\mathfrak{m})$ of $\mathfrak{m}$. $O(\mathfrak{m})$ is obviously closed under addition, subtraction and multiplication, and it contains the unit element of K, i.e. $O(\mathfrak{m})$ is a ring with unit. By Theorem 2, $O(\mathfrak{m})$ is contained in O_K.

Let $\mu \neq 0$ be an element of $\mathfrak{m}$. Then $\mu(O(\mathfrak{m})) \subset \mathfrak{m}$, hence $O(\mathfrak{m}) \subset \mu^{-1}\mathfrak{m}$. Thus $O(\mathfrak{m})$ is a finitely generated module along with $\mu^{-1}\mathfrak{m}$ (A1, Theorem 4). We bring these results together as follows.

Theorem 3. *The order of a module from* $\mathfrak{M}_K$ *different from* $\{0\}$ *is a ring* R *in* O_K *with the following properties:*

(i) $\Gamma \subset R$
(ii) $R \in \mathfrak{M}_K$.

Conversely, each ring R *in* O_K *with properties* (i), (ii) *is its own order, when regarded as a* Γ*-module.* □

Another consequence of Theorem 2 is

Theorem 4. O_K *is a ring.*

Proof. Let α_1 and α_2 be elements of O_K, and let $\mathfrak{a}_1$ and $\mathfrak{a}_2$ be modules $\neq \{0\}$ from $\mathfrak{M}_K$ with $\alpha_1\mathfrak{a}_1 \subset \mathfrak{a}_1$ and $\alpha_2\mathfrak{a}_2 \subset \mathfrak{a}_2$. Then $\alpha_1, \alpha_2 \in O(\mathfrak{a}_1\mathfrak{a}_2)$. □

Theorem 5. *If an element* α *of* K *satisfies an equation*

$$\alpha^m + \alpha_1\alpha^{m-1} + \dots + \alpha_n = 0$$

with coefficients $\alpha_1,\dots,\alpha_m$ *in* O_K, *then* α *is in* O_K.

Proof. Let $\mathfrak{a}_i \neq \{0\}$ be modules from $\mathfrak{M}_K$ with $\alpha_i\mathfrak{a}_i \subset \mathfrak{a}_i$ for $i = 1,\dots,m$. Then $\alpha\mathfrak{m} \subset \mathfrak{m}$ for $\mathfrak{m} = (1,\alpha,\dots,\alpha^{m-1})\mathfrak{a}_1\dots\mathfrak{a}_m$. □

Theorem 6. *Let* $\alpha \in O_K$. *Then the minimal polynomial* f_α *of* α *over* P *has coefficients in* Γ.

Proof. Let N be a splitting field of f_α. Along with α, the conjugates of α in N, i.e. the zeros of f_α in N, are also integral. By Viète's root theorem the coefficients of f_α lie in the ring generated by the zeros of f_α, hence by Theorem 4 they are integral in N, and hence also integral in P, by Theorem 1. □

From Theorem 6 one easily obtains a new proof of Theorem 2 in Chapter 3.

18.2 Lattices in finite extensions of P

We now confine ourselves to the case where K is a finite extension of P. Let n be the degree of K/P. A finitely generated Γ-module $\mathfrak{a}$ in K has a rank which is less than or equal to n (A1, Theorem 5). When $\mathfrak{a}$ has rank n, $\mathfrak{a}$ is called a *lattice* in K.

Let $\mathfrak{a} = (\alpha_1,...,\alpha_n)$ and $\mathfrak{b} = (\beta_1,...,\beta_n)$ be lattices in K with $\mathfrak{a} \subset \mathfrak{b}$. Also, let A be the matrix which transforms the basis $\alpha_1,...,\alpha_n$ to the basis $\beta_1,...,\beta_n$. The elements of A lie in Γ. When one goes to other bases of $\mathfrak{a}$ and $\mathfrak{b}$, det A is multiplied by a unit in Γ. Thus the class of associates of det A in Γ is independent of the choice of bases and we denote it by $[\mathfrak{b}:\mathfrak{a}]$. By A1.4, $[\mathfrak{b}:\mathfrak{a}] = [\mathfrak{b}/\mathfrak{a}]$. In the case $\Gamma = \mathbb{Z}$, $\mathfrak{a}$ is a subgroup of $\mathfrak{b}$ of index $|\det A|$.

By Chapter 17, (7), we have

$$\Delta(\alpha_1,...,\alpha_n) = (\det A)^2\Delta(\beta_1,...,\beta_n). \tag{2}$$

From this it is clear that the class of associates of $\Delta(\alpha_1,...,\alpha_n)$ in Γ is independent of the choice of basis $\alpha_1,...,\alpha_n$ of $\mathfrak{a}$. We denote this class by $\Delta(\mathfrak{a})$. Then it follows from (2) that

$$\Delta(\mathfrak{a}) = [\mathfrak{b}:\mathfrak{a}]^2\Delta(\mathfrak{b}). \tag{3}$$

In the case $\Gamma = \mathbb{Z}$ it follows from (2) that the number $\Delta(\alpha_1,...,\alpha_n)$ is already independent of the choice of basis. In this case we set $\Delta(\mathfrak{a}) = \Delta(\alpha_1,...,\alpha_n)$ and interpret (3) as an equation between numbers. $\Delta(\mathfrak{a})$ is called the *discriminant* of $\mathfrak{a}$.

The most important result of this section is the following.

Theorem 7. O_K *is a lattice in* K.

Proof. Let $\alpha_1,...,\alpha_n$ be a basis of K over P which consists of elements of O_K, and let $\mathfrak{a} := (\alpha_1,...,\alpha_n)$. By definition of the discriminant and Theorem 6 we then have $\Delta(\mathfrak{a}) \subset \Gamma$. If $\mathfrak{a}$ is different from O_K we choose a $\beta \in O_K - \mathfrak{a}$ and set $\mathfrak{b} := (\beta,\alpha_1,...,\alpha_n)$. Then $\mathfrak{b}$ is also a lattice in K (A1, Theorem 4) and by (3), $\Delta(\mathfrak{b})$ is a proper divisor of $\Delta(\mathfrak{a})$. If $\mathfrak{b}$ is different from O_K we continue the process. After finitely many steps we arrive at O_K, since $\Delta(\mathfrak{a})$ has only finitely many divisors (up to associates, A1, Theorem 3). □

18.3 The integers of quadratic fields

The determination of a basis of O_K is generally a difficult problem. In this section we consider the simplest case, where K is a quadratic field, i.e. an extension of $\mathbb{Q}$ of degree 2.

A quadratic field results from $\mathbb{Q}$ by adjunction of a square root $\sqrt{d}$, where one can assume that d is a square-free integer. These conditions uniquely determine d for a given K.

Each number α in $K = \mathbb{Q}(\sqrt{d})$ has the form $(a_1+a_2\sqrt{d})/a$, where a_1,a_2,a are relatively prime numbers in $\mathbb{Z}$. Then α lies in O_K if and only if $2a_1/a$ and $(a_1^2-da_2^2)/a^2$ lie in $\mathbb{Z}$. Suppose this is the case. Then it follows that $4da_2^2/a^2$ also lies in $\mathbb{Z}$. Since a_1,a_2,a are relatively prime, a is a divisor of 2. Suppose, without loss of generality, that $a > 0$.

We now distinguish two cases:

1. $d \equiv 1 \pmod 4$,
2. $d \equiv 2,3 \pmod 4$.

In case 1, $\alpha = a_1 + a_2\sqrt{d}$ or $a_1^2 \equiv a_2^2 \pmod 4$, $a = 2$ and hence $a_1 \equiv a_2 \equiv 1 \pmod 2$. These conditions are obviously also sufficient for α to be integral. It follows that 1 and $\omega := (1+\sqrt{d})/2$ constitute a basis of O_K.

In case 2 it follows from $(a_1^2-da_2^2)/a^2 \in \mathbb{Z}$ that $a = 1$. The numbers 1 and $\omega := \sqrt{d}$ constitute a basis of O_K.

In Section 18.1 we have considered the orders $O(\mathfrak{m})$ of modules $\mathfrak{m}$, and have seen that in the case $\Gamma = \mathbb{Z}$ these may be characterised as rings of algebraic integers which contain $\mathbb{Z}$ and are finitely generated as $\mathbb{Z}$-modules. When such a ring generates the field K we call it an *order* of K. We now want to gain an overview of the orders of quadratic fields.

Let R be an order of $\mathbb{Q}(\sqrt{d})$. There is a natural number b with $b\omega \in R$. The collection of all such numbers b is the set of multiples of a number f determined by R. This set is called the *conductor* of R. We have $R = (1,f\omega)$.

The ring of Gaussian integers, which we have considered in 16.1, is the ring of integers of $\mathbb{Q}(\sqrt{-1})$. As we have seen, unique prime decomposition holds in this ring. In the other orders of $\mathbb{Q}(\sqrt{-1})$, $f\sqrt{-1}$ is a prime element and f^2 has at least two essentially different prime decompositions. The following example shows that in general unique prime decomposition also fails in the maximal order O_K of a quadratic field.

We consider the ring $R = (1,\sqrt{-5})$ of integers of $\mathbb{Q}(\sqrt{-5})$. The number $a+b\sqrt{-5}$ from R has the norm

$$N(a+b\sqrt{-5}) = a^2 + 5b^2 \tag{4}$$

and hence it is a unit in R if and only if $a^2 + 5b^2 = 1$. The only units in R are therefore the numbers ± 1.

The four numbers $3, 7, 1+2\sqrt{-5}, 1-2\sqrt{-5}$ are indecomposable. E.g., if we had $3 = \alpha_1\alpha_2$ where $\alpha_1,\alpha_2 \in R$ were not units, then we should have $N(3) = 9 = N(\alpha_1)N(\alpha_2)$, and hence $N(\alpha_1) = 3$, which is impossible by (4). On the other hand,

$$21 = 3.7 = (1+2\sqrt{-5})(1-2\sqrt{-5}).$$

Thus there are two essentially different decompositions of 21 into products of primes.

In order to arrive at unique prime decomposition, despite this, Kummer introduced "*ideal numbers*", though admittedly only in the case of cyclotomic fields. In his *Supplement* XI, Dedekind pursued this line of thinking for the example $\mathbb{Q}(\sqrt{-5})$, in order to show that difficulties were encountered with arbitrary finite number fields, and hence that another approach to the theory of divisibility should be adopted.

18.4 The ideals of O_K

We now return to the general situation. Let Γ be a euclidean ring, whose field of fractions P has characteristic 0, let K be an extension field of P of degree n and let O_K be the ring of integers of K over Γ.

In questions of divisibility, elements are determined only up to associates. Instead of the class of associates of an $\alpha \in O_K$ we may just as well take the collection αO_K of all multiples of α. For $\alpha_1,\alpha_2 \in O_K$, α_1 is a divisor of α_2 if and only if $\alpha_1 O_K \supset \alpha_2 O_K$.

The collection αO_K is an O_K-module contained in O_K. Dedekind's basic idea is that it is possible to arrive at a satisfactory divisibility theory in O_K by passing from the modules αO_K to *arbitrary* O_K-modules in O_K, which he called *ideals*, following Kummer. The ideals of the form αO_K are called *principal ideals.*

In a euclidean ring all ideals are principal ideals. In general, a ring in which each ideal is principal is called a *principal ideal ring*. One can show that the theorem of unique prime decomposition holds in each principal ideal ring which is also an integral domain.

By an *ideal* of K (relative to O_K) we mean a finitely generated O_K-module in K distinct from $\{0\}$. Each ideal $\neq \{0\}$ of O_K is an ideal of K, i.e. it has a finite set of generators, because this is the case for O_K. Over and above that, we have

Theorem 8. *An ideal* $\mathfrak{A} \neq \{0\}$ *of* K *is a lattice.*

Proof. It suffices to show that there are n elements in $\mathfrak{A}$ which are linearly independent over P. Let $\alpha \neq 0$ be an element of $\mathfrak{A}$, and let $\alpha_1,\dots,\alpha_n$ be linearly independent elements of O_K. Then $\alpha\alpha_1,\dots,\alpha\alpha_n$ are n linearly independent elements of $\mathfrak{A}$. $\square$

We now consider the decomposition of 21 in the field $K = \mathbb{Q}(\sqrt{-5})$ from the standpoint of ideals in O_K. We set

$$\mathfrak{A}_1 = (3,1+2\sqrt{-5})O_K,\ \mathfrak{A}_2 = (3,1-2\sqrt{-5})O_K,$$

$$\mathfrak{B}_1 = (7,1+2\sqrt{-5})O_K,\ \mathfrak{B}_2 = (7,1-2\sqrt{-5})O_K.$$

Then

$$\mathfrak{A}_1\mathfrak{A}_2 = (9,3-6\sqrt{-5},\ 3+6\sqrt{-5},\ 21)O_K = 3O_K.$$

Similarly, one finds

$$\mathfrak{A}_1\mathfrak{B}_1 = (1+2\sqrt{-5})O_K,\ \mathfrak{A}_1\mathfrak{B}_2 = (-4+\sqrt{-5})O_K,$$

$$\mathfrak{A}_2\mathfrak{B}_1 = (-4-\sqrt{-5})O_K,$$

$$\mathfrak{A}_2\mathfrak{B}_2 = (1-2\sqrt{-5})O_K,\ \mathfrak{B}_1\mathfrak{B}_2 = 7O_K$$

and

$$21O_K = \mathfrak{A}_1\mathfrak{A}_2\mathfrak{B}_1\mathfrak{B}_2. \tag{5}$$

The ideals $\mathfrak{A}_1, \mathfrak{A}_2, \mathfrak{B}_1, \mathfrak{B}_2$ are prime ideals (A1.1), and cannot be split into products of ideals in O_K which are different from O_K. Likewise, one can show that (5) is the unique decomposition of $21O_K$ into the product of prime ideals.

We end this section with a few remarks on prime ideals in O_K, where K is an arbitrary finite extension of P.

Theorem 9. *An ideal* $\mathfrak{P} \neq \{0\}$ *of* O_K *is prime if and only if* $\mathfrak{P}$ *is a maximal ideal.*

Proof. An ideal $\mathfrak{P}$ is maximal in O_K if and only if $O_K/\mathfrak{P}$ is a field (A1, Theorem 1). It therefore suffices to show that the integral domain $O_K/\mathfrak{P}$ is a field for a prime ideal $\mathfrak{P} \neq \{0\}$ of O_K.

Along with $\mathfrak{P}$, $\mathfrak{p} := \mathfrak{P} \cap \Gamma$ is also a prime ideal, and $\mathfrak{p}$ is different from $\{0\}$. Because if $\alpha \neq 0$ lies in $\mathfrak{P}$ then $N_{K/P}\alpha \neq 0$ lies in $\mathfrak{p}$, since $\alpha \mid N_{K/P}\alpha$. The inclusion $\Gamma \subset O_K$ induces a monomorphism

$$\phi : \Gamma/\mathfrak{p} \to O_K/\mathfrak{P}. \tag{6}$$

Here, $\Gamma/\mathfrak{p}$ is a field, because Γ is a principal ideal ring. Each $\overline{\alpha} \in O_K/\mathfrak{P}$ satisfies an algebraic equation with coefficients in $\phi(\Gamma/\mathfrak{p})$. When $\overline{\alpha} \neq 0$ we can assume that the absolute coefficient of this equation is non-zero. It follows that $\overline{\alpha}$ has an inverse in $O_K/\mathfrak{P}$. □

It is usual to identify $\Gamma/\mathfrak{p}$ with $\phi(\Gamma/\mathfrak{p})$ and to regard (6) as a field extension. By Theorem 7 this field extension has a finite degree, which is called the *inertia degree of* $\mathfrak{P}$ *(relative to* $\mathfrak{p}$*)*.

An immediate consequence of Theorem 9 is

Theorem 10. *Let* $\mathfrak{P} \neq \{0\}$ *be a prime ideal and let* $\mathfrak{A}$ *be an arbitrary ideal of* O_K. *Then* $\mathfrak{A} \subset \mathfrak{P}$ *or* $\mathfrak{A} + \mathfrak{P} = O_K$. □

We also have

Theorem 11. *When a prime ideal* $\mathfrak{P}$ *contains the product of ideals* $\mathfrak{A}$ *and* $\mathfrak{B}$, *then* $\mathfrak{A}$ *is contained in* $\mathfrak{B}$ *or* $\mathfrak{P}$.

Proof. If neither $\mathfrak{A}$ nor $\mathfrak{B}$ is contained in $\mathfrak{P}$, then by Theorem 10 there are elements π_1, π_2 of $\mathfrak{P}$ and $\alpha \in \mathfrak{A}, \beta \in \mathfrak{B}$ with $1 = \alpha+\pi_1$, $1 = \beta+\pi_2$. By multiplication of these equations one obtains $1 = \alpha\beta + (\alpha\pi_2+\beta\pi_1+\pi_1\pi_2)$. This means $O_K = \mathfrak{A}\mathfrak{B} + \mathfrak{P}$, contrary to the hypothesis $\mathfrak{P} \subset \mathfrak{A}\mathfrak{B}$. □

18.5 The fundamental theorem of ideal theory

The fundamental theorem of ideal theory is the generalisation of the theorem of unique prime decomposition. It reads as follows:

Theorem 12. *Each ideal* $\neq \{0\}$ *of* O_K *is representable as a product of prime ideals. This representation is unique up to the order of factors.*

The proof of this theorem was heavy going for Dedekind. Hilbert felt that Dedekind's proof was inelegant, and gave several new proofs himself. Here we follow an arrangement of the proof, also used by van der Waerden in his textbook of algebra, which can be regarded as a simplification of Dedekind's proof in *Supplement* XI.

We begin with a few lemmas.

Lemma 1. *Each strictly increasing sequence of ideals* $\mathfrak{A}_1 \subsetneq \mathfrak{A}_2 \subsetneq \ldots$ *terminates in finitely many steps.*

Proof. By Theorem 8 we can construct the sequence $[O_K:\mathfrak{A}_1], [O_K:\mathfrak{A}_2], \ldots$ in which each term is a proper divisor of the one before. □

Lemma 2. *For each ideal* $\mathfrak{A}$ *there are prime ideals;* $\mathfrak{P}_1, \ldots, \mathfrak{P}_r$ *with* $\mathfrak{A} \supset \mathfrak{P}_1 \ldots \mathfrak{P}_r$.

Proof. In view of Lemma 1, let

$$\mathfrak{A} = \mathfrak{A}_1 \subsetneq \mathfrak{A}_2 \subsetneq \dots \subsetneq \mathfrak{A}_s \subsetneq O_K$$

be an ascending chain of maximal length for $\mathfrak{A}$. Then we call s the *length of* $\mathfrak{A}$. The ideals of length 1 are precisely the prime ideals. Thus Lemma 2 is true for them. Suppose the lemma is already proved for ideals of length $\leq s$, and let $\mathfrak{A}$ be an ideal of length $s + 1$. Since $\mathfrak{A}$ is not a prime ideal, there are elements β, γ from O_K with $\beta\gamma \in \mathfrak{A}$ and $\beta \notin \mathfrak{A}, \gamma \in \mathfrak{A}$. The ideals $\mathfrak{B} = (\beta,\mathfrak{A})$ and $\mathfrak{C} = (\gamma,\mathfrak{A})$ properly contain $\mathfrak{A}$. Their length is therefore smaller than $s + 1$, hence Lemma 2 is true for $\mathfrak{B}$ and $\mathfrak{C}$ by induction hypothesis. Since $\mathfrak{B}\mathfrak{C} \subset \mathfrak{A}$ also, the assertion follows. □

For an ideal $\mathfrak{A} \neq \{0\}$ in O_K we let $\mathfrak{A}^{-1}$ denote the collection of elements $\beta \in K$ with $\beta\mathfrak{A} \subset O_K$. If $\alpha \in \mathfrak{A}$ then $\alpha\mathfrak{A}^{-1}$ is an ideal of O_K. Therefore $\mathfrak{A}^{-1}$ is a finitely generated O_K-module, i.e. an ideal of K (cf. Section 18.4). One also speaks of *fractional ideals* and calls the ideals of O_K *integral ideals*.

Let $\mathcal{I}_K$ be the set of fractional ideals of K. The following considerations lead to the conclusion that $\mathcal{I}_K$ is a group with respect to multiplication of modules. Obviously the ideal O_K is the identity element of $\mathcal{I}_K$.

Lemma 3. *Let* $\mathfrak{P} \neq \{0\}$ *be a prime ideal. Then there is an element in* $\mathfrak{P}^{-1}$ *which does not lie in* O_K.

Proof. Let γ be a non-zero element of $\mathfrak{P}$. By Lemma 2 there are prime ideals $\mathfrak{P}_1,\dots,\mathfrak{P}_r$ with $\gamma O_K \supset \mathfrak{P}_1 \dots \mathfrak{P}_r$. We can assume that γO_K contains no subproduct of $\mathfrak{P}_1 \dots \mathfrak{P}_r$. Since $\mathfrak{P} \supset \mathfrak{P}_1 \dots \mathfrak{P}_r$, it follows from Theorem 11 that $\mathfrak{P}$ contains one of the prime ideals $\mathfrak{P}_1,\dots,\mathfrak{P}_r$. Suppose $\mathfrak{P} \supset \mathfrak{P}_1$, i.e. $\mathfrak{P} = \mathfrak{P}_1$. Then

$$\gamma O_K \supset \mathfrak{P}\mathfrak{P}_2 \dots \mathfrak{P}_r, \quad \gamma O_K \not\supset \mathfrak{P}_2 \dots \mathfrak{P}_r.$$

Hence there is a $\beta \in \mathfrak{P}_2 \dots \mathfrak{P}_r$ which is not in γO_K, i.e. $\beta/\gamma \notin O_K$. Also $\gamma O_K \supset \beta\mathfrak{P}$, hence $\beta/\gamma \in \mathfrak{P}^{-1}$. □

The following lemma justifies the notation $\mathfrak{P}^{-1}$.

Lemma 4. *Let* $\mathfrak{P} \neq \{0\}$ *be a prime ideal. Then* $\mathfrak{P}\cdot\mathfrak{P}^{-1} = O_K$.

Proof. By definition of $\mathfrak{P}^{-1}$, $O_K \subset \mathfrak{P}^{-1}$. Therefore $\mathfrak{P} \subset \mathfrak{P}\mathfrak{P}^{-1} \subset O_K$. Hence by Theorem 9, $\mathfrak{P}\mathfrak{P}^{-1} = \mathfrak{P}$ or $\mathfrak{P}\mathfrak{P}^{-1} = O_K$. Suppose that $\mathfrak{P}\cdot\mathfrak{P}^{-1} = \mathfrak{P}$. Then $\beta\mathfrak{P} \subset \mathfrak{P}$ for all $\beta \in \mathfrak{P}^{-1}$, i.e. β is an integer, by Theorem 2, contrary to Lemma 3. □

Lemma 5. *Suppose the ideal* $\mathfrak{A} \neq \{0\}$ *is contained in the prime ideal* $\mathfrak{P}$. *Then there is an ideal* $\mathfrak{B}$ *with* $\mathfrak{A} = \mathfrak{P}\mathfrak{B}$ *and* $\mathfrak{A} \subsetneqq \mathfrak{B}$.

Proof. $\mathfrak{B} := \mathfrak{P}^{-1}\mathfrak{A}$ does what is wanted. □

Now we can easily show that each ideal $\mathfrak{A} \neq \{0\}$ of O_K is a product of prime ideals. By Lemma 1, $\mathfrak{A}$ is contained in a prime ideal. By Lemma 5, $\mathfrak{A} = \mathfrak{P}_1\mathfrak{A}_1$ with $\mathfrak{A} \subsetneqq \mathfrak{A}_1$. By continuing the process one obtains a representation $\mathfrak{A} = \mathfrak{P}_1 \ldots \mathfrak{P}_s\mathfrak{A}_s$ and a chain $\mathfrak{A} \subsetneqq \mathfrak{A}_1 \subsetneqq \mathfrak{A}_2 \subsetneqq \ldots \subsetneqq \mathfrak{A}_s$. By Lemma 1 this chain terminates, and we obtain the desired representation for $\mathfrak{A}$.

Let

$$\mathfrak{P}_1 \ldots \mathfrak{P}_s = \mathfrak{Q}_1 \ldots \mathfrak{Q}_r \tag{7}$$

be two decompositions of $\mathfrak{A}$ into the product of prime ideals. Then $\mathfrak{P}_1$ contains the product $\mathfrak{Q}_1 \ldots \mathfrak{Q}_r$ and hence by Theorem 11 it is equal to one of the prime ideals $\mathfrak{Q}_i$, i = 1,...,s. Suppose $\mathfrak{P}_1 = \mathfrak{Q}_1$. Multiplying (7) by $\mathfrak{P}_1^{-1}$ one obtains

$$\mathfrak{P}_2 \ldots \mathfrak{P}_s = \mathfrak{Q}_2 \ldots \mathfrak{Q}_r.$$

By continuing the process one finds that the two decompositions (7) of $\mathfrak{A}$ agree. Hence Theorem 12 is proved. □

18.6 Consequences of the fundamental theorem

It follows easily from Lemma 4 and Theorem 12 that $\mathcal{I}_K$ is a group and that each $\mathfrak{B} \in \mathcal{I}_K$ is uniquely expressible in the form

$$\mathfrak{B} = \prod_{\mathfrak{P}} \mathfrak{P}^{n(\mathfrak{P})},$$

where $\mathfrak{P}$ runs through the prime ideals of O_K and the exponents $n(\mathfrak{P}) = \nu_{\mathfrak{P}}(\mathfrak{B})$ are integers which are almost all 0. The integer $\nu_{\mathfrak{P}}(\mathfrak{B})$ is called the $\mathfrak{P}$-*exponent of* $\mathfrak{B}$.

We note the following generalisation of Lemma 5.

Theorem 13. *Suppose the ideal* $\mathfrak{A} \neq \{0\}$ *is contained in the ideal* $\mathfrak{C}$. *Then there is exactly one ideal* $\mathfrak{B}$ *with* $\mathfrak{A} = \mathfrak{B}\mathfrak{C}$. □

As a generalisation of divisibility of numbers we define *divisibility of ideals of* O_K. We say that an ideal $\mathfrak{A}$ is divisible by the ideal $\mathfrak{B}$ when there is an ideal $\mathfrak{C}$ with $\mathfrak{A} = \mathfrak{B}\mathfrak{C}$ (cf. Section 18.4). By Theorem 13, $\mathfrak{A}$ is divisible by $\mathfrak{B}$ exactly when $\mathfrak{A} \subset \mathfrak{B}$.

On the basis of this divisibility concept we can define the concepts of *greatest common divisor*, $\gcd(\mathfrak{A},\mathfrak{B})$ and *least common multiple* $\mathrm{lcm}(\mathfrak{A},\mathfrak{B})$ in the usual way. By Theorem 13 we have the formulae

$$\gcd(\mathfrak{A},\mathfrak{B}) = \mathfrak{A} + \mathfrak{B}, \quad \mathrm{lcm}(\mathfrak{A},\mathfrak{B}) = \mathfrak{A} \cap \mathfrak{B}. \tag{8}$$

The following theorems relate to this:

Theorem 14. *Let* $\mathfrak{P}_1,\dots,\mathfrak{P}_s$ *be different prime ideals and let* $\mathfrak{A} \neq \{0\}$ *be an arbitrary ideal. Then there is an element* α *in* $\mathfrak{A}$ *which is divisible by none of the ideals* $\mathfrak{A}\mathfrak{P}_1,\dots,\mathfrak{A}\mathfrak{P}_s$.

Proof. Let $\mathfrak{B} = \mathfrak{P}_1\dots\mathfrak{P}_s$ and let α_i be an element of $\mathfrak{A}\mathfrak{B}\mathfrak{P}_i^{-1}$ which does not lie in $\mathfrak{A}\mathfrak{B}$, for $i = 1,\dots,s$. Then $\alpha := \alpha_1 + \dots + \alpha_s$ belongs to $\mathfrak{A}$, but not to $\mathfrak{A}\mathfrak{P}_i$, since otherwise we should have $\alpha_i \in \mathfrak{A}\mathfrak{P}_i$, contrary to the choice of α_i. □

Theorem 15. *Let* $\mathfrak{A}$ *and* $\mathfrak{B}$ *be arbitrary ideals different from* $\{0\}$. *Then there is an* $\alpha \in \mathfrak{A}$ *with*

$$\mathfrak{A}\mathfrak{B} + \alpha O_K = \mathfrak{A}, \quad \mathfrak{A}\mathfrak{B} \cap \alpha O_K = \alpha\mathfrak{B}. \tag{9}$$

Proof. Suppose, without loss of generality, that $\mathfrak{B} \neq O_K$, and let $\mathfrak{P}_1,\dots,\mathfrak{P}_s$ be the different prime ideals of $\mathfrak{B}$. Let α be an element of $\mathfrak{A}$, given by Theorem 14, which is not divisible by $\mathfrak{A}\mathfrak{P}_1,\dots,\mathfrak{A}\mathfrak{P}_s$. Then

$$\gcd(\mathfrak{A}\mathfrak{B},\alpha O_K) = \mathfrak{A}, \quad \mathrm{lcm}(\mathfrak{A}\mathfrak{B},\alpha O_K) = \alpha\mathfrak{B}. \quad \square$$

18.7 The norm of an ideal

An important rôle is played by the *norm of an ideal* $\mathfrak{A}$ of O_K. Its definition depends on Sections 17.6 and 18.2. From them we have, for $0 \neq \alpha \in O_K$,

$$N(\alpha)\Gamma = [O_K : \alpha O_K]$$

where N denotes the norm of K/P. This suggests the definition

$$N(\mathfrak{A}) := [O_K:\mathfrak{A}] \quad \text{for} \quad \mathfrak{A} \neq \{0\}, \quad N(\{0\}) := \{0\}.$$

We have

Lemma 6. *Let* $\mathfrak{A}$ *and* $\mathfrak{B}$ *be ideals different from* $\{0\}$. *Then*

$$N(\mathfrak{B}) = [\mathfrak{A}:\mathfrak{A}\mathfrak{B}].$$

Proof. Let α be an element of $\mathfrak{A}$ satisfying (9). Then $[\mathfrak{A}:\mathfrak{A}\mathfrak{B}] = [(\mathfrak{A}\mathfrak{B}+\alpha O_K)/\mathfrak{A}\mathfrak{B}] = [\alpha O_K/(\mathfrak{A}\mathfrak{B} \cap \alpha O_K)] = [\alpha O_K/\alpha\mathfrak{B}] = [O_K/\mathfrak{B}]$. □

Remark. The proof technique used here, based on Theorem 15, is called *semi-localisation* in modern language. The essence of it is that when one is only interested in the product of a given finite set of prime ideals one can pass from ideals to elements of O_K.

Theorem 16. *For any ideals* $\mathfrak{A}$ *and* $\mathfrak{B}$

$$N(\mathfrak{A}\mathfrak{B}) = N(\mathfrak{A})N(\mathfrak{B}). \tag{10}$$

Proof. Assume without loss of generality that $\mathfrak{A}$ and $\mathfrak{B}$ are different from $\{0\}$. Then by Lemma 6

$$N(\mathfrak{A}\mathfrak{B}) = [O_K:\mathfrak{A}\mathfrak{B}] = [O_K:\mathfrak{A}][\mathfrak{A}:\mathfrak{A}\mathfrak{B}] = N(\mathfrak{A})N(\mathfrak{B}) \text{ (A1, Theorem 9).} \ \square$$

We now specialise to the norm of a prime ideal $\mathfrak{P}$. Let p be a prime element of Γ with

$$\mathfrak{P} \mid pO_K. \tag{11}$$

The principal ideals pO_K and $p\Gamma$ are uniquely determined by (11). By (10), $N(\mathfrak{P})\,|\,N(p)\Gamma = p^n\Gamma$. Hence there is a natural number $f \le n$ with $N(\mathfrak{P}) = p^f\Gamma$.

Theorem 17. f *equals the inertia degree of* $\mathfrak{P}$ (cf. Section 18.4).

Proof. We choose a basis $\omega_1,\dots,\omega_n$ in O_K, so that $\mathfrak{P}$ has a basis of the form $a_1\omega_1,\dots,a_n\omega_n$, with $a_1|\dots|a_n$ (A1, Theorem 4). Then $N(\mathfrak{P}) = a_1\dots a_n\Gamma$, and since $p\omega_i \in \mathfrak{P}$, a_i is not divisible by p^2, for $i = 1,\dots,n$. Hence there is a number s with $a_1\Gamma,\dots,a_s\Gamma = \Gamma$, $a_{s+1}\Gamma,\dots,a_n\Gamma = p\Gamma$. Consequently the classes of $\omega_{s+1},\dots,\omega_n$ form a basis of $O_K/\mathfrak{P}$ over $\Gamma/p\Gamma$, and $N(\mathfrak{P}) = p^{n-s}\Gamma$. □

Let $pO_K = \mathfrak{P}_1^{e_1}\dots\mathfrak{P}_g^{e_g}$ be the prime decomposition of p. By passing to the norm one obtains

$$n = e_1f_1 + \dots + e_gf_g, \tag{12}$$

where f_i is the inertia degree of $\mathfrak{P}_i$. The coefficient e_i is called the *ramification exponent* of $\mathfrak{P}_i$, $i = 1,\dots,g$.

With each ideal $\mathfrak{a}$ of P we associate the ideal of K generated by $\mathfrak{a}$. In this way one obtains a monomorphism of the group $\mathcal{I}_P$ into the group $\mathcal{I}_K$. We identify the ideals from $\mathcal{I}_P$ with their images in $\mathcal{I}_K$. As one easily sees, the norm may be extended in a unique way to a homomorphism of the group $\mathcal{I}_K$ into the group $\mathcal{I}_P$.

Lemma 6 may now be generalised to the following theorem.

Theorem 18. *Suppose* $\mathfrak{A},\mathfrak{B} \in \mathcal{I}_K$ *and* $\mathfrak{A} \subset \mathfrak{B}$. *Then*

$$N(\mathfrak{A}/\mathfrak{B}) = [\mathfrak{B}:\mathfrak{A}].$$

Proof. Let $\alpha \neq 0$ be an element of K with $\alpha\mathfrak{A} \subset O_K$, $\alpha\mathfrak{B} \subset O_K$. Then

$$N(\mathfrak{A}/\mathfrak{B}) = N(\alpha\mathfrak{A}/\alpha\mathfrak{B}) = [\alpha\mathfrak{B}/\alpha\mathfrak{A}] = [\mathfrak{B}/\mathfrak{A}]. \quad \square$$

If $\mathfrak{A}$ is an ideal of O_K then

$$\mathfrak{A}\,|\,N(\mathfrak{A}). \tag{13}$$

18.8 Congruences

We make a few remarks about congruences, mainly as generalisations of theorems from Chapter 1.

Let $\mathfrak{A}$ be an ideal of O_K. As a generalisation of Gauss's definition of congruence,

$$\alpha \equiv \beta \pmod{\mathfrak{A}} \quad \text{for} \quad \alpha,\beta \in O_K$$

means that $\alpha-\beta$ lies in $\mathfrak{A}$. The rules for congruences given in Section 1.1 still hold. We emphasise the following:

Theorem 19. *Let* $\alpha \in O_K$ *and let* $\mathfrak{B}$ *be an ideal relatively prime to* αO_K. *Then there is an element* ξ *in* O_K *with*

$$\alpha\xi \equiv 1 \pmod{\mathfrak{B}}.$$

Proof. This follows from $\alpha O_K + \mathfrak{B} = O_K$ (cf. (8)). □

Calculating with congruences in O_K is the same as calculating in the congruence class ring $O_K/\mathfrak{A}$. The Chinese remainder theorem now runs as follows:

Theorem 20. *Let* $\mathfrak{A}$ *and* $\mathfrak{B}$ *be relatively prime ideals. Then*

$$\psi : \gamma + \mathfrak{A}\mathfrak{B} \mapsto (\gamma + \mathfrak{A}, \gamma + \mathfrak{B}), \; \gamma \in O_K,$$

defines an isomorphism of $O_K/\mathfrak{A}\mathfrak{B}$ *onto* $O_K/\mathfrak{A} + O_K/\mathfrak{B}$.

Proof. Because of (8), ψ is injective. Let α, β be arbitrary elements of O_K. Because of (8), there are elements α_1 of $\mathfrak{A}$ and β_1 of $\mathfrak{B}$ with $\alpha_1 + \beta_1 = 1$. We have $\psi(\alpha\beta_1 + \beta\alpha_1 + \mathfrak{A}\mathfrak{B}) = (\alpha + \mathfrak{A}, \beta + \mathfrak{B})$. □

For the remainder of this section we confine ourselves to the case $\Gamma = \mathbb{Z}$.

Let $\mathfrak{A}$ be an ideal of O_K. The norm $N(\mathfrak{A})$ is now a non-negative integer, the number of elements of $O_K/\mathfrak{A}$.

Let $\Phi(\mathfrak{A})$ be the number of prime congruence classes mod $\mathfrak{A}$. Then for all numbers π in O_K prime to $\mathfrak{A}$ we have

$$\pi^{\Phi(\mathfrak{A})} \equiv 1 \pmod{\mathfrak{A}}. \tag{14}$$

In fact the prime congruence classes mod $\mathfrak{A}$ constitute the group of units of the ring $O_K/\mathfrak{A}$ (A1.1 and Theorem 19). In particular, a prime ideal $\mathfrak{P}$ satisfies the Fermat theorem

$$\alpha^{N(\mathfrak{P})} \equiv \alpha \pmod{\mathfrak{P}} \text{ for all } \alpha \in O_K. \tag{15}$$

$q := N(\mathfrak{P})$ is a power of the prime number p such that $p\mathbb{Z} = \mathfrak{P} \cap \mathbb{Z}$. As in the proof of Theorem 5, Chapter 1, one shows that the multiplicative group of $O_K/\mathfrak{P}$ is cyclic. Let ζ be a generator of this group. The Galois theory developed in Chapter 17 carries over directly to the field extension $O_K/\mathfrak{P}$ over $\mathbb{Z}/p\mathbb{Z}$. This extension is normal, since it is generated by ζ and ζ is a zero of the polynomial $x^{q-1}-1$, whose other zeros are the powers of ζ. Let f be the degree of the field extension of $O_K/\mathfrak{P}$ over $\mathbb{Z}/p\mathbb{Z}$. Then $q = p^f$. By Theorem 3, Chapter 17, the Galois group G of this extension has order f. On the other hand, $\zeta \mapsto \zeta^p$ defines an automorphism τ of G which has order f. G is therefore a cyclic group with generator τ, and it is called the *Frobenius automorphism.*

18.9 Prime ideal decomposition in quadratic fields

As an example we now look more closely at the prime ideal decomposition in the quadratic field $K = \mathbb{Q}(\sqrt{d})$, using the notation of Section 18.3.

Because of (12) there are three possibilities for the prime ideal decomposition of a prime p in O_K:

$$pO_K = \mathfrak{P}_1\mathfrak{P}_2 \text{ with } \mathfrak{P}_1 \neq \mathfrak{P}_2,\ pO_K = \mathfrak{P} \text{ and } pO_K = \mathfrak{P}^2.$$

In the first case p is said to *split*, in the second case to *be inert* and in the third case to *ramify*. The following theorem gives information about which of these three cases occurs for a prime number p.

Theorem 21. *Let* D *be the discriminant of* O_K (Section 18.2). *For* $d \equiv 1 \pmod 4$, $D = d$, *and for* $d \equiv 2$ *or* $3 \pmod 4$, $D = 4d$.

(i) p *ramifies if and only if* p *is a divisor of* D.

(ii) *Suppose* $p \nmid 2D$. *Then* p *splits or is inert according as the Legendre symbol* $(\frac{D}{p})$ *equals* 1 *or* -1 (Section 1.4).

(iii) *Suppose* $2 \nmid D$. *Then* 2 *splits or is inert according as* $D \equiv 1 \pmod 8$ *or* $D \equiv 5 \pmod 8$.

Proof. In what follows, $(\alpha_1,\ldots,\alpha_s)$ denotes the ideal generated by $\alpha_1,\ldots,\alpha_s$. If $p|D$ then $p|d$ or $p = 2$ and $d \equiv 3 \pmod 4$. In the first case $(p) = (p,\sqrt{d})^2$. In the second case $(p) = (p,1+\sqrt{d})^2$.

For $d \nmid 2D$ and $(\frac{D}{p}) = 1$ there is an $x \in \mathbb{Z}$ with $x^2 \equiv D \pmod p$. Then $(p) = (p,x+\sqrt{D})(p,x-\sqrt{D})$ and $(p,x+\sqrt{D}) \neq (p,x-\sqrt{D})$.

For $p \nmid 2D$ and $(\frac{D}{p}) = -1$ there is no $\alpha \in O_K$ with $p|N(\alpha)$ and $p \nmid \alpha$. Hence there is no prime ideal $\mathfrak{P}$ with $p = N(\mathfrak{P})$, and it follows that p is inert.

For $p = 2$, $D \equiv 1 \pmod 8$ we have

$$(2) = (2,\frac{1+\sqrt{d}}{2})(2,\frac{1-\sqrt{d}}{2}) \text{ and } (2,\frac{1+\sqrt{d}}{2}) \neq (2,\frac{1-\sqrt{d}}{2}).$$

For $p = 2$, $D \equiv 5 \pmod 8$ we have the same as in the case $p|2D$ and $(\frac{D}{p}) = -1$. □

Exercises

18.1 (Stickelberger) Let $K/\mathbb{Q}$ be an extension of degree n, and let $\alpha_1,\ldots,\alpha_n$ be integers of K. Show that the discriminant $\Delta(\alpha_1,\ldots,\alpha_n)$ is congruent to 0 or 1 modulo 4. (Hint: Use Chapter 17, (6).)

18.2 Let $K = \mathbb{Q}(\alpha)$, where α is a zero of the irreducible polynomial $f(x) = x^n + a_1x^{n-1} + \ldots + a_n$ with coefficients $a_1,\ldots,a_n$ from $\mathbb{Z}$. Also, let p be a prime number and let $\bar{f}(x) = x^n + \bar{a}_1x^{n-1} + \ldots + \bar{a}_n$ be the corresponding polynomial with coefficients from $\mathbb{Z}/p\mathbb{Z}$. Let

$$\bar{f}(x) = \bar{f}_1(x)\ldots\bar{f}_s(x) \qquad (16)$$

be the decomposition of $f(x)$ in $\mathbb{Z}/p\mathbb{Z}[x]$ into irreducible factors of degrees $n_1,\ldots,n_s$.

a) Show that $\bar{f}(x)$ decomposes into distinct irreducible factors if and only if p is not a divisor of the discriminant $\Delta(\alpha)$.

b) Suppose p is not a divisor of $\Delta(\alpha)$. Show that the ideals;

$$\mathfrak{P}_1 := (p,f_1(\alpha))O_K,\ldots,\mathfrak{P}_s := (p,f_s(\alpha))O_K$$

are distinct prime ideals with degrees of inertia $n_1,\ldots,n_s$.

c) Show that $pO_K = \mathfrak{P}_1\ldots\mathfrak{P}_s$.

d) Let $\Delta(\alpha)$ be square-free. Show that $1,\alpha,\ldots,\alpha^{n-1}$ is a basis of O_K.

18.3 Let α be a zero of the polynomial x^3-x-1 and let $K = \mathbb{Q}(\alpha)$.

a) Show that $1,\alpha,\alpha^2$ is a basis of O_K.

b) Show that $23O_K = \mathfrak{P}_1\mathfrak{P}_2^2$ for distinct prime ideals $\mathfrak{P}_1$, $\mathfrak{P}_2$.

c) Show that $2O_K$, $3O_K$ and $7O_K$ are prime ideals.

18.4 Give examples illustrating the following deep theorem: if p is a prime number different from 23, then the polynomial x^3-x-1 splits into two irreducible factors in $\mathbb{Z}/p\mathbb{Z}[x]$ if and only if -23 is not a quadratic residue modulo p; x^3-x-1 splits into three irreducible factors if and only if $p = \frac{a^2+23b^2}{4}$ with $a,b \in \mathbb{Z}$.

18.5 Let $K = \mathbb{Q}(\sqrt{2},\sqrt{a})$ with a square-free integer $a \equiv 3 \pmod 4$. Show that $1, \sqrt{2}, \sqrt{a}, \sqrt{2a}$ form a basis of O_K.

18.6 Let $\mathfrak{A}$ and $\mathfrak{B}$ be relatively prime ideals in the ring of integers of an algebraic number field. Show that $\Phi(\mathfrak{A}\mathfrak{B}) = \Phi(\mathfrak{A})\Phi(\mathfrak{B})$.

18.7 Let $\mathfrak{P}$ be a prime ideal. Show that $\Phi(\mathfrak{P}^h) = N(\mathfrak{P}^h) - N(\mathfrak{P})^{h-1}$.

18.8 A polynomial $f(x) := x^n + a_1x^{n-1} + \ldots + a_n$ with coefficients in $\mathbb{Z}$ is called an *Eisenstein polynomial* for the prime number p when $a_1,\ldots,a_n$ are divisible by p and p appears in a_n only to the first power. Let α be a zero of $f(x)$.

a) Show that $\mathfrak{P} := (\alpha,p)$ in $K := \mathbb{Q}(\alpha)$ is a prime ideal with $\mathfrak{P}^n = pO_K$.

b) Use a) to show that $f(x)$ is irreducible (*Eisenstein's irreducibility criterion*, Theorem 3, Chapter 3).

18.9 Let K be an imaginary quadratic field with discriminant D_K and let O_K be a euclidean ring.

a) Let γ be an element of minimal height in O_K-{0,1,-1}. Show that each α in O_K is congruent to 0, 1 or -1 (mod γ).

b) Suppose $D_K < -11$. Show that $N(\alpha) > 3$ for all $\alpha \in O_K$-{0,1,-1}.

c) Show that $D_K > -11$ (cf. Chapter 16, Exercise 8).

19. The ideal class group and the group of units

19.1 The finiteness of the class number

In this and the following chapters we confine ourselves to algebraic number fields, more precisely to finite extensions K of $\mathbb{Q}$. As above, O_K denotes the ring of integers of K and $\mathcal{I}_K$ denotes the group of fractional ideals of K. The latter group contains the subgroup $(K^{\times})$ of *fractional principal ideals*, i.e. the O_K-modules of the form $\alpha O_K =: (\alpha)$ for $\alpha \in K^{\times}$. The quotient group $Cl_K =: \mathcal{I}_K/(K^{\times})$ is called the *ideal class group* of K.

The mapping of $K^{\times}$ into $\mathcal{I}_K$ which associates with each $\alpha \in K^{\times}$ the principal ideal (α), has as kernel the group E_K of units of O_K. We therefore have an exact sequence of groups

$$\{1\} \to E_K \to K^{\times} \to \mathcal{I}_K \to Cl_K \to \{1\}. \tag{1}$$

Knowledge of the groups Cl_K and E_K, which we study in this chapter, is necessary for the mastery of arithmetic in K.

The order of Cl_K is a measure of the difference between the arithmetic of O_K and the arithmetic of the principal ideal ring. The following theorem says that this difference is finite. The order of Cl_K is called the *class number of* K.

Theorem 1. *The class number is finite.*

Proof. Let $\omega_1,\ldots,\omega_n$ be a basis of O_K as a $\mathbb{Z}$-module, so that n is the degree of K over $\mathbb{Q}$. Also let $\delta = h_1\omega_1 + \ldots + h_n\omega_n$ be a number of O_K whose coordinates $h_1,\ldots,h_n$ are less than or equal in absolute value to a positive constant K, and for an isomorphism g of K into $\mathbb{C}$ let

$$r_g := |g\omega_1| + \ldots + |g\omega_n|.$$

Then

$$|N(\delta)| = \prod_g |h_1 g\omega_1 + \ldots + h_n g\omega_n| \le k^n \prod_g r_g \text{ (Sect. 17.6),} \tag{2}$$

where the product is taken over all isomorphisms g of K into $\mathbb{C}$. We set $\prod_g r_g =: t$.

Now let $\underset{\sim}{A}$ be a given ideal of O_K and let k be the natural number with

$$k^n \le |N(\underset{\sim}{A})| < (k+1)^n.$$

Among the $(k+1)^n$ numbers of the form $h_1\omega_1 + ... + h_n\omega_n$ with $0 \le h_i \le k$ and $h_i \in \mathbb{Z}$, for $i = 1,...,n$, there are two different ones β, γ whose difference lies in $\underset{\sim}{A}$, because by choice of K the number of congruence classes of O_K mod $\underset{\sim}{A}$ is strictly less than $(k+1)^n$. Hence at least one congruence class must contain two numbers β, γ. Thus the number $\alpha = \beta - \gamma$ lies in $\underset{\sim}{A}$.

Remark. The type of reasoning applied here was used by Dirichlet in his work cited in Section 16.2, and is called *Dirichlet's pigeon-hole principle.*

By (2) and Chapter 18, (10) we have

$$|N(\alpha)| \le N(\underset{\sim}{A})t, \quad N(\alpha\underset{\sim}{A}^{-1}) \le t.$$

Hence the class $\underset{\sim}{A}^{-1}$ contains an integral ideal whose norm is less than or equal to t. Since t is independent of the choice of $\underset{\sim}{A}$ and there are only finitely many ideals whose norm is smaller than a given number, the assertion follows. □

The bound t is too rough for practical computation of the class number. Minkowski derived substantially better bounds. We come back to this problem in Chapter 24.

19.2 Units in quadratic fields

We want to gain an overview of the units in O_K, and first we consider the case of a quadratic field.

As in Section 18.3 let $K = \mathbb{Q}(\sqrt{d})$. A number ε in O_K is a unit if and only if $|N(\varepsilon)| = 1$. One therefore has to determine the integers $\varepsilon = (u+t\sqrt{d})/2$ with

$$|u^2 - t^2d| = 4, \tag{3}$$

where $u \equiv t \pmod 2$ and, in the case $d \equiv 2,3 \pmod 4$, where $u \equiv t \equiv 0 \pmod 2$ as well.

When d is negative, (3) has only the solutions $u = \pm 2$, $t = 0$, except in the cases $d = -1$ and $d = -3$, where there are also the solutions $u = 0$, $t = \pm 2$ resp. $u = \pm 1$, $t = \pm 1$. I.e., the group of units consists only of the two roots of unity, apart from the cases $d = -1$ and $d = -3$, where the group of units consists of the fourth resp. sixth roots of unity.

When d is positive, (3) is called *Pell's equation*. Its complete solution was first determined by Lagrange (*Solution d'un Problème d'Arithmétique, Miscellanea Taurinensia* 4 (1766)). The result is that there is a basic unit ε_0 such that all units ε are uniquely expressible in the form

$$\varepsilon = \pm\varepsilon_0^i,\ i \in \mathbb{Z}.$$

19.3 The structure of the group of units in the orders of algebraic number fields

We now go to the general case. Let $K = \mathbb{Q}(\theta)$ be an algebraic number field of degree n, generated by the integer θ. Dirichlet determined the units of the subring $\mathbb{Z}[\theta]$ of O_K (Section 16.2). We follow Dirichlet's method, but consider more generally an arbitrary order B of K, i.e. a subring of O_K which contains $\mathbb{Z}$ and whose field of fractions equals K. By Theorem 7, Chapter 18, and Theorem 4, A.1, B is a lattice in K. Let E be the group of units of B.

To formulate Dirichlet's result we consider the isomorphisms g of K into $\mathbb{C}$. When the image of g lies in $\mathbb{R}$, g is called real, otherwise complex. When g is complex and j denotes the transition to complex conjugates, $jg \neq g$. Thus the complex isomorphisms always occur in pairs. Let r_1 be the number of real isomorphisms, let r_2 be half the number of complex isomorphisms, and let $r = r_1 + r_2$. By Theorem 2, Chapter 17, $r_1 + 2r_2 = n$.

Theorem 2. *(Dirichlet's unit theorem). There are* $r-1$ *units* $\varepsilon_1,\dots,\varepsilon_{r-1}$ *in* E *and a root of unity* ζ *such that each unit* ε *in* E *is uniquely expressible in the form*

$$\varepsilon = \zeta^i \prod_{K=1}^{r-1} \varepsilon_K^{i_K}, \tag{4}$$

where t *is a nonnegative integer smaller than the order of* ζ *and* $i_1,\dots,i_{r-1}$ *are integers.*

The proof will be carried out in several steps over the next few sections, which also can claim to be of independent interest. We assume $r > 1$.

19.4 The logarithmic components

An important role in the proof is played by the *logarithmic components* of the units of B. To define them we fix a sequence of n isomorphisms $g_1,\dots,g_n$ of K into $\mathbb{C}$ such that $g_1,\dots,g_{r_1}$ are real and $g_{r_1+1},\bar{g}_{r_1+1},\dots,g_r,\bar{g}_r$ are the pairs of conjugate complex isomorphisms. Also set $l_i = 1$ for $i = 1,\dots,r_1$ and $l_i = 2$ for $i = r_1 + 1,\dots,r$. Then the ith logarithmic component $l_i(\alpha)$ of $\alpha \in K^\times$ is equal to $l_i \log|g_i\alpha|$. Also let $\mathbf{l}(\alpha)$ denote the vector $l_1(\alpha),\dots,l_r(\alpha)$ in $\mathbb{R}^r$. The units ε are characterised in B by the condition $|N(\varepsilon)| = 1$. This is obviously equivalent to

$$l_1(\varepsilon) + \dots + l_r(\varepsilon) = 0. \tag{5}$$

We see that $\varepsilon \mapsto \mathbf{l}(\varepsilon)$ is a group homomorphism $\mathbf{l}$ of E into the (r-1)-dimensional subspace U of $\mathbb{R}^r$ defined by the equation $x_1 + \dots + x_r = 0$ for the vectors $(x_1,\dots,x_r)$ of $\mathbb{R}^r$.

To prove Theorem 2 it suffices to show that the kernel of $\mathbf{l}$ is a finite group and that the image of $\mathbf{l}$ is a lattice in U.

19.5 The kernel of l

We first show that Ker $\mathbf{l}$ is a finite group. To do this we need the following

Proposition 1. *Let* c *be a positive constant. Then there are only finitely many numbers* δ *in* B *with* $|g_i\delta| \le c$ *for* $i = 1,\dots,n$.

Proof. Let $\omega_1,\dots,\omega_n$ be a basis for B over $\mathbb{Z}$ and let $\kappa_1,\dots,\kappa_n$ be the complementary basis (Chapter 17, (11)). For

$$\delta = h_1\omega_1 + \dots + h_n\omega_n \text{ with } h_i \in \mathbb{Z}$$

we then have

$$h_i = S(\delta\kappa_i) = g_1\delta g_1\kappa_i + \dots + g_n\delta g_n\kappa_i,$$

$$|h_i| \le c(|g_1\kappa_i| + \dots + |g_n\kappa_i|) \text{ for } i = 1,\dots,n. \quad \square$$

By definition, the kernel of l consists of the units ε with $|g_i\varepsilon| = 1$ for $i = 1,\dots,n$. By Proposition 1 there are only finitely many such units.

19.6 The image of l

In this section we shall show that the group $l(E)$ is finitely generated. Since $l(E)$ lies in the vector space U of dimension $r-1$, it follows that $l(E)$ has rank at most $r-1$ (A1.4).

First we prove a simple lemma.

Lemma 1. *Let* V *be a vector space over* $\mathbb{R}$ *of dimension* s *and let* M *be a subgroup of* V *which is discrete in* V*, i.e., each* $m \in M$ *has a neighbourhood in* V *containing no other points of* M.

Then M *is a finitely generated group.*

Proof. We can assume without loss of generality that the vector space generated by M is equal to V. Let $v_1,\dots,v_s$ be a basis of V consisting of elements of M. It suffices to show that the group M_0 generated by $v_1,\dots,v_s$ has finite index in M. In each coset in M/M_0 there is exactly one representative $v = a_1v_1 + \dots + a_sv_s$ with $0 \le a_s < 1$ for $i = 1,\dots,s$. Since M is discrete, there are only finitely many such v. $\square$

We now show that $l(E)$ is finitely generated. By Lemma 1 it suffices to show that $l(E)$ is discrete in $\mathbb{R}^r$. Let t be any positive number and suppose $l_i(\varepsilon) \le t$ for $i = 1,\dots,r$. Then

$$|g_i\varepsilon| \le e^t \text{ for } i = 1,\dots,r. \tag{6}$$

By Proposition 1 there are only finitely many $\varepsilon \in B$ satisfying (6), so the assertion follows.

19.7 The rank of l(E)

To prove Theorem 2 it remains to show that l(E) has rank r-1. This is the main difficulty. We first prove two more propositions.

Proposition 2. *Let* c_1 *and* c_2 *be positive real numbers and let*

$$u := \max\{|g_i\omega_1| + \ldots + |g_i\omega_n| \,|\, i = 1,\ldots,n\}.$$

Let the set G *of isomorphisms of* K *into* $\mathbb{C}$ *be partitioned into non-empty disjoint subsets* G_1, G_2*, with conjugate complex isomorphisms in the same set.*

Then there is a number γ *from* B *with*

$$|g\gamma| < c_1 \text{ for } g \in G_1, \quad |g\gamma| > c_2 \text{ for } g \in G_2$$

and

$$N(\gamma) < (3u)^n.$$

Proof. We first remark that the function $f(x) := (x+1)^s - x^s - 1$ is positive for $s > 1$ and $x > 0$, because $f(0) = 0$ and $f'(x) > 0$.

Alongside the real isomorphisms in G_1 we consider the following mappings defined for pairs g, g' of conjugate complex isomorphisms in G_1:

$$\delta \mapsto \operatorname{Re} g\delta, \quad \delta \mapsto \operatorname{Im} g\delta \quad \text{for } \delta \in K.$$

In this way we obtain $\mu := [G_1]$ mappings of K into $\mathbb{R}$ which we denote in arbitrary order by $w_1,\ldots,w_\mu$.

Let k be a natural number and let

$$\Omega(k) = \{h_1\omega_1+\ldots+h_n\omega_n \,|\, h_i \in \mathbb{Z},\ 0 \le h_i \le k \text{ for } i = 1,\ldots,n\}.$$

By definition of u and ω_j we have

$$|w_i(\delta)| \leq uk \quad \text{for} \quad \delta \in \Omega(k),\ i = 1,\dots,\mu. \tag{6}$$

Since $n > \mu > 0$ and $k > 0$, it follows from the remark above about the function $f(x)$ that

$$(k+1)^{n/\mu} - k^{n/\mu} > 1.$$

Hence there is a natural number m with

$$(k+1)^{n/\mu} > m > k^{n/\mu},$$

whence

$$(k+1)^n > m^\mu > k^n.$$

We now avail ourselves of the pigeon-hole principle already used in Section 19.1, in the following way: consider the closed interval $[-uk,uk]$, which contains all the values $w_1(\delta),\dots,w_\mu(\delta)$ for the $(k+1)^n$ numbers δ in $\Omega(k)$. Let $d := 2uk/m$, so that

$$d < 2uk^{1-n/\mu}. \tag{7}$$

We divide $[-uk,uk]$ into m subintervals of length d. Interval s, for $s = 1,\dots,m$, is the interval $[-uk + (s-1)d,\ -uk + sd]$. Each of the μ values $w_1(\delta),\dots,w_\mu(\delta)$ falls into one of these intervals, and if a number falls on the boundary of two intervals, it is assigned arbitrarily to one of them. Let the corresponding interval numbers be $s_1,\dots,s_\mu$. We say that δ *corresponds to* the sequence $s_1,\dots,s_\mu$. There are altogether m^μ interval sequences (these are our pigeon holes). Since we have $(k+1)^n$ numbers in $\Omega(k)$ and $(k+1)^n > m^\mu$, there must be two different numbers α, β in $\Omega(k)$ which have the same interval sequence (i.e., which lie in the same pigeon hole). We set $\gamma := \alpha-\beta$.

Then

$$w_i(\gamma) = w_i(\alpha) - w_i(\beta) \leq d \quad \text{for} \quad i = 1,\dots,\mu,$$

hence $|g\gamma| \leq \sqrt{2}d$ for $g \in G_1$, and hence by (7)

$$|g\gamma| < 3uk^{1-n/\mu} \quad \text{for} \quad g \in G_1. \tag{8}$$

We set

$$M_j = \prod_{g\in G_j} |g\gamma| \quad \text{for} \quad j = 1,2.$$

Then $M_1M_2 = |N(\gamma)|$ and by (8)

$$M_1 < (3u)^{\mu}k^{\mu-n}. \tag{9}$$

Since $|g\gamma| \le uk$ for $g \in G$, (9) implies

$$|N(\gamma)| < (3u)^n \tag{10}$$

and

$$M_2 = |N(\gamma)|M_1^{-1} \ge M_1^{-1} > (3u)^{-\mu}k^{n-\mu}.$$

Hence for $g \in G_2$

$$|g\gamma| = M_2 \prod_{h\in G_2-\{g\}} |h\gamma|^{-1} > (3u)^{-\mu}k^{n-\mu}(uk)^{-n+\mu+1} = 3^{-\mu}u^{1-n}k. \tag{11}$$

If one chooses k sufficiently large, then (8), (10) and (11) allow the requirements of Proposition 2 to be met. □

Proposition 3. *There is a unit* ε *in* B *with*

$$|g\varepsilon| < 1 \quad \textit{for} \quad g \in G_1,$$

$$|g\varepsilon| > 1 \quad \textit{for} \quad g \in G_2.$$

Proof. By Proposition 2 there is an infinite sequence $\gamma_1,\gamma_2,\ldots$ of numbers from B with the properties

$$|N(\gamma_i)| < (3u)^n \quad \text{for} \quad i = 1,2,\ldots$$

$$|g\gamma_1| > |g\gamma_2| > \ldots \text{ for } g \in G_1, \tag{12}$$

$$|g\gamma_1| < |g\gamma_2| < \ldots \text{ for } g \in G_2. \tag{13}$$

Since the $|N(\gamma_i)|$ are bounded natural numbers, we can assume without loss of generality that $|N(\gamma_i)|$ has the same value a for all $i = 1,2,\dots$. However, there are only finitely many pairwise non-associated numbers $\gamma \in B$ with $|N(\gamma)| = a$. More precisely, two numbers $\alpha,\beta \in B$ with $|N(\alpha)| = |N(\beta)| = a$ and $\alpha-\beta \in aB$ are associated.

It follows from $\alpha-\beta = a\delta$ for $\delta \in B$ that

$$\frac{\alpha}{\beta} = 1 + \frac{|N(\beta)|}{\beta}\delta \in B, \ \frac{\beta}{\alpha} = 1 - \frac{|N(\alpha)|}{\alpha}\delta \in B.$$

There are exactly a^n classes in B/aB. A system of representatives of these classes in B is given by the elements $a_1\omega_1 + \dots + a_n\omega_n$ with $0 \le a_i < a$ for $i = 1,\dots,n$.

It follows that there are associates γ_1,γ_2 in B satisfying (12), (13). Then $\varepsilon = \gamma_2\gamma_1^{-1}$ does what is wanted. □

Now we are in a position to give units $\varepsilon_1,\dots,\varepsilon_{r-1}$ such that $l(\varepsilon_1),\dots,l(\varepsilon_{r-1})$ are linearly independent vectors in $\mathbb{R}^r$. Proposition 3 allows us to take ε_i, for $i = 1,\dots,r-1$, to be a unit with

$$|g_i\varepsilon_i| > 1, \ |g_j\varepsilon_i| < 1 \ \text{ for } j = 1,\dots,r, \ j \ne i$$

i.e.

$$l_i(\varepsilon_i) > 0, \ l_j(\varepsilon_i) < 0 \ \text{ for } j = 1,\dots,r, \ j \ne i.$$

This implies

$$\sum_{K=1}^{r-1} l_K(\varepsilon_i) = -l_r(\varepsilon_i) > 0.$$

We have to show that the determinant of the matrix $(l_j(\varepsilon_i))_{i,j=1,\dots,r-1}$ is non-zero. This results from the following

Lemma 2. (*Minkowski*). *Let* (a_{ij}) *be an s-rowed matrix of real numbers with*

$$a_{ij} \le 0 \ \textit{for} \ i \ne j, \ \sum_{k=1}^{s} a_{ik} > 0 \ \textit{for} \ i = 1,\dots,s.$$

Then $\det(a_{ij}) \ne 0$.

Proof. Let $x_1,\dots,x_s$ be real numbers with

$$\sum_{k=1}^{s} a_{ik}x_k = 0 \quad \text{for} \quad i = 1,\dots,s. \tag{14}$$

We choose a j with $|x_j| \geq |x_i|$ for $i = 1,\dots,s$. Suppose, without loss of generality, that $x_j \geq 0$. Then

$$0 = \sum_{k=1}^{s} a_{jk}x_k \geq a_{jj}x_j + \sum_{k\neq j} a_{jk}x_j = \left(\sum_{k=1}^{s} a_{jk}\right)x_j \geq 0$$

and hence $x_j = 0$. It follows that the system of equations (14) has only the zero solution. □

A set $\{\varepsilon_1,\dots,\varepsilon_{r-1}\}$ of units from B is called a *fundamental system of units* of B when the vectors $l(\varepsilon_1),\dots,l(\varepsilon_{r-1})$ generate the $\mathbb{Z}$-module $l(E)$. The number $R(\varepsilon_1,\dots,\varepsilon_{r-1}) := |\det(l_j(\varepsilon_j))_{i,j=1,\dots,r-1}|$ is independent, by (5), of the choice and ordering of the isomorphisms $g_1,\dots,g_{r-1}$. Also, $R(\varepsilon_1,\dots,\varepsilon_{r-1})$ has the same value for all fundamental units. This common value is called the *regulator* of B. In the case $r = 1$ the regulator is set equal to 1.

19.8 Dirichlet's unit theorem as an assertion about diophantic equations

Theorem 2 is easily interpreted as a theorem about the solution of diophantine equations. Let $x_1,\dots,x_n$ be indeterminates and let

$$N(x_1\omega_1 + \dots + x_n\omega_n) = \prod_{i=1}^{n} (x_1g_i\omega_1 + \dots + x_ng_i\omega_n).$$

This is an nth degree form in $x_1,\dots,x_n$ with integral coefficients. The solutions of the equation

$$N(x_1\omega_1 + \dots + x_n\omega_n) = \pm 1 \tag{15}$$

with integral coefficients are in one-to-one correspondence with the units of B. Theorem 2 therefore gives an overview of the solutions of the diophantine equation (15).

More generally, one can regard the theory of algebraic numbers, as we have developed it in Chapters 18 and 19, as encoding information about ordinary integers in a way suitable for formulating and proving the laws of ordinary integers in the clearest possible manner. (See also Chapter 21.)

Exercises

19.1 Let θ be an algebraic integer with $|g\theta| = 1$ for all conjugates $g\theta$ of θ. Show that θ is a root of unity.

19.2 Let K be a finite extension of $\mathbb{Q}$ and let S be a finite set of prime ideals of K. An element $\alpha \in K$ is called an *S-unit* when only prime ideals from S appear in the prime decomposition of α. Show that the group of all S-units of K is finitely generated and has rank $[S] + r_1 + r_2 - 1$.

19.3 Let $K = \mathbb{Q}(\sqrt{d})$ be a real quadratic field and let $\varepsilon = \frac{x+y\sqrt{d}}{2}$ be a unit of K. Show that ε is greater than 1 if and only if $x > 0$ and $y > 0$.

19.4 Let $\varepsilon_1 = \frac{x_1+y_1\sqrt{d}}{2}$ and $\varepsilon_2 = \frac{x_2+y_2\sqrt{d}}{2}$ be two units of $\mathbb{Q}(\sqrt{d})$ with $1 < \varepsilon_1 < \varepsilon_2$. Show that $x_2 > x_1$ and $y_2 \geq y_1$, where the case $y_2 = y_1$ occurs only for $\varepsilon_1 = \frac{1+\sqrt{5}}{2}$, $\varepsilon_2 = \frac{3+\sqrt{5}}{2}$.

19.5 Let $d \neq 5$. By Exercise 4 the integer $\frac{x+y\sqrt{d}}{2} > 1$ is a fundamental unit if and only if $|x^2-y^2d| = 4$ holds with minimal y. Use this to compute the fundamental units of $\mathbb{Q}(\sqrt{d})$ for $d = 2,3,5,6,7,10,13,14,15,17$.

19.6 Suppose d contains a prime factor $p \equiv 3 \pmod 4$. Show that the norm of each unit in $\mathbb{Q}(\sqrt{d})$ equals 1.

19.7 Let p be a prime number with $p \equiv 1 \pmod 4$. Show that the norm of the fundamental unit ε_0 of $K = \mathbb{Q}(\sqrt{p})$ equals -1. (Hint: Prove first that each $\alpha \in K$ with $N(\alpha) = 1$ is representable in the form $\alpha = \beta/\sigma\beta$ with $\beta \in K$, where σ is the non-trivial automorphism of K, and then investigate the principal ideal (β) in a representation $\varepsilon_0 = \beta/\sigma\beta$.)

20. The Dedekind ζ-function

20.1 Definition of the Dedekind ζ-function

In this chapter the notations of Chapter 19 become worthwhile: K denotes an algebraic number field of degree n, the number of real isomorphisms of K into $\mathbb{C}$ is denoted by r_1, half the number of complex isomorphisms of K into $\mathbb{C}$ is denoted by r_2, $r := r_1 + r_2$. Also, R denotes the regulator, h the class number, $D := \Delta(O_K)$ the discriminant of O_K and w the number of roots of unity in K.

Dirichlet found a closed expression for the class number of quadratic forms (Chapter 2) by analytic methods, and Kummer did something similar for cyclotomic fields. Dedekind generalised part of these considerations to an arbitrary algebraic number field K and introduced the following function $\zeta_K(s)$ which is known as the *Dedekind ζ-function*. For real $s > 1$

$$\zeta_K(s) := \Sigma N(\mathfrak{A})^{-s}$$

where the sum is taken over all ideals $\mathfrak{A} \neq \{0\}$ of O_K.

The convergence of this series for $s > 1$, as well as the limit formula

$$\lim_{s\to 1}(s-1)\zeta_K(s) = \frac{2^r\pi^{n-r}Rh}{w\sqrt{|D|}} \qquad (1)$$

results from the following two theorems.

Theorem 1. *(Dedekind). Let* $T(t)$ *be the number of ideals of* O_K *with norm less than or equal to* t. *Then*

$$\lim_{t\to\infty} \frac{T(t)}{t} = \frac{2^r\pi^{n-r}Rh}{w\sqrt{|D|}}. \qquad (2)$$

Theorem 2. *(Dirichlet). Let* $a_1, a_2, \ldots$ *be a sequence of complex numbers and let*

$$T(t) := \sum_{m\leq t} a_m.$$

Suppose the limit $\lim_{t\to\infty} T(t)/t$ *exists and equals* a. *Then the series*

$$\xi(s) := \sum_{m=1}^{\infty} a_m m^{-s}$$

converges for $s > 1$ *and*

$$\lim_{s\to 1}(s-1)\xi(s) = a. \tag{3}$$

20.2 Preliminaries to the proof of Theorem 1

In what follows we use the term *ideal class* of K to mean the set of integral ideals in a class of Cl_K.

To prove Theorem 1 it suffices to prove the following Theorem 3.

Theorem 3. *Let* C *be an ideal class of* $\mathcal{I}_K$ *and let* T(C,t) *be the number of ideals* $\mathfrak{A}$ *in* C *with* $N(\mathfrak{A}) \le t$. *Then*

$$\lim_{t\to\infty} \frac{T(C,t)}{t} = \frac{2^r \pi^{n-r} R}{w\sqrt{|D|}}. \tag{4}$$

Proof. We choose an arbitrary ideal $\mathfrak{B}$ in C^1. Then for $\mathfrak{A} \in C$ with $N(\mathfrak{A}) \le t$ the product $\mathfrak{A}\mathfrak{B}$ is a principal ideal (α) with $\alpha \in \mathfrak{B} - \{0\}$ and

$$|N(\alpha)| = N(\mathfrak{A})N(\mathfrak{B}) \le tN(\mathfrak{B}). \tag{5}$$

Conversely, the ideal $\mathfrak{A} = (\alpha)\mathfrak{B}^{-1}$ with $N(\mathfrak{A}) \le t$ belongs to (α) with $\alpha \in \mathfrak{B} - \{0\}$ and $|N(\alpha)| \le tN(\mathfrak{B})$. Thus we have

$$T(C,t) = [\{(\alpha) | \alpha \in \mathfrak{B} - \{0\}, |N(\alpha)| \le tN(\mathfrak{B})\}].$$

We now want to distinguish a generator in each principal ideal. To do this we fix a system $\zeta, \varepsilon_1, \ldots, \varepsilon_{r-1}$ of generators of E according to Theorem 2, Chapter 19, and for $\alpha \in K^\times$ we consider the system of equations

$$l_i(\varepsilon_1)x_1 + \ldots + l_i(\varepsilon_{r-1})x_{r-1} + l_i x_r = l_i(\alpha), \quad i = 1,\ldots,r. \tag{6}$$

The determinant of this system is equal to $nR \neq 0$. Hence for each $\alpha \in K^{\times}$ there is a unique solution $x_1(\alpha),\dots,x_r(\alpha)$. By addition of the equations (6) one obtains

$$nx_r(\alpha) = \log|N(\alpha)|. \tag{7}$$

For two numbers $\alpha,\beta \in K^{\times}$,

$$x_i(\alpha\beta) = x_i(\alpha) + x_i(\beta), \quad i = 1,\dots,r. \tag{8}$$

Since, obviously, $x_i(\varepsilon_j) = \delta_{ij}$ for $i,j = 1,\dots,r-1$ there is a number α' in the set αE with

$$0 \leq x_i(\alpha') < 1 \quad \text{for} \quad i = 1,\dots,r-1. \tag{9}$$

The number α' is called *reduced* (relative to $\varepsilon_1,\dots,\varepsilon_{r-1}$). By Section 19.5, α' is uniquely determined by (9) up to multiplication by a power of ζ.

Let $\omega_1,\dots,\omega_n$ be a basis of $\mathfrak{B}$. Then $wT(C,t)$ equals the number of n-tuples $(a_1,\dots,a_n) \in \mathbb{Z}^n$ such that $\alpha' = a_1\omega_1 + \dots + a_n\omega_n$ satisfies condition (9) and the condition

$$|N(\alpha')| \leq tN(\mathfrak{B}). \tag{10}$$

20.3 Reduction to a volume calculation[1)]

We now want to interpret the number $wT(C,t)$ as the number of points of the lattice $\mathbb{Z}^n$ in $\mathbb{R}^n$ inside a region $V_t \subset \mathbb{R}^n$. To do this we introduce the r functions

$$f_j(\xi) := l_j\log|\xi_1 g_j\omega_2 + \dots + \xi_n g_j\omega_n|, \quad j = 1,\dots,r,$$

of $\xi = (\xi_1,\dots,\xi_n) \in \mathbb{R}^n$. We define V_t to be the set of all points $\xi \in \mathbb{R}^n$ which satisfy the conditions

$$f_j(\xi) \neq \infty \quad \text{for} \quad j = 1,\dots,r,$$

[1)] We recommend that the reader first try to understand the content of Sections 20.3 and 20.4 in the case $n = 2$.

$$\sum_{j=1}^{r} f_j(\xi) \leq \log t + \log N(\mathfrak{B}) \tag{11}$$

and

$$0 \leq x_i(\xi) < 1 \quad \text{for} \quad i = 1,\ldots,r\text{-}1, \tag{12}$$

where

$$x_1 = x_1(\xi),\ldots,x_{r-1} = x_{r-1}(\xi),\ x_r = \frac{1}{n}\sum_{j=1}^{r} f_j(\xi) \tag{13}$$

is the unique solution of the system (6) of linear equations with $l_j(\alpha)$ replaced by $f_j(\xi)$.

Then $wT(C,t)$ equals the number of lattice points in V_t.

Finally we define the positive number δ by $\delta^{-n} = tN(\mathfrak{B})$ and consider the linear mapping $\xi \mapsto \delta\xi$ of $\mathbb{R}^n$ onto itself. With $\eta = \delta\xi$ we get $f_j(\eta) = f_j(\xi) + l_j \log \delta$ and $x_i(\eta) = x_i(\xi)$.

V_t becomes the region V consisting of all $\eta \in \mathbb{R}^n$ with $f_j(\eta) \neq \infty$ for $j = 1,\ldots,r$, $\sum_{j=1}^{r} f_j(\eta) \leq 0$ and

$$0 \leq x_i(\eta) < 1 \quad \text{for} \quad i = 1,\ldots,r\text{-}1,$$

where

$$x_1 = x_1(\eta),\ldots,x_{r-1} = x_{r-1}(\eta),\ x_r = \frac{1}{n}\sum_{j=1}^{r} f_j(\eta)$$

is the unique solution of the system of equations

$$l_j(\varepsilon_1)x_1 + \ldots + l_j(\varepsilon_{r-1})x_{r-1} + l_j x_r = f_j(\eta), \quad j = 1,\ldots,r.$$

$\mathbb{Z}^n$ becomes the lattice $\delta\mathbb{Z}^n$. The conditions for V do not depend on δ.

The number $wT(C,t)$ is now the number of points of $\delta\mathbb{Z}^n$ which lie in V. Since $\delta \to 0$ as $t \to \infty$, by counting these lattice points we can determine the volume $I(V)$ of V, assuming V has a volume, which we shall show to be the case in the next section. Thus we have

$$I(V) = \lim_{\delta\to 0} wT(C,t)\delta^n = \lim_{t\to\infty} \frac{wT(C,t)}{N(\mathfrak{B})t},$$

since δ^n is the volume of a fundamental cell of the lattice $\delta\mathbb{Z}^n$.

20.4 Volume calculation

To prove Theorem 3 it now suffices to show that

$$I(V) = \frac{2^r \pi^{n-r} R}{N(\mathfrak{B})\sqrt{|D|}}. \tag{14}$$

This is done by passing to new coordinates.

First we perform the linear transformations

$$\zeta_i = \sum_{j=1}^{n} \eta_j g_i \omega_j \quad \text{for} \quad i = 1,\ldots,r_1,$$

$$\zeta_i = \sum_{j=1}^{n} \eta_j \mathrm{Re}\, g_i \omega_j,\ \zeta_{i+r_2} = \sum_{j=1}^{n} \eta_j \mathrm{Im}\, g_{i+r_2} \omega_j \ \text{for}\ i = r_1 + 1,\ldots,r_1+r_2.$$

Then by Chapter 17, (6) and Chapter 18, (3)

$$I(V) = \int_V d\eta_1 \ldots d\eta_n = \int_{V_1} \frac{2^{r_2} d\zeta_1 \ldots d\zeta_n}{N(\mathfrak{B})\sqrt{|D|}}. \tag{15}$$

Here V_1 consists of all $\zeta = (\zeta_1,\ldots,\zeta_n)$ with $\zeta_j \neq 0$ for $j = 1,\ldots,r_1$, $\zeta_j^2 + \zeta_{j+r_2}^2 \neq 0$ for $j = r_1 + 1,\ldots,r$,

$$\sum_{j=1}^{r_1} \log|\zeta_j| + \sum_{j=r_1+1}^{r} \log(\zeta_j^2 + \zeta_{j+r_2}^2) \leq 0$$

and

$$0 \leq x_i(\zeta) < 1 \quad \text{for} \quad i = 1,\ldots,r\text{-}1$$

where

$$x_1 = x_1(\zeta),\ldots,x_{r-1} = x_{r-1}(\zeta),\ x_r = \frac{1}{n}\left[\sum_{j=1}^{r_1} \log|\zeta_j| + \sum_{j=r_1+1}^{r} \log(\zeta_j^2 + \zeta_{j+r_2}^2)\right]$$

is the unique solution of the system of equations

$$l_j(\varepsilon_1)x_1 + \dots + l_j(\varepsilon_{r-1})x_{r-1} + l_j x_r = \begin{cases} \log|\zeta_j|, & j = 1,\dots,r_1 \\ \log(\zeta_j^2+\zeta_{j+r_2}^2), & j = r_1+1,\dots,r. \end{cases}$$

V_1 is symmetric under the reflections

$$\sigma_j : \zeta_j \mapsto -\zeta_j,\ \zeta_i \mapsto \zeta_i \quad \text{for } i \neq j.$$

Hence if V_2 is the subregion of V_1 with the additional conditions $\zeta_j \geq 0$ for $j = 1,\dots,r_1$ we obviously have

$$\int_{V_1} d\zeta_1 \dots d\zeta_n = 2^{r_1} \int_{V_2} d\zeta_1 \dots d\zeta_n. \tag{16}$$

We now introduce the coordinate transformation

$$\zeta_j = \exp(\theta_j) \quad \text{for } j = 1,\dots,r_1$$

$$\zeta_j = \exp(\theta_j/2)\cos\theta_{j+r_2},\ \zeta_{j+r_2} = \exp(\theta_j/2)\sin\theta_{j+r_2} \quad \text{for } j = r_1+1,\dots,r,$$

with $0 \leq \theta_{j+r_2} < 2\pi$ for $j = r_1+1,\dots,r$. Then

$$\int_{V_2} d\zeta_1 \dots d\zeta_n = \frac{1}{2^{r_2}} \int_{V_3} \exp(\theta_1 + \dots + \theta_r) d\theta_1 \dots d\theta_n. \tag{17}$$

Here V_3 consists of all $\theta = (\theta_1,\dots,\theta_n)$ with

$$\sum_{j=1}^{r} \theta_j \leq 0, \tag{18}$$

$0 \leq \theta_j < 2\pi$ for $j = r+1,\dots,n$ and

$$0 \leq x_i(\theta) < 1 \quad \text{for } i = 1,\dots,r-1, \tag{19}$$

where

$$x_1 = x_1(\theta),\dots,x_{r-1} = x_{r-1}(\theta),\ x_r = \frac{1}{n}\sum_{j=1}^{r} \theta_j$$

is the unique solution of the system of equations

$$l_j(\varepsilon_1)x_1 + \dots + l_j(\varepsilon_{r-1})x_{r-1} + l_j x_r = \theta_j,\ j = 1,\dots,r. \tag{20}$$

The integrals over $d\theta_{r+1},\dots,d\theta_n$ may now be split off. We obtain

$$\int_{V_2} d\zeta_1\dots d\zeta_n = \pi^{r_2}\int_{V_4} \exp(\theta_1+\dots+\theta_r)d\theta_1\dots d\theta_r, \tag{21}$$

where V_4 consists of the $\theta = (\theta_1,\dots,\theta_r)$ which satisfy conditions (18) and (19).

Finally we go to the coordinates $x_1,\dots,x_r$ with the help of (20). The absolute value of the determinant of (20) is Rn. Therefore

$$\int_{V_3} \exp(\theta_1+\dots+\theta_r)d\theta_1\dots d\theta_r = Rn\int \exp(nx_r)dx_1\dots dx_r, \tag{22}$$

with $0 \le x_i < 1$ for $i = 1,\dots,r-1$, $x_r \le 0$, hence

$$\int \exp(nx)dx_1\dots dx_r = \int_{-\infty}^{0} \exp(nx_r)dx_r = \frac{1}{n}. \tag{23}$$

The value (14) now follows from (15) to (23), hence Theorem 3 is proved.□

20.5 Proof of Theorem 2

As in Section 6.3 we write the partial sums of $\xi(s)$ in the form

$$\sum_{m=1}^{h} a_m m^{-s} = T(1)(1^{-s}-2^{-s}) + T(2)(2^{-s}-3^{-s}) + \dots$$
$$+ T(h-1)((h-1)^{-s}-h^{-s}) + \frac{T(h)}{h}h^{1-s}.$$

By hypothesis

$$\lim_{h\to\infty} \frac{T(h)}{h}h^{1-s} = 0 \quad \text{for} \quad s > 1,$$

hence

$$\xi(s) = \sum_{h=1}^{\infty} T(h)(h^{-s}-(h+1)^{-s}).$$

Since

$$h^{-s} - (h+1)^{-s} = s\int_h^{h+1} \frac{dx}{x^{s+1}}$$

we also get

$$\xi(s) = s\int_1^{\infty} \frac{T(t)dt}{t^{s+1}}. \tag{24}$$

Since $\frac{T(t)}{t}$ is bounded by hypothesis and $\int_1^{\infty} \frac{dt}{t^s} = \frac{1}{s-1}$, the integral in (24) converges. This proves the first part of Theorem 2.

In order to carry out the passage to the limit as $s \to 1$ we write $\xi(s)$ in the form

$$\xi(s) = \left(\frac{s}{s-1}\right)a + s\int_1^{\infty} \left(\frac{T(t)}{t} - a\right)\frac{dt}{t^s}. \tag{25}$$

This means (3) is equivalent to

$$\lim_{s\to 1}(s-1)\int_1^{\infty} \left(\frac{T(t)}{t} - a\right)\frac{dt}{t^s} = 0.$$

To prove the latter, we choose, for a given $\varepsilon > 0$, a number u in the interval of integration such that

$$\left|\frac{T(t)}{t} - a\right| \le \varepsilon \quad \text{for} \quad t \ge u.$$

Then

$$\left|(s-1)\int_u^{\infty} \left(\frac{T(t)}{t} - a\right)\frac{dt}{t^s}\right| \le (s-1)\varepsilon\int_u^{\infty} \frac{dt}{t^s} \le \varepsilon.$$

Let c be a number with

$$\left|\frac{T(t)}{t} - a\right| \le c \quad \text{for} \quad 1 \le t \le u.$$

Then

$$\left|(s-1)\int_1^{u} \left(\frac{T(t)}{t} - a\right)\frac{dt}{t^s}\right| \le c\left(1 - \frac{1}{u^{s-1}}\right).$$

Keeping u fixed, we now choose s so near to 1 that

$$c\left(1 - \frac{1}{u^{s-1}}\right) \le \varepsilon.$$

Altogether we have

$$\left|(s-1)\int_1^{\infty}\left[\frac{T(t)}{t}a\right]\frac{dt}{t^s}\right| \leq 2\varepsilon. \qquad \square$$

20.6 Application

As in the case $K = \mathbb{Q}$ (see Section 6.3) one shows that

$$\zeta_K(s) = \prod_{\mathfrak{P}} \frac{1}{1-N(\mathfrak{P})^{-s}} \quad \text{for} \quad s > 1,$$

where the product is taken over all prime ideals $\mathfrak{P}$ of K. It follows from (1) that

$$\sum_{\mathfrak{P}} \frac{1}{N(\mathfrak{P})} \tag{26}$$

is a divergent series. Since the prime ideals whose degree is different from 1 make a convergent contribution to (26), one gets

Theorem 4. *In each number field* K *there are infinitely many prime ideals of first degree.* $\square$

In Section 25.4 we shall use (1) in order to supply the proof of Theorem 4, Chapter 6 which is still outstanding.

Exercises

20.1 Show that the Dedekind ζ-function, as a function of the complex variable s, admits continuation to a regular function $\zeta_K(s)$ for $\mathrm{Re}\, s > 1$.

20.2 Let K be a quadratic field with discriminant D.

a) Show that there is a uniquely determined quadratic character χ modulo D (cf. Section 6.3) with the following properties:

$\chi(p) = 1$ for all prime numbers p which split into two distinct prime factors in $K : (p) = \mathfrak{P}\mathfrak{P}'$;

$\chi(p) = -1$ for all prime numbers which are inert in K:
$(p) = \mathfrak{P}$;

$\chi(p) = 0$ for all prime numbers p which ramify in K:
$(p) = \mathfrak{P}^2$ (cf. Section 18.9).

χ is called the character associated with K.

b) Show that χ is a primitive character (Exercise 6.7).

c) Show that for real $s > 1$

$$\zeta_K(s) = \zeta(s)L(s,\chi)$$

and

$$\lim_{s\to 1}(s-1)\zeta_K(s) = L(1,\chi).$$

d) Show that the class number h of K satisfies

$$h = \begin{cases} \dfrac{\sqrt{D}}{2\log\varepsilon}L(1,\chi) & \text{for } D > 0 \\ \dfrac{w\sqrt{|D|}}{2\pi}L(1,\chi) & \text{for } D < 0, \end{cases}$$

where $\varepsilon > 1$ is the fundamental unit of K and w is the number of roots of unity in K. For $D = -4$ one obtains the Leibniz series (Section 5.6).

20.3 The Gauss sum $\tau_a(\chi)$ was defined in Exercise 6.3.

a) Show that the character χ of the quadratic field K with discriminant D satisfies the equation $\tau_1(\chi)^2 = D$.

More precisely, $\tau_1(\chi) = \sqrt{D}$ for $D > 0$ and $\tau_1(\chi) = i\sqrt{|D|}$ for $D < 0$ (cf. Section 3.9).

b) Use Exercise 6.7 to show

$$h = -\frac{1}{\log\varepsilon}\sum_{m=1}^{[D/2]}\chi(m)\log\sin\frac{\pi m}{D} \quad \text{for } D > 0$$

and

$$h = -\frac{1}{|D|}\sum_{m=1}^{|D|}\chi(m)m \quad \text{for } D < -4.$$

21. Quadratic forms and quadratic fields

21.1 Modules in quadratic fields

Dedekind realised that one can give a clear presentation of Gauss's theory of quadratic forms (Chapter 2), and particularly the composition of classes of forms, when one goes from forms to modules in quadratic fields. Here we consider the modules first and then give the connection between modules and forms.

Let $K = \mathbb{Q}(\sqrt{d})$ be a quadratic field (notation as in Section 18.3). In what follows a module is always a subgroup of K^+ generated by a basis of K over $\mathbb{Q}$. For $\alpha \in K$, α' denotes the conjugate of α.

Theorem 1. *Let* $\mathfrak{m}$ *be a module in* K. *There is a positive rational number* m *such that* $\mathfrak{m}$ *has the form* $\mathfrak{m} = (m, m\gamma)$ *with* $\gamma \in K$. *The number* m *is uniquely determined by* $\mathfrak{m}$.

Proof. Let $\mathfrak{m} = (\alpha, \beta)$ and take $x, y \in \mathbb{Q}$ with

$$1 = x\alpha + y\beta. \tag{1}$$

There is a positive rational number m such that mx, my are integral and relatively prime.

We set $mx =: u$ and $my =: v$. Let r, s be rational numbers with $us - vr = 1$ and let $\gamma := \frac{r\alpha + s\beta}{m}$. Then $m, m\gamma \in \mathfrak{m}$:

$$m = u\alpha + v\beta, \qquad m\gamma = r\alpha + s\beta. \tag{2}$$

If we view (2) as a system of equations for α and β, then we see that α and β are integral linear combinations of m and $m\gamma$, because the associated determinant $us - vr$ is equal to 1.

The rational numbers in $\mathfrak{m}$ are the integer multiples of m. Hence m is uniquely determined by $\mathfrak{m}$. □

The number γ is a zero of an irreducible quadratic polynomial:

$$a\gamma^2 + b\gamma + c = 0. \tag{3}$$

The coefficients a,b,c are uniquely determined when we assume that they have no common divisor and that a is positive.

The number $a\gamma$ is an algebraic integer and hence, by Section 18.3, of the form

$$a\gamma = h + k\omega = \frac{-b+k\sqrt{D}}{2}, \tag{4}$$

where h and k are integers, D is the discriminant of O_K and

$$k^2D = b^2 - 4ac. \tag{5}$$

By passing from γ to $-\gamma$, if necessary, one arranges that $k > 0$. A generator of K with $k > 0$ is called *admissible.*

In Section 18.1 we have defined the order $O(\mathfrak{m})$ as the ring of all numbers $\delta \in K$ with $\delta\mathfrak{m} \subset \mathfrak{m}$. We have

Theorem 2. *The order of* $\mathfrak{m}$ *is* $(1,a\gamma) = (1,k\omega)$.

Remark. Thus by Section 18.3 the number k equals the conductor of $O(\mathfrak{m})$, for admissible γ.

Proof of Theorem 2. Since, for $\alpha \neq 0$, the modules $\mathfrak{m}$ and $\alpha\mathfrak{m}$ have the same order, it suffices to consider $\mathfrak{m}(1,\gamma)$. Now $\delta \in O(\mathfrak{m})$ means that δ and $\delta\gamma$ lie in $\mathfrak{m}$. Therefore δ has the form $\delta = x + y\gamma$ with $x,y \in \mathbb{Z}$ and $y\gamma^2 \in \mathfrak{m}$. By (3), $y\gamma^2 = ya^{-1}(b\gamma-c)$. Consequently yb and yc are divisible by a. Since a,b,c are relatively prime, $a|y$. On the other hand, obviously $a\gamma \in O(\mathfrak{m})$ and hence $O(\mathfrak{m}) = (1,a\gamma)$. □

We carry over the concept of norm of an ideal (Section 18.7) to an arbitrary module $\mathfrak{m} = (\alpha,\beta)$: let $O(\mathfrak{m}) = (\alpha_1,\beta_1)$ and let A be the matrix which transforms α_1,β_1 to α,β. Then $|\det \mathbf{A}|$ is independent of the choice of bases α,β in $\mathfrak{m}$ and α_1,β_1 in $O(\mathfrak{m})$. We set

$$N(\mathfrak{m}) := |\det \mathbf{A}|.$$

By Theorems 1 and 2

$$N(\mathfrak{m}) = \frac{m^2}{a}. \tag{6}$$

Now we set

$$\mathfrak{m}' := \{\alpha' \mid \alpha \in \mathfrak{m}\}.$$

Then $\mathfrak{m}'$ has the same order and the same norm as $\mathfrak{m}$.

Now we consider the multiplication of modules (Section 18.1). Let R be an order in K and let $\mathfrak{m}$ be a module with $O(\mathfrak{m}) = R$. Then $\mathfrak{m}R = \mathfrak{m}$.

Theorem 3. $\mathfrak{m}\mathfrak{m}' = N(\mathfrak{m})R$.

Proof. By Theorem 2 and (6)

$$\mathfrak{m}\mathfrak{m}' = (m, m\gamma') = m^2(1,\gamma,\gamma',\gamma\gamma') = \frac{m^2}{a}(a, a\gamma, b, c) = N(\mathfrak{m})R,$$

since the greatest common divisor of a,b,c is 1. □

Now let $\mathfrak{m}$ and $\mathfrak{m}_1$ be modules with the same order R. Obviously

$$(\mathfrak{m}\mathfrak{m}_1)' = \mathfrak{m}'\mathfrak{m}_1'$$

and therefore

$$\mathfrak{m}\mathfrak{m}_1(\mathfrak{m}\mathfrak{m}_1)' = N(\mathfrak{m}\mathfrak{m}_1)O(\mathfrak{m}\mathfrak{m}_1) = \mathfrak{m}\mathfrak{m}'\mathfrak{m}_1\mathfrak{m}_1' = N(\mathfrak{m})N(\mathfrak{m}_1)R. \tag{7}$$

This implies

$$O(\mathfrak{m}\mathfrak{m}_1) = R \tag{8}$$

and

$$N(\mathfrak{m}\mathfrak{m}_1) = N(\mathfrak{m})N(\mathfrak{m}_1). \tag{9}$$

Theorem 4. *The collection* $\mathfrak{M}(R)$ *of modules with a given order* R *constitutes a group under multiplication.*

Proof. Equation (8) says that $\mathfrak{M}(R)$ is closed under multiplication. By Theorem 3, $(1/N(\mathfrak{m}))\mathfrak{m}'$ is the inverse of $\mathfrak{m}$. □

Two modules $\mathfrak{m}_1, \mathfrak{m}_2$ from $\mathfrak{M}(R)$ are called *similar in the strict sense* when there is an $\alpha \in K$ with $\mathfrak{m}_2 = \alpha\mathfrak{m}_1$ and $N(\alpha) > 0$.

We denote the group of all $\alpha \in K^\times$ with $N(\alpha) > 0$ by A. For $D < 0$, $A = K^\times$; for $D > 0$, A is a subgroup of index 2 in $K^\times$. The class of modules which are similar in the strict sense are the elements of the quotient group $\mathfrak{M}(R)/\{\alpha R \mid \alpha \in A\}$. The group $R^\times$ of units of R has finite index in $E = O_K^\times$ (Chapter 19, Theorem 2, A1, Theorem 4), which we denote by e_f. Let n_f be the index of $AR^\times$ in $K^\times$. One easily sees that

$n_f = 1$ for $D < 0$,
$n_f = 1$ for $D > 0$ when there is an $\varepsilon \in R^\times$ with $N(\varepsilon) = -1$,
$n_f = 2$ for $D > 0$ when $N(\varepsilon) = 1$ for all $\varepsilon \in R^\times$.

Also let

$$Cl(R) = \mathfrak{M}(R)/\{\alpha R \mid \alpha \in K^\times\},\ Cl_0(R) = \mathfrak{M}(R)/\{\alpha R \mid \alpha \in A\}.$$

$Cl_0(R)$ is the group of interest for the theory of quadratic forms. In the comparison of classes of modules and forms the following lemma plays a leading rôle.

Lemma 1. *Let* γ *and* γ_1 *be admissible numbers from* K. *The modules* $(1,\gamma)$ *and* $(1,\gamma_1)$ *are similar in the strict sense if and only if there is a matrix* $A := \begin{pmatrix} k & l \\ m & n \end{pmatrix}$ *with integer coefficients and* $\det A = 1$ *such that*

$$\gamma_1 = \frac{k\gamma + l}{m\gamma + n}. \tag{10}$$

Proof. Let $(1,\gamma)$ and $(1,\gamma_1)$ be similar in the strict sense. Then there is an $\alpha \in K^\times$ with $N(\alpha) > 0$ and a matrix $A = \begin{pmatrix} k & l \\ m & n \end{pmatrix}$ with $\det A = \pm 1$ such that

$$\begin{pmatrix} \gamma_1 \\ 1 \end{pmatrix} = A \begin{pmatrix} a\gamma \\ a \end{pmatrix}. \tag{11}$$

It follows that

$$\begin{pmatrix} \gamma_1 & \gamma_1' \\ 1 & 1 \end{pmatrix} = A \begin{pmatrix} \alpha\gamma & \alpha'\gamma' \\ \alpha & \alpha' \end{pmatrix}$$

and

$$\gamma_1 - \gamma_1' = (\det A) N(\alpha)(\gamma - \gamma'). \tag{12}$$

Since γ and γ_1 are special, it follows that $\det A = 1$. Also, the desired relation (10) follows from (11).

Conversely, suppose (10) is satisfied. Then (11) holds with $\alpha = \frac{1}{m\gamma+n}$. It follows from (12) that $N(\alpha) > 0$. □

21.2 Comparison with the ideal group

We want to compare the structure of $\mathfrak{M}(R)$ with that already known for $\mathfrak{M}(O_K)$.

The correspondence $\mathfrak{m} \mapsto \mathfrak{m}O_K$ defines a homomorphism ϕ of $\mathfrak{M}(R)$ into $\mathfrak{m}(O_K)$.

Theorem 5. $\operatorname{Im} \phi = \mathfrak{M}(O_K)$.

Proof. Since each principal ideal lies in the image of ϕ, it suffices to show that all prime ideals of O_K with degree of inertia 1 lie in the image of ϕ. Let $\mathfrak{P}$ be such a prime ideal and let f be the conductor of the order R. Let $f = p^i f_1$ with $p = N(\mathfrak{P})$, $p \nmid f_1$. Then $\mathfrak{P}$ may be generated, as an O_K-ideal, by p^{i+1} and an element $\alpha \in O_K$ of the form $\alpha = x + f_1\omega$ with $x \in \mathbb{Z}$, where $\mathfrak{P}$ goes into α exactly to the first power (Chapter 18, Theorem 21). By Theorem 2 the module (p^{i+1},α) has order $R = (1,f\omega)$ and $\phi(p^{i+1},\alpha) = \mathfrak{P}$. □

Theorem 6. *The kernel of* ϕ *consists of all modules* $\mathfrak{m}$ *of the form* $\mathfrak{m} = (f,f\omega,\xi)$, *where* f *is the conductor of* R *and* ξ *is an arbitrary element of* O_K *relatively prime to* f.

We first give a characterisation of the conductor of R.

Lemma 2. Let $\mathfrak{m}$ be a module in $\mathfrak{M}(R)$ with $\mathfrak{m}O_K = O_K$. Then the conductor of R is the smallest natural number f with $fO_K \subset \mathfrak{m}$.

Proof. By definition, the conductor of R is the smallest natural number h with $h\omega\mathfrak{m} \subset \mathfrak{m}$. Since $h\mathfrak{m} \subset \mathfrak{m}$ and $\mathfrak{m}O_K = O_K$, this condition is equivalent to $hO_K \subset \mathfrak{m}$. □

Proof of Theorem 6. Let $\mathfrak{m}$ be a module from $\mathfrak{M}(R)$ with $\phi\mathfrak{m} = O_K$. Then $\mathfrak{m}O_K = O_K$. Hence by Lemma 2 the conductor of R equals the smallest natural number f with

$fO_K \subset \mathfrak{m}$. It follows that the group $\mathfrak{m}/fO_K$ is cyclic. Let ξ represent a generator of this group in $\mathfrak{m}$. Then $\mathfrak{m} = (f, f\omega, \xi)$ and ξ is relatively prime to f.

On the other hand, one easily shows that each module of the form $(f, f\omega, \xi)$, with an $\xi \in O_K$ which is relatively prime to f, belongs to Ker ϕ. □

Obviously $(f, f\omega, \xi_1) = (f, f\omega, \xi_2)$ for $\xi_1 \equiv \xi_2 \pmod f$. Moreover, one easily sees that ξ_1, ξ_2 relatively prime to f satisfy the relation

$$(f, f\omega, \xi_1)(f, f\omega, \xi_2) = (f, f\omega, \xi_1\xi_2).$$

The correspondence $\xi \mapsto (f, f\omega, \xi)$ therefore defines an epimorphism ψ of $(O_K/fO_L)^\times$ onto Ker ϕ.

Theorem 7. *The kernel of* ψ *consists of the classes of* (O_K/fO_K) *which have representatives in* $\mathbb{Z}$.

Proof. The class ξ belongs to Ker ψ if and only if $(f, f\omega, \xi) = R$. That means that there is a $b \in \mathbb{Z}$ with $1 - b\xi \in fO_K$. □

We can summarise the foregoing results on the structure of $\mathfrak{M}(R)$ in the following exact sequence:

$$\{1\} \to (\mathbb{Z}/f\mathbb{Z})^\times \to (O_K/fO_K)^\times \to \mathfrak{M}(R) \underset{\phi}{\to} \mathfrak{M}(O_K) \to \{1\}, \tag{13}$$

We also want to express the order of $Cl_0(R)$ in terms of the order of $Cl(O_K)$, i.e. in terms of the ideal class number h of K.

Theorem 8. *Let* $\Phi(f) = [(O_K/fO_K)^\times]$, $\phi(f) = [(\mathbb{Z}/f\mathbb{Z})^\times]$. *Then*

$$[Cl_0(R)] = \frac{n_f \Phi(f)}{e_f \phi(f)} h. \tag{14}$$

Proof. First, one sees easily that

$$[Cl_0(R)] = n_f [Cl(R)]. \tag{15}$$

Next we consider the natural homomorphism $\overline{\phi}$ of $Cl(R)$ into $Cl(O_K)$. By Theorem 5, $\overline{\phi}$ is surjective, and hence

$$[Cl(R)] = h[\mathrm{Ker}\ \overline{\phi}]. \tag{16}$$

From the diagram

$$\begin{array}{ccccc} E/R^{\times} & \to & K^{\times}/R^{\times} & \to & K^{\times}/E \\ & & \downarrow & & \downarrow \\ \mathrm{Ker}\ \phi & \to & \mathfrak{M}(R) & \underset{\phi}{\to} & \mathfrak{M}(O_K) \\ & & \downarrow & & \downarrow \\ \mathrm{Ker}\ \overline{\phi} & \to & Cl(R) & \underset{\overline{\phi}}{\to} & Cl(O_K) \end{array}$$

one derives the exact sequence

$$\{1\} \to E/R^{\times} \to \mathrm{Ker}\ \phi \to \mathrm{Ker}\ \overline{\phi} \to \{1\},$$

from which it follows that

$$[\mathrm{Ker}\ \overline{\phi}] = [\mathrm{Ker}\ \phi]/e_f. \tag{17}$$

One obtains

$$[\mathrm{Ker}\ \phi] = \Phi(f)/\phi(f) \tag{18}$$

from (13), and (14) now follows from (15) to (18). □

21.3 Forms and modules

We now come to the crux of this chapter, the connection between quadratic forms and modules.

We confine ourselves once again to primitive forms (a,b,c) (Section 2.5). In the case $\sigma = 2$ we go from the form $ax^2 + 2bxy + cy^2$ to the form $(a/2)x^2 + bxy + (c/2)y^2$, whose coefficients are relatively prime. By a *primitive form* we now mean an indecomposable form $ax^2 + bxy + cy^2$ whose coefficients a,b,c are relatively prime integers. We denote this form by [a,b,c]. As discriminant of this form we now take the number $b^2 - 4ac$. The form decomposes (into the product of two linear forms) if and only if its discriminant is a square (Chapter 2, Theorem 1). In addition, we assume that $a > 0$.

Let γ be an admissible number of the quadratic field K. We associate γ, on the one hand, with the class of the module $(1,\gamma)$ in $Cl_0(R)$, where R is the order of $(1,\gamma)$, and on the other hand with the class of property equivalent forms which includes the

$$a(x-\gamma y)(x-\gamma' y) = ax^2 + bxy + cy^2. \tag{19}$$

Here the coefficients a,b,c are determined by (3), so that $a > 0$ and a,b,c are relatively prime, i.e. the form (19) is primitive. It follows easily from Lemma 1 that this gives a one-to-one correspondence between the elements of $Cl_0(R)$ and the classes of primitive forms which include forms [a,b,c] with $a > 0$ whose discriminant equals the discriminant $\Delta(1,a\gamma)$ of R (Theorem 2). We call this correspondence the *canonical correspondence* between classes of modules and classes of forms.

Multiplication of classes in $Cl_0(R)$ corresponds, under the canonical correspondence, to multiplication of classes of forms. This is Gauss's composition of forms. The reader will easily verify that the special cases of composition given in Section 2.9 indeed correspond to multiplication in $Cl_0(R)$.

Exercises

21.1 Let R be an order in an imaginary quadratic field, let $D < 0$ be the discriminant of R, and let $(1,\gamma)$ be a module in $\mathfrak{M}(R)$. The number γ and the module $(1,\gamma)$ are called *reduced* when the following holds: $\operatorname{Im}\gamma > 0$, $-\frac{1}{2} < \operatorname{Re}\gamma \le \frac{1}{2}$, $|\gamma| > 1$ in case $-\frac{1}{2} < \operatorname{Re}\gamma < 0$, and $|\gamma| \ge 1$ in case $0 \le \operatorname{Re}\gamma \le \frac{1}{2}$. Use Proposition 4, Chapter 13 to show that in each class of similar modules $(1,\gamma)$ there is exactly one reduced module (cf. also Section 13.4).

21.2 With the notation of (3), let $\gamma = (-b + i\sqrt{|D|})/2a$. Show that γ is reduced if and only if the following conditions are satisfied: $-a \le b < a$, $c \ge a$ in case $b \le 0$, and $c > a$ in case $b > 0$.

21.3 Compare reduced modules with reduced forms (Section 2.6).

21.4 Compute the class numbers of the imaginary quadratic fields with discriminant $D \ge -47$.

21.5 Let $F_1 = [a_1,b_1,c_1]$ and $F_2 = [a_2,b_2,c_2]$ be primitive forms with the same discriminant and suppose $a_1 > 0$, $a_2 > 0$. Suppose that F_1 represents the number n_1 and F_2 represents the number n_2. Show that the forms from the class which results from composition of the classes of F_1 and F_2 represent the number n_1n_2.

22. The different and the discriminant

22.1 Relative extensions

Let K be an algebraic number field of degree n over $\mathbb{Q}$. A prime number p is called *unramified* in K when all ramification indices $e_1,\ldots,e_g$ equal 1 in the prime ideal decomposition $pO_K = \mathfrak{P}_1^{e_1}\ldots\mathfrak{P}_g^{e_g}$ (Section 18.7), otherwise ramified. The motivation for this terminology comes from the theory of function fields, which we take up in Chapter 23.

It is an important result of Dedekind that there are always only finitely many primes which are ramified in K. A more precise description of this situation was given by Dedekind in his work "*Über die Diskriminanten endlicher Körper*" (*Abh. Königl. Gesellsch. Wiss. Göttingen* 29 (1882)), developing the theory of differents and discriminants which we present in this chapter.

In the introduction to this work, Dedekind says that all theorems remain valid, with easy modifications, when one proceeds from an arbitrary ground field K in place of $\mathbb{Q}$. He promises to carry out this *important generalisation* in a later work. However, he never came back to it. In fact, the *relative theory* was first presented by Hilbert (*Grundzüge einer Theorie des Galoisschen Zahlkörpers, Nachr. Königl. Gesellsch. Wiss. Göttingen* 1894).

As we have already done in Chapter 18, we draw function fields in one indeterminate into our considerations, and hence proceed from the following assumptions: Γ is a euclidean ring with field of fractions P of characteristic 0, K is a finite extension of P, and F is an intermediate field of K/P.

By definition, $O_F = O_K \cap F$ (Section 18.1). We associate with a fractional ideal $\mathfrak{a}$ of F the fractional ideal $\iota(\mathfrak{a}) := \mathfrak{a}O_K$ of K. In this way one obtains a homomorphism ι of the group $\mathcal{I}_F$ of fractional ideals of F into $\mathcal{I}_K$.

The homomorphism ι is a monomorphism. Because it follows from $\iota(\mathfrak{a}) = O_K$ that $\mathfrak{a}$ is integral. Also, $\iota(\mathfrak{a}^{-1}) = \iota(\mathfrak{a})^{-1} = O_K$. Hence $\mathfrak{a}^{-1}$ is also integral, i.e. $\mathfrak{a} = O_F$. In what follows we identify $\mathfrak{a}$ with $\iota(\mathfrak{a})$.

We shall now define the *relative norm* $N_{K/F}\mathfrak{A}$ of an ideal $\mathfrak{A}$ of O_F. This should be compatible with the relative norm $N_{K/F}\alpha$ of an element α of K:

$$N_{K/F}(\alpha O_K) = (N_{K/F}\alpha)O_F. \tag{1}$$

We proceed as follows: in accordance with Theorem 14, Chapter 18, we choose a $\beta \in \mathfrak{A}$ such that, for all prime ideals $\mathfrak{P}$ which divide $\mathfrak{a} = \mathfrak{A} \cap O_F$, β does not lie in $\mathfrak{A}\mathfrak{P}$. Let

$$(N_{K/F}\beta)O_F = \prod_{\mathfrak{p}} \mathfrak{p}^{b_{\mathfrak{p}}},$$

where the product is taken over all prime ideals $\mathfrak{p}$ of O_F. Then we define

$$N_{K/F}\mathfrak{A} = \prod_{\mathfrak{p} | \mathfrak{a}} \mathfrak{p}^{b_{\mathfrak{p}}}.$$

This definition is independent of the choice of β. For if β' is another element of $\mathfrak{A}$ with the desired properties, then β/β' is divisible by no prime ideal which divides $\mathfrak{a}$. Thus to prove the claim it suffices to show the following proposition.

Proposition 1. *Let* $\gamma \neq 0$ *be an element of* K *which is relatively prime to the ideal* $\mathfrak{a}$ *of* O_F. *Then* $N_{K/F}\gamma$ *is relatively prime to* $\mathfrak{a}$.

Proof. Let N be a normal extension of F which contains K (Section 17.2). The Galois group G of N/F acts in a natural way on $\mathcal{I}_N$: for $\tau \in G$ and $\mathfrak{A} \in \mathcal{I}_n$ we set $\tau\mathfrak{A} = \{\tau\alpha | \alpha \in \mathfrak{A}\}$. Then τ is an automorphism of the group $\mathcal{I}_N$ which leaves the subgroup $\mathcal{I}_F$ fixed. It follows that the conjugates of γ likewise have no prime divisor in common with $\mathfrak{a}$. Hence $N_{K/F}\gamma$ is also relatively prime to $\mathfrak{a}$. □

The definition of the relative norm extends in an obvious way to fractional ideals. Now one easily proves the basic properties of the relative norm, which are the following

a) $N_{K/F}$ is a homomorphism of $\mathcal{I}_K$ into $\mathcal{I}_F$ which sends $\mathfrak{a} \in \mathcal{I}_F$ to $\mathfrak{a}^n$.

b) With notation as in the proof of Proposition 1,

$$N_{K/F}\mathfrak{A} = \prod_{\tau} \tau\mathfrak{A} \quad \text{for } \mathfrak{A} \in \mathcal{I}_K, \tag{2}$$

where the product is taken over all conjugates τ of K/F.

c) For $F = P$, $N_{K/F}$ coincides with the absolute norm defined in Section 18.7.

d) For a tower of fields $F \subset L \subset K$, $N_{K/F} = N_{L/F}N_{K/L}$.

e) For all ideals; $\mathfrak{A}$, $\mathfrak{B}$ of O_K, $\mathfrak{A}/\mathfrak{B}$ implies $N_{K/F}\mathfrak{A} | N_{K/F}\mathfrak{B}$.

We now consider the relations between a prime ideal $\mathfrak{P}$ of O_K and the underlying prime ideal $\mathfrak{p} = \mathfrak{P} \cap O_F$ of O_F.

We define the *ramification index* $e = e_{K/F}(\mathfrak{P})$ to be the exponent of the highest power of $\mathfrak{P}$ which divides $\mathfrak{p}$. Also, let the *inertia degree* $f = f_{K/F}(\mathfrak{P})$ be defined by $N_{K/F}\mathfrak{P} = \mathfrak{p}^f$. (Since $\mathfrak{P} | \mathfrak{p}$ we have $N_{K/F}\mathfrak{P} | N_{K/F}\mathfrak{p} = \mathfrak{p}^n$, i.e. $N_{K/F}\mathfrak{P}$ is a power of $\mathfrak{p}$.)

For a tower of fields $F \subset L \subset K$ we have

$$e_{K/F}(\mathfrak{P})L = e_{K/L}(\mathfrak{P})e_{L/F}(\mathfrak{P}_L) \tag{3}$$

and

$$f_{K/F}(\mathfrak{P}) = f_{K/L}(\mathfrak{P})f_{L/F}(\mathfrak{P}_L), \tag{4}$$

where $\mathfrak{P}_L = \mathfrak{P} \cap O_L$ is the prime ideal of O_L underlying $\mathfrak{P}$.

It follows from (4), applied to the tower of fields $P \subset F \subset K$, and from Theorem 18, Chapter 18, that the field extension $O_K/\mathfrak{P}$ over $O_F/\mathfrak{p}$ has degree $f_{K/F}(\mathfrak{P})$.

The relations (12) and (13) from Section 18.7 now carry over easily to the field extension K/F.

22.2 Complementary modules

Let M be an O_F-module in K. We define the complementary module M^* of M by

$$M^* := \{\alpha \in K | Tr_{K/F}(\alpha M) \subset O_F\}.$$

(For a subset H of K, $Tr_{K/F}H$ denotes the set of traces (Section 17.6) of elements of H.)

M^* is again an O_F-module. If $\omega_1,\ldots,\omega_n$ is a basis of K/F and $\kappa_1,\ldots,\kappa_n$ is the associated complementary basis (Chapter 17 (11)), then the complementary module of $(\omega_1,\ldots,\omega_n)O_F$ is $(\kappa_1,\ldots,\kappa_n)O_F$. In particular we have

Theorem 1. *Let* $\alpha \in O_K$ *be a generator of the field extension* K/F. *Then the complementary module of* $O_F[\alpha]$ *equals* $D_{K/F}(\alpha)^{-1}O_F[\alpha]$.

Proof. In accordance with Theorem 10, Chapter 17, one has to show

$$(\beta_0,\ldots,\beta_{n-1})O_F = (1,\alpha,\ldots,\alpha^{n-1})O_F \tag{5}$$

with the $\beta_0,\ldots,\beta_{n-1}$ appearing there. By definition, the characteristic polynomial $\chi_\alpha(x)$ of α over K/F has coefficients in O_F. Hence one can easily compute that β_i is a polynomial of degree $n-i$ in α with coefficients from O_F and leading coefficient 1, for $i = 0,\ldots,n-1$. This implies (5). □

We now consider the case where M is in fact an O_K-module, i.e. M is in $\mathcal{I}_K$. Then M^* is also in $\mathcal{I}_K$. Moreover, we have

Theorem 2. (i) $(O_K^*)^{-1}$ *is an integral ideal.*
(ii) *Let* $\mathfrak{A} \in \mathcal{I}_K$. *Then* $\mathfrak{A}^* = O_K^*\mathfrak{A}^{-1}$, $\mathfrak{A}^{**} = \mathfrak{A}$.

Proof. Since O_K is obviously contained in O_K^*, we have $(O_K^*)^{-1} \subset O_K$, i.e. $(O_K^*)^{-1}$ is an integral ideal.

To prove (ii) we note that a $\beta \in K$ lies in $\mathfrak{A}^*$ if and only if $Tr_{K/F}(\beta\mathfrak{A}O_K) \subset O_F$. This is equivalent to $\beta\mathfrak{A} \subset O_K^*$, i.e. $\beta \in O_K^*\mathfrak{A}^{-1}$. Therefore $\mathfrak{A}^* = O_K^*\mathfrak{A}^{-1}$. If one replaces $\mathfrak{A}$ here by $\mathfrak{A}^*$, then one obtains $\mathfrak{A}^{**} = O_K^*\mathfrak{A}^{*-1} = \mathfrak{A}$. □

The ideal $(O_K^*)^{-1}$ is called the *different* of K/F and is denoted by $\mathfrak{D}_{K/F}$.

Theorem 3 *(Different tower theorem). If* $F \subset L \subset K$ *is a tower of fields, then*

$$\mathfrak{D}_{K/F} = \mathfrak{D}_{K/L}\mathfrak{D}_{L/F}. \tag{6}$$

Proof. Equation (6) is equivalent to

$$\mathfrak{D}_{K/F}^{-1} = \mathfrak{D}_{K/L}^{-1}\mathfrak{D}_{L/F}^{-1}, \tag{7}$$

which follows from the trace formula

$$\mathrm{Tr}_{K/F}(\beta) = \mathrm{Tr}_{L/F}(\mathrm{Tr}_{K/L}\beta) \quad \text{for} \quad \beta \in K, \tag{8}$$

which is an easy consequence of the principles of Galois theory (Chapter 17):

$$\begin{aligned} \mathfrak{D}_{K/F} &= \{\beta \mid \mathrm{Tr}_{K/F}(\beta O_K) \subset O_F\} \\ &= \{\beta \mid \mathrm{Tr}_{L/F}(\mathrm{Tr}_{K/L}(\beta O_K)) \subset O_F\} \\ &= \{\beta \mid \mathrm{Tr}_{K/L}(\beta O_K) \subset \mathfrak{D}^{-1}_{L/F}\} \\ &= \{\beta \mid \mathrm{Tr}_{K/L}(\beta \mathfrak{D}_{L/F}) \subset O_F\} \\ &= \{\beta \mid \beta \in \mathfrak{D}^{-1}_{L/F}\mathfrak{D}^{-1}_{K/L}\}. \quad \square \end{aligned}$$

We define the *discriminant* $\Delta_{K/F}$ *of* K/F as the norm of the different:

$$\Delta_{K/F} := N_{L/F}(\mathfrak{D}_{K/F}).$$

One obtains the connection with the discriminant definition in Section 18.2 through the following theorem:

Theorem 4 *(First fundamental theorem of Dedekind). Let* $\omega_1,\dots,\omega_n$ *be a basis of* O_K *over* Γ. *Then*

$$\Delta_{K/P} = \Delta(\omega_1,\dots,\omega_n)\Gamma.$$

Proof. The complementary basis $\kappa_1,\dots,\kappa_n$ of $\omega_1,\dots,\omega_n$ is a basis of $\mathfrak{D}^{-1}_{K/P}$ over Γ. Hence by Chapter 18, (3) and Chapter 18, Theorem 18

$$\begin{aligned} \Delta(\omega_1,\dots,\omega_n)\Gamma &= [O^*_K/O_K]^2\Delta(\kappa_1,\dots,\kappa_n)\Gamma \\ &= N_{K/P}(\mathfrak{D}_{K/P})^2\Delta(\kappa_1,\dots,\kappa_n)\Gamma. \end{aligned} \tag{9}$$

On the other hand, with the notation of 17.6 we have the matrix equation

$$(g_j\omega_i)_{i,j}(g_j\kappa_k)_{j,k} = (\mathrm{Tr}(\omega_j\kappa_k)) = (\delta_{ik}),$$

whence

$$\Delta(\omega_1,\dots,\omega_n)\Delta(\kappa_1,\dots,\kappa_n) = 1. \tag{10}$$

The assertion follows from (9) and (10). □

22.3 The second fundamental theorem of Dedekind

We now come to a deeper theorem, which relates the different to the element differents relative to K/F. To prove it we have to assume that, for all prime ideals $\mathfrak{P}$ of O_K, the congruence class field extension $O_K/\mathfrak{P}$ over $O_F/\mathfrak{p}$, where $\mathfrak{p} = \mathfrak{P} \cap O_F$, is generated by one element. This condition is satisfied in the two cases of particular interest to us, where $\Gamma = \mathbb{Z}$ or $\Gamma = F_0[x]$ is the polynomial ring over a field F_0 of characteristic 0. In the first case $O_K/\mathfrak{P}$ is a finite field, generated by the single generator of its cyclic multiplicative group. In the second case the congruence class field has characteristic 0, hence there is a generator by the theorem of the primitive element (Section 7.5). In what follows we need even stricter hypotheses. Hence we assume in this section that the congruence class field is finite or has characteristic 0.

Theorem 5. *(Second fundamental theorem of Dedekind). $\mathfrak{D}_{K/F}$ is the greatest common divisor of the differents $D_{K/F}(\alpha)$ for $\alpha \in O_K$.*

Proof. By Theorem 1

$$D_{K/F}(\alpha)^{-1}O_K \supset D_{K/F}(\alpha)^{-1}O_F[\alpha] = (O_F[\alpha])^* \supset O_K^* = \mathfrak{D}_{K/F}^{-1} \tag{11}$$

and hence

$$D_{K/F}(\alpha) \in \mathfrak{D}_{K/F}.$$

It therefore remains to show that, for each prime ideal $\mathfrak{P}$, there is an $\alpha \in O_K$ such that the powers of $\mathfrak{P}$ which divide $D_{K/F}(\alpha)$ at most equal the power of $\mathfrak{P}$ which divides $\mathfrak{D}_{K/F}$. To do this we need a few auxiliary facts about arithmetic in $R := O_F[\alpha]$ for $\alpha \in O_K$ with $F(\alpha) = K$.

We define the *conductor* f_R of R to be the largest ideal of O_K contained in R.

Proposition 2. $f_R = D_{K/F}(\alpha)\mathfrak{D}_{K/F}^{-1}$.

Proof. Because of (11), we have $D_{K/F}(\alpha)\mathfrak{D}_{K/F}^{-1} \subset O_F(\alpha)$. On the other hand, by Theorem 1

$$Tr_{K/F}(f_R D_{K/F}(\alpha)^{-1}) \subset O_F.$$

This implies $D_{K/F}(\alpha)^{-1} \in f^*_K = f_R^{-1} D_{K/F}^{-1}$, so

$$f_R \subset D_{K/F}(\alpha) \mathfrak{P}_{K/F}^{-1}. \qquad \square$$

Proposition 3. *Let* $\mathfrak{P}$ *be a prime ideal of* O_K *and suppose*

$$\mathfrak{p} = \mathfrak{P} \cap O_F = \mathfrak{P}^e \mathfrak{A} \quad \textit{with} \quad \mathfrak{P} \nmid \mathfrak{A}.$$

Also suppose $\alpha \in \mathfrak{A}$ *is a generator of* K/F *with* $\mathfrak{P} | \alpha$.

If the natural homomorphism

$$R/(R \cap \mathfrak{P}^m) \to O_K/\mathfrak{P}^m$$

is an isomorphism for all $m = 1,2,...,$ *then* $\mathfrak{P} \nmid f_R$.

Proof. Let $a \in O_F$ be such that

$$N_{K/F}(D_{K/F}(\alpha)) | \mathfrak{p}^k a, \quad \mathfrak{p} \nmid a.$$

Then, by hypothesis, for each $\beta \in O_K$ there is a $\rho \in R$ with

$$\gamma := \beta - \rho \equiv 0 \pmod{\mathfrak{P}^{ek}}.$$

By Theorem 1 we have

$$R = D_{K/F}(\alpha) R^* \supset D_{K/F}(\alpha) O_K \supset N_{K/F}(D_{K/F}(\alpha)) O_K \supset \mathfrak{p}^k a O_K \supset \gamma \alpha^k a O_K,$$

so

$$\beta \alpha^k a = \gamma \alpha^k a + \rho \alpha^k a \in R.$$

Since β is an arbitrary element of O_K, it follows that $\alpha^k a O_K \subset R$. Therefore f_R is a divisor of $\alpha^k a$, which is not divisible by $\mathfrak{P}$. $\square$

To prove Theorem 5 it now remains only to show that for each $\mathfrak{P}$ there is an α which satisfies the conditions of Proposition 3:

Proposition 4. *Let* $\mathfrak{P}$ *be a prime ideal of* O_K *and let* $\mathfrak{A}$ *be an ideal of* O_K *such that* $\mathfrak{P} \nmid \mathfrak{A}$. *Then there is an* $\alpha \in \mathfrak{A}$, *with* $\mathfrak{P} \nmid \alpha$, *such that for each natural number* m *the order* $O_F[\alpha]$ *contains a complete system of residues of* O_K *modulo* $\mathfrak{P}^m$.

Proof. Let $\mathfrak{p} := \mathfrak{P} \cap O_F$. By the hypotheses of this section, the extension $O_K/\mathfrak{P}$ of $O_F/\mathfrak{p}$ is generated by a single element ζ with $\zeta \in O_K$. Let $\phi(t)$ be a polynomial from $O_K[t]$ such that $\overline{\phi}(t) \in (O_F/\mathfrak{p})[t]$ is the minimal polynomial of ζ.

If $\phi(\zeta) \in \mathfrak{P}^2$ we take an element $\pi \in \mathfrak{P} - \mathfrak{P}^2$. Since

$$\phi(\zeta+\pi) \equiv \phi(\zeta) + \phi'(\zeta)\pi \pmod{\mathfrak{P}^2},\ \phi'(\zeta) \notin \mathfrak{P},$$

we have $\phi(\zeta+\pi) \notin \mathfrak{P}^2$. We can therefore assume $\phi(\zeta) \notin \mathfrak{P}^2$. By the Chinese remainder theorem there is an $\alpha \in O_K$ with $\alpha \equiv 0 \pmod{\mathfrak{A}}$ and $\alpha \equiv \zeta \pmod{\mathfrak{P}^2}$. Such an α satisfies the requirements of Proposition 4: let $\xi \in O_K$. Then there is a polynomial $\psi(t) \in O_F[t]$ with

$$\xi \equiv \psi(\zeta) \pmod{\mathfrak{P}}$$

and consequently

$$\xi \equiv \psi(\alpha) \pmod{\mathfrak{P}}.$$

This proves the assertion for $m = 1$. For $m > 1$, the fact that $\phi(\alpha) \equiv \phi(\zeta) \pmod{\mathfrak{P}^2}$ is used to obtain a complete residue system of O_K modulo $\mathfrak{P}^m$ in the form

$$\xi_0 + \xi_1\phi(\alpha) + \dots + \xi_{m-1}\phi(\alpha)^{m-1},$$

where $\xi_0, \xi_1, \dots, \xi_{m-1}$ each run through a complete system of residues of O_F modulo $\mathfrak{P}$. □

The third main theorem on the different is the following *Dedekind different theorem*:

Theorem 6. *Let* $\mathfrak{P}$ *be a prime ideal of* O_K *and suppose* $\mathfrak{p} := \mathfrak{P} \cap O_F$, $v_{\mathfrak{P}}(\mathfrak{p}) =: e$. *Also suppose* p *is the characteristic of the field* $O_K/\mathfrak{P}$. *Then*

$$v_{\mathfrak{P}}(\mathfrak{D}_{K/F}) = e - 1 \text{ if } p \nmid e,$$

and

$$v_{\mathfrak{P}}(\mathfrak{D}_{K/F}) > e - 1 \text{ if } p \mid e. \qquad \blacksquare$$

We prove Theorem 6 in Section 25.2 with the help of Theorem 5.

By taking norms, one obtains assertions about the discriminant from Theorem 3 and Theorem 6.

Theorem 7 *(Discriminant tower theorem). If* $F \subset L \subset K$ *is a tower of fields with* $m := [K:L]$ *then*

$$\Delta_{K/F} = N_{L/F}(\Delta_{K/M})\Delta^m_{L/F}. \qquad \square$$

Theorem 8 *(Dedekind discriminant theorem). Let* $\mathfrak{p}$ *be a prime ideal of* O_F *and suppose* $\mathfrak{p} = \mathfrak{P}_1^{e_1} \dots \mathfrak{P}_g^{e_g}$ *is the prime decomposition of* $\mathfrak{p}$ *in* O_K. *Also suppose* p *is the characteristic of the field* $O_F/\mathfrak{p}$. *Then*

$$v_{\mathfrak{p}}(\Delta_{K/F}) = (e_1-1)f_1 + \dots + (e_g-1)f_g \quad \text{if } p \nmid e_i \text{ for } i = 1,\dots,g,$$

$$v_{\mathfrak{p}}(\Delta_{K/F}) > (e_1-1)f_1 + \dots + (e_g-1)f_g \quad \text{if } p \mid e_i \text{ for some } i.$$

Here, f_i *is the degree of inertia of* $\mathfrak{P}_i$ *for* $i = 1,\dots,g$. $\square$

As an immediate consequence of Theorem 8 one obtains

Theorem 9. *A prime ideal* $\mathfrak{p}$ *of* O_F *is ramified in* K/F *if and only if* $\mathfrak{p}$ *is a divisor of the discriminant of* K/F. $\square$

Exercises

22.1 Let K/F be a normal extension of number fields with Galois group S_3. Show that K contains no prime ideal whose degree over F is 6.

22.2 Let L and K be any finite extensions of an algebraic number field F.

a) Let $\alpha \in O_K$. Show that $D_{LK/L}(\alpha) \mid D_{K/F}(\alpha)$.

b) Show that $\mathfrak{D}_{LK/L} \mid \mathfrak{D}_{K/F}$.

c) Let $\mathfrak{P}$ be a prime ideal of KL which is ramified in KL/L. Show that $\mathfrak{P} \cap K$ is ramified in K/F.

d) Let $\mathfrak{p}$ be a prime ideal of F which is unramified in L/F and K/F. Show that $\mathfrak{p}$ is unramified in KL/F.

22.3 Let F be a quadratic field with discriminant D. Then D is called a *prime discriminant* when only one prime number divides D.

a) Show that each prime discriminant is of the form -4, 8, -8 or $(-1)^{(p-1)/2}p$ for a prime number $p \neq 2$.

b) Show that the discriminant D of any quadratic field is uniquely representable as the product of pairwise relatively prime prime discriminants $D_1,\ldots,D_s$.

c) Show that all prime ideals; of $\mathbb{Q}(\sqrt{D_1},\ldots,\sqrt{D_s})$ are unramified over $\mathbb{Q}(\sqrt{D})$.

22.4 Let p be a prime number with $p \equiv 5 \pmod 8$. Show that the class numbers of the fields $\mathbb{Q}(\sqrt{2p})$ and $\mathbb{Q}(\sqrt{-2p})$ are divisible by 2.

23. Theory of algebraic functions of one variable

23.1 Algebraic function fields

In Chapter 11 we have treated the theory of meromorphic functions on closed Riemann surfaces. In particular, we have seen in Section 11.7 that these functions are algebraic functions of complex variable z. However, we had to assume the Riemann existence theorem for differentials, which was not rigorously proved by Riemann. Weierstrass pointed this out in 1870. This prompted Dedekind and Weber to give the theory a new foundation. In their joint work in *J. reine angew. Math.* 92 (1882) they called the theory of algebraic functions of one variable "*one of the main results created by Riemann*". In their work they give an independent construction "*from a simple but at the same time rigorous and completely general standpoint*". Their starting point is the unique decomposition of polynomials into irreducible factors, which we have already carried out in Chapter 18.

In the light of Chapter 11, Theorem 13, we base our considerations on a field L which is a finite extension of the field $\mathbb{C}(\xi)$ of rational functions in an indeterminate ξ.

Corresponding to Section 18.1, $\alpha \in L$ is called an *entire function* of ξ when α is integral over $\mathbb{C}[\xi]$, i.e. when α satisfies an equation

$$\alpha^m + a_1\alpha^{m-1} + \ldots + a_m = 0 \quad \text{with} \quad a_1,\ldots,a_m \in \mathbb{C}[\xi].$$

Each element η in $L - \mathbb{C}$ is called a *variable*. The indeterminate ξ satisfies an algebraic equation with coefficients in $\mathbb{C}(\eta)$. Therefore $L/\mathbb{C}(\eta)$ is a finite extension, and η plays the same rôle as ξ with respect to L. The elements of $\mathbb{C}$ are called *constants.*

We want to point out a special feature of arithmetic in L immediately. Let O_ξ be the ring of entire functions of ξ in L and let $\mathfrak{P}$ be a prime ideal of O_ξ with $\mathfrak{p} = \mathfrak{P} \cap \mathbb{C}[\xi]$. Then by the fundamental theorem of algebra $\mathfrak{p} = (\xi - c)\mathbb{C}[\xi]$ with a unique $c \in \mathbb{C}$ and $\mathbb{C}[\xi]/\mathfrak{p} = \mathbb{C}$. Since $O_\xi/\mathfrak{P}$ is a finite extension of $\mathbb{C}[\xi]/\mathfrak{p}$, we also have $O_\xi/\mathfrak{P} = \mathbb{C}$, i.e., $\mathfrak{P}$ has inertia degree 1.

23.2 The Riemann surface

The field L is initially a purely algebraic object for us. We can view it as the residue class field of an irreducible polynomial with coefficients in $\mathbb{C}(\xi)$ (A1.5). We now want to interpret the elements of L as functions. The functions can take the value ∞ as well as complex values, i.e. they can have poles. Computations with ∞ involve the following:

For $c,d \in \mathbb{C}$, $d \neq 0$, one has

$$\infty/c = \infty,\ c \pm \infty = \infty,\ c/\infty = 0,\ d.\infty = \infty,\ \infty\cdot\infty = \infty.$$

The expressions $\infty \pm \infty$, $0.\infty$, $0/0$, ∞/∞ are regarded as meaningless.

We now come to the definition of a point P of L. It is a mapping of L onto $\mathbb{C} \cup \{\infty\}$ with the following properties:

(i) If $c \in \mathbb{C}$ then $P(c) = c$

(ii) If $\alpha,\beta \in L$ then

$$P(\alpha+\beta) = P(\alpha) + P(\beta),\ P(\alpha-\beta) = P(\alpha) - P(\beta),$$

$$P(\alpha\beta) = P(\alpha)P(\beta),\ P(\alpha/\beta) = P(\alpha)/P(\beta),$$

provided both sides of the equations are defined.

The collection of points of L is called the *Riemann surface associated with* L. It is denoted by $\mathcal{F}(L)$. When L is given as the field $K(\mathcal{F})$ of meromorphic functions on a Riemann surface $\mathcal{F}$, in the sense of Section 11.7, then $P \in \mathcal{F}$ gives a point of $K(\mathcal{F})$ in the sense of the above definition as the function $f \mapsto f(P)$ where f ranges over $K(\mathcal{F})$, and all points of $K(\mathcal{F})$ are obtained in this way.

We first establish the basic connection between points of L and prime ideals in L.

Suppose $P \in \mathcal{F}(L)$ and let η be a variable of L with $P(\eta) \neq \infty$. Such an η always exists, because if $P(\eta) = \infty$ then $P(1/\eta) = 0$. Then by (i) and (ii) all functions in $\mathbb{C}[\eta]$ have a finite value. Also, if $\alpha \neq 0$ is a polynomial function of η, then there is an equation

$$1 = a_1(1/\alpha) + \dots + a_m(1/\alpha)^m \quad \text{with} \quad a_1,\dots,a_m \in \mathbb{C}[\eta].$$

It follows that $P(1/\alpha) \neq 0$ and hence $P(\alpha) \neq \infty$. Therefore $P(\alpha) \neq \infty$ for all $\alpha \in O_\eta$, i.e. P is a ring homomorphism of O_η onto $\mathbb{C}$. Let $\mathfrak{P}$ be the kernel of this homomorphism. By Theorem 1 of A1, $\mathfrak{P}$ is a prime ideal of O_η.

Conversely, each prime ideal $\mathfrak{P} \neq \{0\}$ of O_η is associated with a point $P \in \mathcal{F}(L)$ with $P(\eta) \neq \infty$; on O_η, P is defined by $O_\eta \mapsto O_\eta/\mathfrak{P} = \mathbb{C}$. By Theorem 14 of Chapter 18, each $\alpha \in L$ is expressible in the form $\alpha = \beta/\gamma$ with $\beta,\gamma \in O_\eta$ and $\mathfrak{P} \nmid \gcd(\beta,\gamma)$. Then $P(\alpha) := P(\beta)/P(\gamma)$ defines $P(\alpha)$ independently of the choice of β,γ. The P constructed in this way satisfies conditions (i) and (ii).

This establishes a one-to-one correspondence between the points P of L with $P(\eta) \neq \infty$ and the prime ideals different from $\{0\}$ in the ring O_η.

Now let P be a point with $P(\eta) = \infty$. Then $P(1/\eta) = 0$, and P corresponds to a prime ideal $\mathfrak{P}$ in $O_{1/\eta}$ with $1/\eta \in \mathfrak{P}$. There are finitely many prime ideals of $O_{1/\eta}$ with this property: the prime divisors of $1/\eta$. In order to obtain all points of L, one has to add the points corresponding to the latter prime ideals to the points corresponding to the prime ideals of O_η.

We apply these considerations to the rational function field $\mathbb{C}(\eta)$ in particular: the points of the Riemann surface of $\mathbb{C}(\eta)$ correspond to the prime ideals of $\mathbb{C}[\eta]$, which correspond in turn to the linear polynomials $\eta - c$, i.e. to the complex numbers c, with an additional "infinite" point corresponding to $1/\eta$. Thus the Riemann surface of $\mathbb{C}(\eta)$ is the completed complex plane, which we denote by $\mathcal{F}_\eta$.

By restriction of the point $P \in \mathcal{F}(L)$ to $\mathbb{C}(\eta)$ one obtains a point of $\mathcal{F}_\eta$. Under the above identification of $\mathcal{F}_\eta$ with $\mathbb{C} \cup \{\infty\}$, this point equals $P(\eta)$. Thus the Riemann surface $\mathcal{F}(L)$ is spread over the completed complex plane.

23.3 The order of a function at a point

Let P be a point of L. We want to associate with each $\alpha \neq 0$ in L an integer $v_P(\alpha)$, the order of α at P.

When $P(\alpha) \neq \infty$ we define $v_P(\alpha)$ to be the smallest integer m with the property that $P(\beta^m/\alpha)$ is finite for all $\beta \in L$ with $P(\beta) = 0$. When $P(\alpha) = \infty$ we set $v_P(\alpha) := v_P(1/\alpha)$. Let η be a variable in L with $P(\eta) \neq \infty$ and let $\mathfrak{P}$ be the prime ideal in O_η associated with P. Then $v_P(\alpha)$ equals the power to which $\mathfrak{P}$ divides α. This immediately implies

$$v_P(\alpha_1\alpha_2) = v_P(\alpha_1) + v_P(\alpha_2) \quad \text{for } \alpha_1,\alpha_2 \in L^\times \tag{1}$$

and, when we also set $v_P(0) = \infty$,

$$v_P(\alpha_1+\alpha_2) \geq \min\{v_P(\alpha_1),v_P(\alpha_2)\} \quad \text{for } \alpha_1,\alpha_2 \in L \tag{2}$$

Corresponding to the terminology in Section 11.6, a *divisor* A of L is a linear combination $n_1P_1 + ... + n_sP_s$ of points of L with integral coefficients $n_1,...,n_s$. The *degree* $\deg A$ is defined by $\deg A := n_1 + ... + n_s$. For an $\alpha \neq 0$ in L there are only finitely many $P \in \mathcal{F}(L)$ with $v_P(\alpha) \neq 0$. We can therefore associate α with the divisor $(\alpha) := \Sigma v_P(\alpha)P$, where the sum is taken over all $P \in \mathcal{F}(L)$. Let m be a non-negative integer. We say that $\alpha \in L^\times$ has an *m-tuple zero* resp. an *m-tuple pole* at P when $v_P(\alpha) = m$ resp. $v_P(\alpha) = m$. We say that α takes the value $c \in \mathbb{C}$ *m-tuply* at P when $v_P(\alpha\text{-}c) = m$. When α has an m-tuple pole at P we also say that α takes the value ∞ m-tuply at P.

Theorem 1. *Let* η *be an arbitrary variable of* L, *and let the degree of* L *over* $\mathbb{C}(\eta)$ *be* n. *Then* η *takes each value from* $\mathbb{C} \cup \{\infty\}$ *exactly* n *times. The degree of the divisor associated with* η *is* 0.

Proof. Suppose $c \in \mathbb{C}$.. Then

$$(\eta\text{-}c)O_\eta = \Pi\mathfrak{P}^{v_P(\eta\text{-}c)}$$

where the product is taken over all prime ideals $\mathfrak{P}$ of O_η and P denotes *the point of* $\mathcal{F}(L)$ *associated with* $\mathfrak{P}$. By Chapter 18, (2), $\Sigma v_P(\eta\text{-}c) = n$. This gives the first assertion of Theorem 1 for $c \in \mathbb{C}$.

Now η takes the value ∞ *m-tuply* at a point P if and only if $1/\eta$ has an m-tuple zero at P. Hence the case $c = \infty$ is reduced to the case $c = 0$ by passing from η to $1/\eta$.

The degree of (η) equals the number of zeros of η minus the number of poles of η, i.e. 0. □

The divisor

$$(\eta)_0 := \sum_{v_P(\eta)>0} v_P(\eta)P \quad \text{resp.} \quad (\eta)_\infty := -\sum_{v_P(\eta)<0} v_P(\eta)P$$

is called the *zero divisor* resp. *pole divisor* of η.

Departing somewhat from Chapters 10 and 11, we understand a *local uniformising function* π of P to be an arbitrary element of L with $v_P(\pi) = 1$. An element $L^\times$ with $v_P(\alpha) = v$ may be written in the form $\alpha = c\pi^v + \alpha_1\pi^{v+1}$, where $c \in \mathbb{C}$ and $v_P(\alpha_1) \geq 0$.

We also consider $\mathcal{F}(L)$ as a covering of $\mathcal{F}_\eta$. A point $P \in \mathcal{F}(L)$ is called *branched* over $\mathcal{F}_\eta$ (or over η) when the value $P(\eta)$ is taken multiply. When $P(\eta)$ is taken e_P-tuply, e_P is called the *ramification index* of P relative to $\mathcal{F}_\eta$.

When $P(\eta)$ is finite, then P is branched over $\mathcal{F}_\eta$ if and only if the prime ideal O_η corresponding to P is ramified, i.e. when it divides the different of $O_\eta/\mathbb{C}[\eta]$ (Theorem 6, Chapter 22). Thus there are only finitely many branch points P in $\mathcal{F}(L)$.

The divisor

$$V_\eta = \Sigma(e_P-1)P,$$

where the sum is taken over all branch points of $\mathcal{F}(L)$ over $\mathcal{F}_\eta$, is called the *ramification divisor* of η.

23.4 Normal bases

The main goal of this chapter is to give a new proof of the Riemann-Roch theorem (Theorem 9, Chapter 11) by means of the Dedekind ideal theory (Chapter 18, Chapter 22). The present section contains some considerations which assist towards this goal.

For a divisor $A = \Sigma n_P P$ and a variable η of L, $(A)_\eta$ denotes the O_η-module

$$\{\alpha \in L \mid v_P(\alpha) \geq n_P \text{ for all } P \in \mathcal{F}(L) \text{ with } P(\eta) \neq \infty\}.$$

We want to construct special bases of $\mathfrak{A} := (-A)_\eta$ as $\mathbb{C}[\eta]$-modules, which will be called *normal bases.*

Each element α of $\mathfrak{A}$ is converted to an element of $\mathfrak{A}' := (-A)_{1/\eta}$ by multiplication by a sufficiently high power $(1/\eta)^e$ of $1/\eta$. For $\alpha \neq 0$ the exponents e with this property are bounded from below independently of α. The smallest such e for α is called the *exponent* of α. Also, let r_1 be the minimum of the exponents for $\alpha \in \mathfrak{A}$, $\alpha \neq 0$.

For each $r \in \mathbb{Z}$ we define $\mathfrak{A}(r)$ to be the $\mathbb{C}$-vector space of all elements of $\mathfrak{A}$ with exponent $\leq r$. For $r < r_1$ we have $\mathfrak{A}(r) = \{0\}$.

The $\mathbb{C}$-vector space $\mathfrak{A}/\eta\mathfrak{A}$ has dimension $n = [L:\mathbb{C}(\eta)]$ (A1, Theorem 4). We also have

Proposition 1. *Let* $\lambda_1,...,\lambda_n \in \mathfrak{A}$ *be a system of representatives of a basis* $\bar{\lambda}_1,...,\bar{\lambda}_n$ *of* $\mathfrak{A}/\eta\mathfrak{A}$. *Then* $\lambda_1,...,\lambda_n$ *is a basis for* L *over* $\mathbb{C}(\eta)$.

Proof. Suppose that the elements $\lambda_1,...,\lambda_n$ are linearly independent over $\mathbb{C}(\eta)$. Then there are $a_1,...,a_n \in \mathbb{C}[\eta]$, not all divisible by η, such that

$$a_1\lambda_1 + ... + a_n\lambda_n = 0.$$

It follows that the classes $\bar{\lambda}_1,...,\bar{\lambda}_n$ are linearly independent over $\mathbb{C}$. □

We now choose $\lambda_1,...,\lambda_n$ corresponding to the filtration of $\mathfrak{A}/\eta\mathfrak{A}$ given by $\bar{\mathfrak{A}}(r) := (\mathfrak{A}(r) + \eta\mathfrak{A})/\eta\mathfrak{A}$. That is, let $\lambda_1,...,\lambda_{s_1}$ be a system of representatives of a basis of $\bar{\mathfrak{A}}(r_1)$ in $\mathfrak{A}(r_1)$, let $\lambda_{s_1+1},...,\lambda_{s_2}$ be a system of representatives of a basis of $\bar{\mathfrak{A}}(r_{s_1+1})$ in $\mathfrak{A}(r_{s_1+1})$, where r_{s_1+1} is the smallest integer with $r_{s_1+1} > r_1$ and $\bar{\mathfrak{A}}(r_{s_1+1}) \supsetneq \bar{\mathfrak{A}}(r_1)$, etc.. A basis $\lambda_1,...,\lambda_n$ of $L/\mathbb{C}(\eta)$ constructed in this way is called a *normal basis*. The numbers

$$r_1 = r_2 = \dots = r_{s_1},\ r_{s_1+1} = r_{s_1+2} = \dots = r_{s_2}, \dots, r_n$$

are the exponents corresponding to $\lambda_1, \dots, \lambda_n$. They are independent of the choice of normal basis.

Proposition 2. *Each normal basis is a basis for* $\mathfrak{A}$ *as a* $\mathbb{C}[\eta]$*-module.*

Proof. Each $\alpha \in \mathfrak{A}$ may be represented in the form

$$\alpha = \frac{a_1}{a}\lambda_1 + \dots + \frac{a_n}{a}\lambda_n \tag{3}$$

with $a_1, \dots, a_n,\ a \in \mathbb{C}[\eta]$ and $\gcd(a_1, \dots, a_n, a) = 1$. If $\lambda_1, \dots, \lambda_n$ were not a basis of $\mathfrak{A}$, there would be an $\alpha \in \mathfrak{A}$ with non-constant a. Suppose then that $\eta - c$ is a divisor of a. It follows from (3) that there is an $\alpha' \in \mathfrak{A}$ with

$$(\eta-c)\alpha' = c_1\lambda_1 + \dots + c_n\lambda_n,\quad c_1, \dots, c_n \in \mathbb{C},$$

where not all $c_1, \dots, c_n$ are zero. Let s be the maximal index for which $c_s \neq 0$. Then the exponent of α' is smaller than r_s. In fact

$$v_P(\alpha'/\eta^{r_s - 1}) = v_P(\alpha') \quad \text{for} \quad P(\eta) \neq 0, \infty$$

and

$$v_P(\alpha'/\eta^{r_s - 1}) \leq \min\left\{v_P\left(\frac{\eta c_1}{\eta - c}\frac{\lambda_1}{\eta^{r_s}}\right), \dots, v_P\left(\frac{\eta c_s}{\eta - c}\frac{\lambda_s}{\eta^{r_s}}\right)\right\}$$

$$\geq -n_P \quad \text{for} \quad P(\eta) = \infty.$$

Therefore α, and consequently λ_s, is congruent to a linear combination of elements $\lambda_1, \dots, \lambda_{s-1} (\text{mod } \eta\mathfrak{A})$, contrary to the choice of λ_s. □

Proposition 3. *The functions*

$$\lambda_1' = \lambda_1/\eta^{r_1}, \dots, \lambda_n' = \lambda_n/\eta^{r_n}$$

in $\mathfrak{A}'$ *constitute a normal basis of* $\mathfrak{A}'$.

Proof. This follows immediately from the equation

$$\mathfrak{A}'(r) = \eta^{-r}\mathfrak{A}(r) \quad \text{for} \quad r \in \mathbb{Z}. \qquad \square$$

Proposition 4. *Let* $\tilde{\lambda}_1,\dots,\tilde{\lambda}_n$ *be a basis for* $\mathfrak{A}$ *as a* $\mathbb{C}[\eta]$*-module and let* $\eta^{-\tilde{r}_1}\tilde{\lambda}_1,\dots,\eta^{-\tilde{r}_n}\tilde{\lambda}_n$ *be a basis for* $\mathfrak{A}'$ *as a* $\mathbb{C}[1/\eta]$*-module with* $\tilde{r}_1 \le \tilde{r}_2 \le \dots \le \tilde{r}_n$. *Then* $\tilde{r}_i = r_i$ *for* $i = 1,\dots,n$.

Proof. Let (c_{ij}) be the matrix which transforms $\tilde{\lambda}_1,\dots,\tilde{\lambda}_n$ into the normal basis $\lambda_1,\dots,\lambda_n$. Then

$$(c'_{ij}) := (c_{ij}\eta^{\tilde{r}_j - r_j}) \tag{4}$$

converts $\eta^{-\tilde{r}_1}\tilde{\lambda}_1,\dots,\eta^{-\tilde{r}_n}\tilde{\lambda}_n$ to $\eta^{-r_1}\lambda_1,\dots,\eta^{-r_n}\lambda_n$. One concludes from this that if $r_i < \tilde{r}_j$ for a pair i,j then $c_{ij} = 0$. If there is an i with $r_i < \tilde{r}_i$, then we also have $r_k < \tilde{r}_\ell$ for $k \le i \le \ell$. But then $\det(c_{ij}) = 0$, contrary to hypothesis. Hence $r_i \ge \tilde{r}_i$ for all i.

It also follows from (4) that

$$\det(c'_{ij}) = \eta^{(\tilde{r}_1+\dots+\tilde{r}_n - r_1 - \dots - r_n)}\det(c_{ij})$$

and, since $\det(c'_{ij})$ resp. $\det(c_{ij})$ is a unit in $\mathbb{C}[1/\eta]$ resp. $\mathbb{C}[\eta]$, $\tilde{r}_1 + \dots + \tilde{r}_n = r_1 + \dots + r_n$. $\square$

Proposition 5. *Let* v_η *be the degree of the branch divisor* V_η *of* η. *Then*

$$v_\eta = 2(r_1+\dots+r_n) + 2\deg A.$$

Proof. By the Dedekind discriminant theorem (Theorem 8, Chapter 22)

$$v_\eta = -v_{1/\eta}(\Delta(O_\eta)) + v_{1/\eta}(\Delta(O_{1/\eta})).$$

(One notes that $v_{1/\eta}(a) = -m$ for a polynomial a of degree m in η.)

By Propositions 2 and 3

$$\Delta(\mathfrak{A}') = \eta^{(-2r_1-\dots-2r_n)}\Delta(\mathfrak{A}).$$

Moreover, (Chapter 18, (3))

$$\Delta(\mathfrak{A}) = \Delta(O_\eta)N(\mathfrak{A})^2, \quad \Delta(\mathfrak{A}') = \Delta(O_{1/\eta})N(\mathfrak{A}')^2,$$

$$N(\mathfrak{A}) = \prod_{P(\eta)\neq\infty}(\eta-P(\eta))^{-n_P}, \quad N(\mathfrak{A}') = \prod_{P(\eta)\neq 0}(1/\eta-1/P(\eta))^{-n_P}.$$

It follows that

$$\begin{aligned} v_\eta &= v_{1/\eta}(\Delta(\mathfrak{A})^{-1}N(\mathfrak{A})^2\Delta(\mathfrak{A}')N(\mathfrak{A}')^{-2}) \\ &= 2(r_1+\dots+r_n) + 2\deg A. \end{aligned}$$ □

23.5 The function space associated with a divisor

Let $A = \Sigma n_P P$ be a divisor of L. We associate with it the vector space

$$D(A) := \{\alpha \in L \mid v_P(\alpha) \geq -n_P \text{ for } P \in \mathcal{F}(L)\} = (-A)_\eta \cap (-A)_{1/\eta}.$$

Theorem 2. *Let* $\lambda_1,\dots,\lambda_n$ *be a normal basis of* $(-A)_\eta$, *and let* $r_1,\dots,r_n$ *be the associated exponents. Then the dimension* $d(A)$ *of* $D(A)$ *as a* $\mathbb{C}$*-vector space equals* $\Sigma(1-r_i)$, *where the sum is taken over all* $r_1,\dots,r_n$ *with* $r_i \leq 0$.

Proof. By Proposition 3, $\alpha = a_1\lambda_1 + \dots + a_n\lambda_n \in (-A)_\eta$ with $a_i \in \mathbb{C}[\eta]$ for $i = 1,\dots,n$ belongs to $D(A)$ if and only if $a_i\eta^{r_i}$ belongs to $\mathbb{C}[1/\eta]$ for $i = 1,\dots,n$. This is the case precisely when $-r_i \geq \deg a_i$. □

23.6 Differentials

In this section we want to introduce the concept of the *differential*. First we define the *differential quotient* $d\alpha/d\eta$ for an $\alpha \in L$ and a variable $\eta \in L$.

Let $f(x,y) \in \mathbb{C}[x,y]$ be the irreducible polynomial with $f(\alpha,\eta) = 0$. Let f_x resp f_y be the partial derivatives of f with respect to x resp. y. In accordance with the rules of differential calculus we define

$$\frac{d\alpha}{d\eta} := -\frac{f_y(\alpha,\eta)}{f_x(\alpha,\eta)}. \tag{5}$$

Proposition 6. *There are only finitely many pairs* c, d *of complex numbers with* $f(c,d) = 0$, $f_x(c,d) = 0$.

Proof. Let

$$f(x,y) = a_0(y)x^h + a_1(y)x^{h-1} + \dots + a_h(y)$$

where $a_i(y) \in \mathbb{C}[y]$, $a_0(y) \neq 0$, and let d be a complex number with $a_0(d) \neq 0$. The polynomials $f(x,d)$ and $f_x(x,d)$ have a common zero if and only if $f(x,d)$ has a multiple zero (Theorem 12, A1). This is the case if and only if the discriminant of $f(x,d)$ equals 0 (Section 7.3). The discriminant of $f(x,d)$ is a rational function of d, and hence has only finitely many zeros. □

By Proposition 6, the following holds for almost all points P of $\mathcal{F}(L)$:

$$P\left[\frac{d\alpha}{d\eta}\right] = -\frac{f_y(P(\alpha),P(\eta))}{f_x(P(\alpha),P(\eta))}. \tag{6}$$

Theorem 3. *For almost all points* P *of* $\mathcal{F}(L)$

$$P\left[\frac{d\alpha}{d\eta}\right] = P\left[\frac{\alpha-P(\alpha)}{\eta-P(\eta)}\right]. \tag{7}$$

Proof. Let $P(\alpha), P(\eta) \neq \infty$. The Taylor expansion of $f(x,y)$ at $P(\alpha), P(\eta)$ has the form

$$f(x,y) = (x-P(\alpha))f_x(P(\alpha),P(\eta)) + (y-P(\eta))f_y(P(\alpha),P(\eta)) + \dots .$$

We set $x = \alpha$, $y = \eta$ and divide by $\eta - P(\eta)$. Then

$$0 = \frac{\alpha-P(\alpha)}{\eta-P(\eta)}f_x(P(\alpha),P(\eta)) + f_y(P(\alpha),P(\eta)) + \dots . \tag{8}$$

When $(\alpha-P(\alpha))/(\eta-P(\eta))$ has a finite value, which is the case for almost all P, one obtains (7) from (6) and (8). □

Two functions from L which have the same value at infinitely many points are equal (Theorem 1). Then with the help of Theorem 3 one can easily derive the rules for calculation with differential quotients. Here we mention only the chain rule:

Theorem 4. *Suppose* $\alpha \in L$ *and let* ζ,η *be variables of* L. *Then*

$$\frac{d\alpha}{d\xi} = \frac{d\alpha}{d\eta}\frac{d\eta}{d\xi}. \quad \square$$

Now we define the *differential* $d\alpha$ to be the mapping of L-$\mathbb{C}$ into L which takes the value $d\alpha/d\eta$ for $\eta \in L$-$\mathbb{C}$. By Theorem 4, two differentials are equal when they take the same value for a single variable. Similarly, $\beta d\alpha$ is the mapping $\eta \mapsto \beta d\alpha/d\eta$ for $\beta \in L$. Theorem 3 yields the following rules:

For $\alpha,\beta \in L$:

$$\begin{aligned} d(\alpha\pm\beta) &= d\alpha \pm d\beta \\ d(\alpha\beta) &= \alpha d\beta + \beta d\alpha \\ d(\alpha/\beta) &= (\beta d\alpha-\alpha d\beta)/\beta^2, \text{ provided } \beta \neq 0. \end{aligned}$$

We now want to determine the divisor of $d\xi/d\eta$ for two variables $\xi,\eta \in L$, and we first prove the following

Proposition 7. *Let* $P \in \mathscr{F}(L)$ *and let* α *be an element of* L *with* $v_P(\alpha) \geq 0$. *Then* $v_P(d\alpha/d\pi) \geq 0$ *for each local uniformising function* π *of* P.

Proof. In the ring

$$O_P := \{\beta \in L \mid v_P(\beta) \geq 0\}$$

the elements γ with $v_P(\gamma) \geq h$ constitute an ideal, which we shall denote by P^h.

Let $f(x,y)$ be the irreducible polynomial with $f(\alpha,\pi) = 0$. One easily proves, by induction on h, that α may be expressed in the form

$$\alpha \equiv c_0 + c_1\pi + \ldots + c_h\pi^h (\text{mod } P^{h+1}) \text{ with } c_0,c_1,\ldots,c_h \in \mathbb{C}. \tag{9}$$

We set

$$\alpha_h := c_0 + c_1\pi + \dots + c_h\pi^h.$$

Then

$$\frac{d\alpha_h}{d\pi} = c_1 + \dots + hc_h\pi^{h-1} \in O_P,$$

$$f(\alpha_h,\pi) \equiv f(\alpha,\pi) \equiv 0 \pmod{P^{h+1}}$$

and

$$\frac{df(\alpha_h,\pi)}{d\pi} \equiv 0 \pmod{P^h},$$

since $f(\alpha_h,\pi)$ is a polynomial in π with coefficients from $\mathbb{C}$.

Moreover,

$$0 \equiv \frac{df(\alpha_h,\pi)}{d\pi} \equiv f_x(\alpha_h,\pi)\frac{d\alpha_h}{d\pi} + f_y(\alpha_h,\pi)$$

$$\equiv f_x(\alpha,\pi)\frac{d\alpha_h}{d\pi} + f_y(\alpha,\pi) \pmod{P^h}.$$

When $k := \nu_P(f_x(\alpha,\pi))$ we set $h = k$ and find that $f_y(\alpha,\pi)$ is divisible by $f_x(\alpha,\pi)$ in O_P. Therefore

$$\frac{d\alpha}{d\pi} = -\frac{f_y(\alpha,\pi)}{f_x(\alpha,\pi)}$$

lies in O_P. □

Remark. Equation (9) serves as a substitute for power series expansions in the theory of analytic functions.

The divisor of $d\xi/d\eta$ is described by the following theorem:

Theorem 5. Let ξ and η be arbitrary variables of L. Then

$$(d\xi/d\eta) = V_\xi - V_\eta - 2(\xi)_\infty + 2(\eta)_\infty.$$

Proof. Let P be a point of $\mathscr{F}(L)$ and let π be a local uniformising function of P. We set $\xi_0 = P(\xi)$ when $P(\xi)$ is finite, and $\xi_0 = 0$ when $P(\xi) = \infty$, and similarly for η.

$\xi - \xi_0$ resp. $\eta - \eta_0$ may be written in the form

$$\xi - \xi_0 = \alpha\pi^r \text{ resp. } \eta - \eta_0 = \beta\pi^s,$$

where $P(\alpha)$ and $P(\beta)$ are finite and non-zero. Then

$$d\xi/d\pi = \alpha r\pi^{r-1} + \pi^r d\alpha/d\pi.$$

By Proposition 7, $v_P(d\alpha/d\pi) \geq 0$. Therefore

$$v_P(d\xi/d\pi) = r-1. \tag{10}$$

Similarly, one shows that $v_P(d\eta/d\pi) = s-1$. Hence, by Theorem 4, $v_P(d\xi/d\eta) = r-s$. By definition, $|r|$ resp. $|s|$ is the ramification index of P relative to ξ resp. η. □

Theorem 5 suggests defining the divisor $(d\xi)$ of $d\xi$ by

$$(d\xi) := V_\xi - 2(\xi)_\infty.$$

By (10), this definition agrees with the definition given in Section 11.6. More generally, for $\beta \in L$ we set

$$(\beta d\xi) := (\beta) + V_\xi - 2(\xi)_\infty.$$

As in Section 11.4, we define the *divisor class group* to be the quotient of the group of all divisors of L by the group $(L^\times)$ of principal divisors. The divisors of differentials all lie in the same class, the canonical class of L. For each divisor A, $d(A)$ and $\deg A$ depend only on the class in which A lies.

Let $n_\eta := [L:\mathbb{C}(\eta)]$. By Theorem 1, $\deg(\eta)_\infty = n_\eta$. Hence by Theorem 5

$$g_L := \frac{1}{2}v_\eta - n_\eta + 1 \tag{11}$$

is a number independent of η. According to the Riemann genus formula (Exercise 10.7), g_L is the genus of $\mathcal{F}(L)$. Hence we can also write (11) in the form

$$\deg(d\eta) = 2g_L - 2$$

known from Chapter 11, (18).

23.7 The Riemann-Roch theorem

Theorem 6. *(Riemann-Roch). Let* A *be an arbitrary divisor of* L, *let* C *be a divisor from the canonical class and let* g_L *be the genus of* L. *Then*

$$d(A) = \deg A - g_L + 1 + d(C-A).$$

Proof. Let η be a variable of L, let $n = n_\eta$ and let $\lambda_1,\ldots,\lambda_\eta$ be a normal basis of $(-A)_\eta$ with exponents $r_1,\ldots,r_n$. By Theorem 2 and Proposition 5

$$d(A) = \sum_{r_i \le 0} (1-r_i) = n_\eta - \sum_{i=1}^{n} r_i + \sum_{r_i \ge 1} (r_i-1)$$

$$= n_\eta - \frac{1}{2}v_\eta + \deg A + \sum_{r_i \ge 2} (r_i-1)$$

$$= 1 - g_L + \deg A + \sum_{r_i \ge 2} (r_i-1).$$

Hence to prove Theorem 6 it suffices to show

$$d(C-A) = \sum_{r_i \ge 2} (r_i-1).$$

Let $\lambda'_1,\ldots,\lambda'_n$ be a normal basis of $(-C+A)_\eta$ with exponents $r'_1,\ldots,r'_n$. Then we have to show

$$\sum_{r'_i \le 0} (1-r'_i) = \sum_{r_i \ge 2} (r_i-1).$$

This certainly holds if

$$r'_i + r_{n-i} = 2 \quad \text{for } i = 1,\ldots,n \tag{12}$$

We show the latter with the help of the theory of complementary ideals (Section 22.2).

We choose $C = (d\eta) = V_\eta - 2(\eta)_\infty$. Then $(C)_\eta = (V_\eta)_\eta = \mathfrak{D}_\eta$, where $\mathfrak{D}_\eta$ denotes the different of $O_\eta/\mathbb{C}[\eta]$ (Theorem 8, Chapter 22), and therefore

$$(-C+A)_\eta = (C)_\eta^{-1}(A)_\eta = (-A)^*_\eta. \tag{13}$$

Similarly

$$\eta^2(-C+A)_{1/\eta} = (-(d(1/\eta)) + A)_{1/\eta} = (-A)^*_{1/\eta}. \tag{14}$$

Let $\kappa_1,\dots,\kappa_n$ be the complementary basis of $\lambda_1,\dots,\lambda_n$. Then $\eta^{r_1}\kappa_1,\dots,\eta^{r_n}\kappa_n$ is the complementary basis of $\eta^{-r_1}\lambda_1,\dots,\eta^{-r_n}\lambda_n$, i.e. $\eta^{r_1}\kappa_1,\dots,\eta^{r_n}\kappa_n$ is a basis of $(-A)^*_{1/\eta}$. Because of (13) resp. (14), $\kappa_1,\dots,\kappa_n$ resp. $\eta^{r_1-2}\kappa_1,\dots,\eta^{r_n-2}\kappa_n$ is a basis of $(-C+A)_\eta$ resp. $(-C+A)_{1/\eta}$. This, together with Proposition 4, implies (12). □

Exercises

23.1 (Hurwitz) Let F be a function field with field of constants $\mathbb{C}$ and let L be a finite extension of F. Also let e_P be the ramification index of the point P of L over F. Show that the genera g_L and g_F of L and F satisfy the relation

$$2g_L - 2 = [L:F](2g_F-2) + \sum_P (e_P-1)$$

where the sum is taken over all points P of L at which L is branched over F.

23.2 Let F_0 be a field of characteristic 0 and let L be a finite extension of the function field $F_0(\xi)$ in the indeterminate ξ. An exponent v of L is a surjective homomorphism of $L^\times$ onto $\mathbb{Z}^+$ which maps F_0 onto $\{0\}$ and satisfies the inequality

$$v(\alpha+\beta) \geq \min\{v(\alpha),v(\beta)\} \quad \text{for } \alpha,\beta \in L^\times.$$

We set $v(0) = \infty$.

a) Suppose $\alpha \in L^\times$ is algebraic over F_0. Show that $v(\alpha) = 0$.

b) The field of elements of L which are algebraic over F_0 is called the *field of constants* of L. Show that the ring $O_v = \{\alpha \in L \mid v(\alpha) \geq 0\}$ has a single prime ideal $\mathfrak{P}_v$ and that $O_v/\mathfrak{P}_v$ is a finite extension of the field of constants of L.

c) Let $F_0 = \mathbb{C}$. Show that there is a one-to-one correspondence between the exponents and the points of L.

23.3 Generalise the results of Chapter 23 to the case, considered in Exercise 23.2, of a function field with an arbitrary field of constants of characteristic 0, where one defines the Riemann surface of L as the set of exponents of L.

24. The geometry of numbers

Minkowski recognised that the geometric concept of *convexity* of a point set in n-dimensional euclidean space can be used in number theory with surprising success. He first presented his ideas in the work "*Über positive quadratische Formen und über die kettenbruchähnlichen Algorithmen*" in the *J. reine angew. Math.* 107 (1891). In this chapter we confine ourselves to their applications to the arithmetic of algebraic number fields.

24.1 The lattice point theorem

We first recall a few basic concepts. A subset M of $\mathbb{R}^n$ is called *convex* when it contains the line segment connecting any two of its points P_1, P_2 (the points of the line segment are those of the form $\lambda_1 P_1 + \lambda_2 P_2$ with $\lambda_1 + \lambda_2 = 1$ and $\lambda_1, \lambda_2 \geq 0$). M is called *centrally symmetric* when -P lies in M along with P. A lattice in $\mathbb{R}^n$ is a subgroup of $\mathbb{R}^n$ of rank n (Section A1.4). Let $e_1,...,e_n$ be a basis of a lattice G. The *fundamental region* of G (with respect to the basis $e_1,...,e_n$) is the point set

$$\{\lambda_1 e_1 + ... + \lambda_n e_n \mid 0 \leq \lambda_i \leq 1 \text{ for } i = 1,...,n\}.$$

The volume of the fundamental region equals $|\det(\alpha_{ij})|$, where $\alpha_{i1},...,\alpha_{in}$ are the coordinates of e_i. The volume of the fundamental region is independent of the choice of basis $e_1,...,e_n$ of G.

Minkowski proved that a convex set always has a volume. In the case of the sets that interest us here we can compute the volume directly.

Theorem 1. *(Minkowski's lattice point theorem). Let* G *be a lattice in* $\mathbb{R}^n$ *and let* g *be the volume of a fundamental region of* G. *Also let* M *be a centrally symmetric convex closed set in* $\mathbb{R}^n$ *with midpoint* 0 *and volume* I(M) = m, *where* $m \geq 2^n g$. *Then* M *contains at least one point of* G *other than* 0.

Proof. Let $e_1,...,e_n$ be a basis of G and let G_0 be the associated fundamental region. Then

$$\bigcup_{x \in G} (x+G_0) = \mathbb{R}^n.$$

Similarly

$$\bigcup_{x\in G} (2x+2G_0) = \mathbb{R}^n.$$

We set $M_x = M \cap (2x+2G_0)$. Then

$$M = \bigcup_{x\in G} M_x,$$

and the closed set $-2x + M_x \subset 2G_0$ satisfies

$$\sum_{x\in G} I(-2x+M_x) = \sum_{x\in G} I(M_x) = m \geq 2^n g = I(2G_0).$$

Therefore there are points $x_1 \neq x_2$ in G with

$$(-2x_1+M_{x_1}) \cap (-2x_2+M_{x_2}) \neq \emptyset.$$

Let x_0 lie in this intersection, so

$$x_0 = -2x_1 + y_1 = -2x_2 + y_2 \quad \text{with} \quad y_1,y_2 \in M.$$

Since M is convex and has midpoint 0,

$$\tfrac{1}{2}(y_1-y_2) = x_2 - x_1 \neq 0 \quad \text{lies in} \quad M \cap G. \qquad \square$$

24.2 Application to the ideals of an algebraic number field

Let K be an algebraic number field of degree n over $\mathbb{Q}$. As in Section 19.4 we let $g_1,\ldots,g_{r_1}$ denote the real isomorphisms, and $g_{r_1+1},\bar{g}_{r_1+1},\ldots,g_r,\bar{g}_r$ the pairs of conjugate complex isomorphisms,of K into $\mathbb{C}$. We obtain an embedding of K in $\mathbb{R}^n$ by the mapping

$$\alpha \mapsto (g_1\alpha,\ldots,g_{r_1}\alpha,\mathrm{Re}\, g_{r_1+1}\alpha,\ldots,\mathrm{Re}\, g_r\alpha,\ldots,\mathrm{Re}\, g_r\alpha,\mathrm{Im}\, g_{r_1+1}\alpha,\mathrm{Im}\, g_r\alpha)$$

for $\alpha \in K$. We denote the coordinates of α under this embedding by $\alpha^{(1)},\ldots,\alpha^{(n)}$.

Theorem 2. *Let* $\mathfrak{A}$ *be an ideal of* K. *Under the given embedding* $\mathfrak{A}$ *is a lattice in* $\mathbb{R}^n$. *The volume of a fundamental region of this lattice equals* $2^{-r_2}N(\mathfrak{A})\sqrt{|D|}$, *where* r_2 *is half the number of complex conjugates of* K *and* D *is the discriminant of* O_K.

Proof. It suffices to prove that a basis $\alpha_1,\ldots,\alpha_n$ of $\mathfrak{A}$ satisfies

$$|\det(\alpha_i^{(j)})| = 2^{-r_2}N(\mathfrak{A})\sqrt{|D|}. \tag{1}$$

Equation (1) follows from Chapter 17, (6), Chapter 18, (3), and Section 18.7. □

We want to apply the Minkowski lattice point theorem to $\mathfrak{A}$, viewed as a lattice in $\mathbb{R}^n$. Our first goal is to find an element α of $\mathfrak{A}$ of smallest possible norm value $|N(\alpha)|$. The value of the norm of $\xi \in K$ is given by

$$|N(\xi)| = \prod_{\nu=1}^{n} |g_\nu \xi| = \prod_{\nu=1}^{r_1} |\xi^{(\nu)}| \prod_{\nu=r_1+1}^{r} |(\xi^{(\nu)}+i\xi^{(\nu+r_2)})|^2.$$

Similarly, if x is a point of $\mathbb{R}^n$ with coordinates $\lambda_1,\ldots,\lambda_n$ the number

$$N(x) := \prod_{\nu=1}^{n} |\lambda_\nu| \prod_{\nu=r_1+1}^{r} |(\lambda_\nu+i\lambda_{\nu+r_2})|^2$$

is called the *norm* of x with respect to K. Also let

$$N := \{x \in \mathbb{R}^n \mid |N(x)| \le 1\}.$$

Theorem 3. *Let* M *be a convex, closed and centrally symmetric set in* N *with midpoint* 0. *Then in each ideal* $\mathfrak{A}$ *of* K *there is an element* $\alpha \neq 0$ *with*

$$|N(\alpha)| \le \frac{2^r}{I(M)} N(\mathfrak{A})\sqrt{|D|}.$$

Proof. We apply Theorem 1 to the case $G = t\mathfrak{A}$, where $t > 0$ is determined by

$$I(M) = (2t)^n N(\mathfrak{A})\sqrt{|D|}2^{-r_2}. \tag{2}$$

Then the hypotheses of Theorem 1 are satisfied. Hence there is a $t\alpha \neq 0$ with $\alpha \in \mathfrak{A}$ and $t\alpha \in M$ and consequently

$$|N(\alpha)|t^n \le 1. \tag{3}$$

It follows from (2) and (3) that α satisfies the conditions of Theorem 3. □

Now it only remains to make a convenient choice of M. We set

$$M_s = \{x \in \mathbb{R}^n \,\Big|\, \sum_{\nu=1}^{r_1} |\lambda_\nu| + 2 \sum_{\nu=r_1+1}^{r} |\lambda_\nu + i\lambda_{\nu+r_2}| \le s\}.$$

Since the geometric mean is less than or equal to the arithmetic mean we have

$$|N(x)| \le \left[\frac{1}{2}\left(\sum_{\nu=1}^{r_1} |\lambda_\nu| + 2 \sum_{\nu=r_1+1}^{r} |\lambda_\nu + i\lambda_{\nu+r_2}|\right)\right]^n$$

and hence $M_n \subset N$. Obviously M_n is centrally symmetric and convex.

Theorem 4. $I(M_s) = \dfrac{2^{r_1-r_2}\pi^{r_2} s^n}{n!}$.

Proof. For $n = 1$ and $n = 2$ one derives the assertion from Fig. 15 which follows

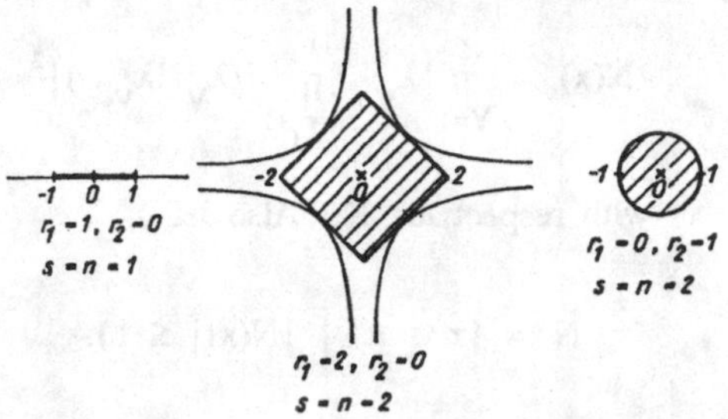

Fig. 15

We prove the theorem by induction on n, and suppose it already proved for natural numbers $< n$. Then with the increase of r_1 in passing from $n-1$ to n we have

$$I(M_s) = 2\int_0^s I(M_{s-\sigma})d\sigma = 2\int_0^s \frac{2^{r_1-1-r_2}\pi^{r_2}}{(n-1)!}(s-\sigma)^{n-1}d\sigma$$

and by increasing r_2 by passing from $n-2$ to n

$$I(M_s) = \int_0^{s/2} I(M_{s-2\sigma})(2\pi\sigma)d\sigma$$

$$= \int_0^{s/2} \frac{2^{r_1-r_2+1}\pi^{r_1-1}}{(n-2)!}(s-2\sigma)^{n-2}(2\pi\sigma)d\sigma.$$

It is easy to see that these integrals have the desired value. □

Because of Theorem 3 and Theorem 4 we have

Theorem 5. *Let* K *be an algebraic number field with* r_1 *real conjugates and* r_2 *pairs of complex conjugates,* $r_1 + 2r_2 = n$. *Also let* $\mathfrak{A}$ *be an ideal of* K. *Then there is an element* $\alpha \neq 0$ *in* K *with*

$$|N(\alpha)| \leq \left(\frac{4}{\pi}\right)^{r_2} \frac{n!}{n^n} N(\mathfrak{A})\sqrt{|D|}. \qquad \square$$

This has two important consequences:

Theorem 6. *(Minkowski discriminant theorem). Let* K *be an algebraic number field of degree* $n > 1$. *Then the absolute value of the discriminant* D *of* O_K *is greater than* 1. *More precisely*

$$|D| \geq \left(\frac{\pi}{4}\right)^{2r_2} \frac{n^{2n}}{n!^2}. \tag{4}$$

Proof. We take O_K as the ring $\mathfrak{A}$ in Theorem 5. Then for $\alpha \in O_K$, $\alpha \neq 0$, we have in any case $|N(\alpha)| \geq 1$ and therefore

$$1 \leq \left(\frac{4}{\pi}\right)^{r_2} \frac{n!}{n^n}\sqrt{|D|}.$$

One obtains $|D| > 1$ from this with the help of Stirling's formula (Chapter 9, Theorem 4). □

Theorem 7. *Each ideal class of* K *contains an integral ideal* $\mathfrak{B}$ *with*

$$N(\mathfrak{B}) \leq \left(\frac{4}{\pi}\right)^{r_2} \frac{n!}{n^n}\sqrt{|D|}.$$

Proof. By Theorem 5 it suffices to take, in the class of $\mathfrak{A}^{-1}$, the integral ideal $\mathfrak{B} := \alpha\mathfrak{A}^{-1}$ □

It follows from Theorem 7 that there are only finitely many ideal classes, as we have already proved in Section 19.1.

Theorem 7 can be used for effective computation of the class number of a number field K when one knows the discriminant of K together with something about the process of decomposition into primes in K and the group of units of K. The case of imaginary quadratic fields is particularly simple. However, in this case the reduction theory of quadratic forms (Section 2.6) yields the best method. Here we consider a cubic field as an example, in which one can manage without computation:

Let $K = \mathbb{Q}(\alpha)$ with $\alpha^3 + \alpha^2 - 2\alpha - 1 = 0$. K is the cubic subfield of the cyclotomic field $\mathbb{Q}(\zeta_7)$, $\alpha = \zeta_7 + \zeta_7^{-1}$ (Section 7.6). In Section 25.3 we shall prove that $7 = \mathfrak{p}^3$ is the only ramified prime number in K. Therefore $|D| = 7^2$ (Theorem 8, Chapter 22). Since K has only real conjugates, we have to investigate integral ideals $\mathfrak{B}$ with $N(\mathfrak{B}) \leq \frac{14}{9}$. It follows that the class number of K equals 1.

Exercises

24.1 Show that the quadratic fields with discriminant 5, 8, 12, 13, 17, 21, 24, 28, 29, 33, 37 have class number 1.

24.2 Show that the field $\mathbb{Q}(\sqrt{10})$ has class number 2.

24.3 Let $K := \mathbb{Q}(\alpha)$, where α is a zero of the polynomial $f(x) := x^3 + a_2x + 1$ with $a_2 = \pm 1$. Show that O_K is a principal ideal domain.

24.4 Let $K := \mathbb{Q}(\alpha)$, where α is a zero of the polynomial $f(x) := x^3 + a_2x + 2$ with $a_2 = \pm 2, \pm 4$.

a) Show that $1, \alpha, \alpha^2$ is a basis of O_K.

b) Show that O_K is a principal ideal domain.

24.5 Let $K_1, K_2, \ldots$ be an infinite sequence of number fields whose degrees $[K_n : \mathbb{Q}]$ tend to infinity as n increases. Show that the absolute values of the associated discriminants likewise tend to infinity.

24.6 (Hermite) Show, with the help of Theorem 1, that there are only finitely many number fields with bounded discriminant.

25. Normal extensions of algebraic number- and function fields

In this chapter we complete the synthesis of Galois theory (Chapter 17) and the theory of algebraic number- and function fields (Chapters 18 to 22) in the theory of normal extensions of such fields. The beginnings of this theory go back to Dedekind. Its later construction is due to Hilbert, who in 1896 wrote a report (the "*Zahlbericht*") on the current state of algebraic number theory for the Deutsche Mathematiker-Vereinigung, with many new results and new proofs. In the foreword to this report he wrote:

"*The theory of number fields is like an edifice of wonderful beauty and harmony; it seems to me that the most richly furnished part of this edifice is the theory of abelian and relatively abelian fields*".

By an "abelian field" Hilbert meant a normal extension of the field $\mathbb{Q}$ of rational numbers with abelian Galois group, and by a "relatively abelian field" he meant such an extension of an arbitrary algebraic number field. The theory of abelian fields stands at the centre of this chapter. The theory of relatively abelian fields was just beginning in 1896. It will be treated in the second volume of this book.

25.1 Normal extensions

We proceed from the same hypotheses as in Section 22.1: Γ is a euclidean ring with field of fractions P of characteristic 0, K is a finite extension of P and F is an intermediate field of K/P. In addition we assume that K/F is normal and that the residue class fields of Γ modulo prime ideals are finite or of characteristic 0. The latter condition guarantees us the applicability of Galois theory (Section 18.8) for normal extensions of residue class fields.

The Galois group G of K/F acts on the group $\mathcal{I}_K$ of fractional ideals of K. For $\mathfrak{A} \in \mathcal{I}_K$ and $g \in G$ one defines

$$g\mathfrak{A} := \{g\alpha \mid \alpha \in \mathfrak{A}\}.$$

Then $g\mathfrak{A}$ is obviously an ideal also, and

$$g(\mathfrak{A}\mathfrak{B}) = g\mathfrak{A}\cdot g\mathfrak{B} \quad \text{for} \quad \mathfrak{A},\mathfrak{B} \in \mathcal{I}_K$$

and

$$gh(\mathfrak{A}) = g(h\mathfrak{A}) \quad \text{for} \quad g,h \in G.$$

It follows from $N_{K/F}\alpha = \prod_{g\in G} g\alpha$ for $\alpha \in K^\times$ and the definition of $N_{K/F}\mathfrak{A}$ that

$$N_{K/F}\mathfrak{A} = \prod_{g\in G} g\mathfrak{A}.$$

In particular, for a prime ideal $\mathfrak{P}$ of K and $\mathfrak{p} = \mathfrak{P} \cap O_F$,

$$N_{K/F}\mathfrak{P} = \mathfrak{p}^f = \prod_{g\in G} g\mathfrak{P}.$$

The set of $g \in G$ with $g\mathfrak{P} = \mathfrak{P}$ constitutes a subgroup $Z_{\mathfrak{P}}$ of G, called the *decomposition group* of $\mathfrak{P}$. Let R be a system of coset representatives for $G/Z_{\mathfrak{P}}$ with $1 \in R$ and let z be the order of $Z_{\mathfrak{P}}$. Then

$$\mathfrak{p}^f = \prod_{g\in R} (g\mathfrak{P})^z,$$

where the prime ideals of $g\mathfrak{P}$ for $g \in R$ are all distinct. Therefore f is a divisor of z and $e = z/f$ resp. f are the common ramification index resp. inertia degree of the $g\mathfrak{P}$. Thus we have the following:

Theorem 1. *Let K/F be a normal extension with Galois group G and let $\mathfrak{P}$ be a prime ideal of K with ramification index e and inertia degree f. The prime divisors $\mathfrak{p} = \mathfrak{P} \cap O_{\mathfrak{P}}$ are the prime ideals $g\mathfrak{P}$ for $g \in G$. They all have the same ramification index e and inertia degree f. The decomposition group $Z_{\mathfrak{P}}$ of $\mathfrak{P}$ has order* ef. □

The fixed field $F_{\mathfrak{P}}$ of $Z_{\mathfrak{P}}$ is called the *decomposition field* of $\mathfrak{P}$.

Theorem 2. *With notation the same as in Theorem* 1, *$\mathfrak{P}$ has ramification index* e *over $F_{\mathfrak{P}}$, and inertia degree* f. *The prime ideal $\mathfrak{P} \cap O_{F_{\mathfrak{P}}}$ of $O_{F_{\mathfrak{P}}}$ has ramification index* 1 *over* F *and inertia degree* 1.

Proof. Since

$$N_{K/F_{\mathfrak{P}}}\mathfrak{P} = \mathfrak{P}^{ef},$$

this follows from Chapter 22, (3), (4) and Chapter 18, (12). □

The nth *ramification group* V_n *of* $\mathfrak{P}$, for $n = 0,1,...$ is defined by

$$V_n = \{t \in Z_{\mathfrak{P}} \mid t\alpha \equiv \alpha (\text{mod } \mathfrak{P}^{n+1}) \text{ for all } \alpha \in O_K\}.$$

The *zero-th ramification group* is also known as the *inertia group*, and the associated fixed field T is known as the *inertia field.* As one easily sees, the ramification groups $V_0, V_1, ...$ constitute a decreasing sequence of normal subgroups of $Z_{\mathfrak{P}}$.

V_0 is defined to be the largest subgroup of $Z_{\mathfrak{P}}$ which acts trivially on the residue class field $O_K/\mathfrak{P}$. We can therefore identify Z_K/V_0 with a subgroup of the Galois group $G_{\mathfrak{P}}$ of the normal extension $O_{\mathfrak{P}}/\mathfrak{P}$ over $O_F/\mathfrak{p}$.

The simplest facts about the ramification groups are collected together in the following theorem.

Theorem 3. *With the same notation as in Theorem* 1,

(i) $Z_{\mathfrak{P}}/V_0 = G_{\mathfrak{P}}$. *In particular, the order of* $Z_{\mathfrak{P}}/V_0$ *equals* f.

(ii) *When* K *is an algebraic number field, there is exactly one* $\bar{t} \in Z_{\mathfrak{P}}/V_0$ *with*

$$t\alpha \equiv \alpha^{N(\mathfrak{p})} (\text{mod } \mathfrak{P}) \text{ for all } \alpha \in O_K;$$

$\bar{t}$ *is a generator of* $Z_{\mathfrak{P}}/V_0$ *and is known as the Frobenius automorphism of* $\mathfrak{P}$ *over* K/F Section 18.8).

(iii) *Let* M *be an arbitrary intermediate field of* K/F *and let* H = G(K/M). *Then the decomposition group resp.* n^{th} *ramification group of* $\mathfrak{P}$ *over* M *is* $Z_{\mathfrak{P}} \cap H$ *resp.* $V_n \cap H$.

(iv) *In* K/T, $\mathfrak{P}$ *has ramification index* $e := [K:T]$ *and in* $T/F_{\mathfrak{P}}$, $\mathfrak{P} \cap O_T$ *has degree of inertia* $f = [T:F_{\mathfrak{P}}]$.

(v) *There are monomorphisms* $V_0/V_1 \to (O_K/\mathfrak{P})^{\times}$ *and* $V_n/V_{n+1} \to (O_K/\mathfrak{P})^{\times}$ *for* $n = 1,2,...$.

(vi) *For sufficiently large* n, $V_n = \{1\}$.

Proof. (i) By Theorem 2 we can assume, without loss of generality, that $F_{\mathfrak{P}} = F$. Then we have to show that each automorphism $\sigma \in G_{\mathfrak{P}}$ is induced by an automorphism of G. Let ζ be a generator of the extension $O_K/\mathfrak{P}$ over $O_F/\mathfrak{p}$ and let $\bar{f}_0$ be the corresponding minimal polynomial. Also let ζ be a representative of ζ in O_K. Then ζ has a minimal polynomial f_1 over F with coefficients in O_F. Let $\bar{f}_1$ be the corresponding polynomial with coefficients in $O_F/\mathfrak{p}$. The polynomials $\bar{f}_0$ and $\bar{f}_1$ have the zero ζ in common. Therefore $\bar{f}_0$ is a divisor of $\bar{f}_1$ (Theorem 11, A1). Thus $\sigma\zeta$ is a zero of $\bar{f}_1$. The assertion now follows from Theorem 3, Chapter 17.

(ii) See Section 18.8.

(iii) This follows immediately from the definition.

(iv) The decomposition group and the inertia group of $\mathfrak{P}$ over T equal V_0 by (iii). It follows, by application of (i) to the extension K/T, that $\mathfrak{P}$ has inertia degree 1 over T. Assertion (iv) now follows from Chapter 22, (3), (4).

(v) Let π be an element of O_K with $\pi \in \mathfrak{P}$ and $\pi \notin \mathfrak{P}^2$. We first show that the following equation holds for $n = 1,2,\ldots$.

$$V_n = \{t \in V_0 \,|\, t\pi \equiv \pi (\mathrm{mod}\ \mathfrak{P}^{n+1})\} \tag{1}$$

Let S be a system of representatives of $O_K/\mathfrak{P}$ in O_K. Then for each $\alpha \in O_K$ there is a unique representation

$$\alpha \equiv \sum_{i=0}^{n} \alpha_i \pi^i (\mathrm{mod}\ \mathfrak{P}^{n+1}),$$

where the $\alpha_0,\ldots,\alpha_n$ are from S. By (iv) we can take the system R of representatives in O_T. Then for $t \in V_0$ with $t\pi \equiv \pi(\mathrm{mod}\ \mathfrak{P}^{n+1})$ we have

$$t\alpha \equiv \sum_{i=0}^{n} \alpha_i (t\pi)^i \equiv \sum_{i=0}^{n} \alpha_i \pi^i \equiv \alpha (\mathrm{mod}\ \mathfrak{P}^{n+1}),$$

whence (1) is proved.

Let $t \in V_0$. Then there is a unique class $\bar{\alpha}_t \in (O_K/\mathfrak{P})^{\times}$ with $\alpha_t \in T$ and

$$t\pi \equiv \pi\alpha_t (\mathrm{mod}\ \mathfrak{P}^2).$$

The map $t \mapsto \bar{\alpha}_t$ induces a monomorphism of V_0/V_1 into $(O_K/\mathfrak{P})^{\times}$.

Let $t \in V_n$, $n = 1,2,\ldots$. Then there is a unique class $\bar{\beta}_t \in O_K/\mathfrak{P}$ with $\beta_t \in T$ and

$$t\pi \equiv \pi + \beta_t \pi^{n+1} (\mathrm{mod}\ \mathfrak{P}^{n+2}). \tag{2}$$

The map $t \mapsto \beta_t$ induces the desired monomorphism of V_n/V_{n+1} into $(O_K/\mathfrak{P})^{\times}$.

(vi) follows from (1): because $t\pi - \pi \in \mathfrak{P}^n$ for all natural numbers n implies $t\pi = \pi$. The extension $T(\pi)/T$ has ramification index e. Hence $T(\pi) = K$ by (iv), and $t\pi = \pi$, for $t \in V_0$, implies $t = 1$. □

25.2 Proof of the Dedekind different theorem

We use the ramification group to prove the Dedekind different theorem (Theorem 6, Chapter 22).

First we consider the case of a normal extension K/F.

Theorem 4. *Notation is the same as in Theorem* 1 *and we set* $[V_n] =: v_n$. *Then* $\mathfrak{P}$ *divides the different* $\mathfrak{D}$ *with exponent*

$$\sum_{n=0}^{\infty} (v_n - 1) = \sum_{n=0}^{\infty} (n+1)(v_n - v_{n+1}). \tag{3}$$

Proof. We use the notation of 25.1. It follows easily from (2) that, for sufficiently large n, the power to which $\mathfrak{P}$ divides the element different $D_{K/T}(\alpha) = \prod_{t \in V_0 - \{1\}} (\alpha - t\alpha)$ is at least as large, for each $\alpha \in O_K$, as the power to which $\mathfrak{P}$ divides $D_{K/T}(\pi)$. By definition of the ramification group, the exponent h to which $\mathfrak{P}$ divides $D_{K/T}(\pi)$ is given precisely by (3). Hence by Theorem 5, Chapter 22, $\mathfrak{P}^h$ is the power of $\mathfrak{P}$ which divides $\mathfrak{D}_{K/T}$. Thus by the different tower theorem (Theorem 3, Chapter 22) it suffices to show that the differents $\mathfrak{D}_{T/F_{\mathfrak{P}}}$ and $\mathfrak{D}_{F_{\mathfrak{P}}/F}$ are prime to $\mathfrak{P}$.

Let ζ be a generator of the extension $O_T/\mathfrak{P} \cap O_T$ and let ζ be a representative of ζ in O_T. Then $\mathfrak{P}$ is not a divisor $\zeta - t\zeta$ for $t \in Z_{\mathfrak{P}} - V_0$, i.e. $\mathfrak{P} \nmid D_{T/F_{\mathfrak{P}}}(\zeta)$ and therefore $\mathfrak{P} \nmid \mathfrak{D}_{T/F_{\mathfrak{P}}}$.

It remains to show $\mathfrak{P} \nmid \mathfrak{D}_{F_{\mathfrak{P}}/F}$. Let $\mathfrak{p}_1 := \mathfrak{P} \cap O_{F_{\mathfrak{P}}}$ and let $\mathfrak{p} = \mathfrak{p}_1\mathfrak{a}$ with $\mathfrak{a} \in \mathcal{I}_F$. By Theorem 2, $\mathfrak{a}$ is prime to $\mathfrak{p}_1$. Hence there is a $\pi_1 \in \mathfrak{p}_1$ with $\mathfrak{q} \nmid \mathfrak{p}_1$ for all prime ideals $\mathfrak{q}$ of O_F which divide $\mathfrak{a}$ (Theorem 14, Chapter 18). We now consider $\mathfrak{p}_1, \mathfrak{q}, \mathfrak{a}$ as ideals of O_K. Since

$$\mathfrak{p} = \prod_{g \in R} g\mathfrak{P}^e = \prod_{g \in R} g\mathfrak{p}_1,$$

$g\pi_1$ is prime to $\mathfrak{p}_1$ for $g \in R - \{1\}$. Hence $D_{F_{\mathfrak{P}}/F}(\pi_1) = \prod_{g \in R - \{1\}} (\pi_1 - g\pi_1)$ is also prime to $\mathfrak{p}_1$. □

We now come to the proof of Theorem 6, Chapter 22, which we formulate as follows in the notation of this chapter:

Let M/F *be an arbitrary finite extension and let* $\mathfrak{p}_M$ *be a prime ideal of* O_M *with ramification index* e_1. *Also let* p *be the characteristic of* $O_M/\mathfrak{p}_M$. *Then* $\mathfrak{p}_M^{e_1 - 1} \mid \mathfrak{D}_{M/F}$, *and* $\mathfrak{p}_M^{e_1} \mid \mathfrak{D}_{M/F}$ *if and only if* $p \mid e_1$.

Proof. Let $K \supset M$ be a normal extension of F. We use the above notation with respect to K/F. Also let $H = G(K/M)$ and let $v'_n = [V_n \cap H]$, $v_0 = e'$. By the different tower theorem

$$\mathfrak{D}_{M/F} = \mathfrak{D}_{K/F}\mathfrak{D}_{K/M}^{-1},$$

and by Theorem 4, $\mathfrak{P}$ divides $\mathfrak{D}$ to the exponent

$$\sum_{n=0}^{\infty} (v_n - v'_n) \geq e - e'. \tag{4}$$

It follows that $\mathfrak{p}_M^{e_1 - 1} = \mathfrak{P}^{e-e'} \mid \mathfrak{D}_{M/F}$.

The equality sign holds in (4) if and only if $V_n \cap H = V_n$ for $n \geq 1$. By Theorem 3, (v), this means that p is not a divisor of e_1. □

25.3 Cyclotomic fields

Let m be a natural number and let ζ_m be a *primitive* m^{th} *root of unity*, i.e. a complex number with $\zeta_m^m = 1$ and $\zeta_m^i \neq 1$ for $1 \leq i < m$. Gauss considered the case of a prime number m (Chapter 3). In this section we want to treat the general case. By a *cyclotomic field* one means a number field of the form $\mathbb{Q}(\zeta_m)$ for a certain m.

We denote the ring of integers of $\mathbb{Q}(\zeta_m)$ by O_m.

We first consider the case where $m = l^h$ is a power of prime number l. Then ζ_m is a zero of the polynomial

$$f_m(x) := (x^m-1)(x^{m/l}-1)^{-1} = x^{m-m/l} + x^{m-2m/l} + \dots + x^{m/l} + 1.$$

One obtains all zeros of this polynomial in the form ζ_m^r, as r runs through a prime residue system R mod m. It follows that $\mathbb{Q}(\zeta_m)$ is a normal extension of $\mathbb{Q}$.

Theorem 5. (i) $\mathbb{Q}(\zeta_m)$ *has degree* $\phi(m) = l^{h-1}(l-1)$.
(ii) l *has only one prime divisor* $\mathfrak{l} = (1-\zeta_m)O_m$ *in* O_m *and is fully ramified:* $lO_m = \mathfrak{l}^{\phi(m)}$.

Proof. By Chapter 18, (12) it suffices to prove the second assertion. We have

$$f_m(x) = \prod_{r\in R} (x-\zeta_m^r)$$

and therefore

$$l = f_m(1) = \prod_{r\in R} (1-\zeta_m^r).$$

Suppose $s \in \mathbb{Z}$ with $rs \equiv 1 \pmod m$. Then

$$(1-\zeta_m^r)(1-\zeta_m)^{-1} \in O_m \text{ and } (1-\zeta_m)(1-\zeta_m^r)^{-1} = (1-\zeta_m^{rs})(1-\zeta_m^r)^{-1} \in O_m,$$

i.e. $1 - \zeta_m$ and $1 - \zeta_m^r$ differ only by a unit of O_m. □

Theorem 6. *The numbers* $1, \zeta_m, \dots, \zeta_m^{\phi(m)-1}$ *constitute a basis of* O_m *over* $\mathbb{Z}$.

Proof. It suffices to show that $\Delta(\zeta_m) = \Delta(1,\zeta_m,\dots,\zeta_m^{\phi(m)-1})$ equals the discriminant of $O_m/\mathbb{Z}$ (Section 18.2). By Section 17.6 we have

$$\Delta(\zeta_m) = \pm N(D(\zeta_m)) = \pm N(\prod_{r\in R-\{1\}} (\zeta_m-\zeta_m^r)) = \pm N(D(\zeta_m-1)), \tag{5}$$

where $D(\alpha)$ denotes the element different of α. (Applying Theorem 5 to the divisor of m). On the other hand, it follows from the proof of Theorem 4 that the power of $\mathfrak{l}$ left in $D(\zeta_m-1)$ is precisely that to which it divides the different of $\mathbb{Q}(\zeta_m)/\mathbb{Q}$. It now follows, from (5) and Theorem 4, Chapter 22, that $\Delta(\zeta_m)$ equals the discriminant of $O_m/\mathbb{Z}$. □

Since l is the only prime divisor of $\Delta(\zeta_m)$, Theorem 8 of Chapter 22 yields the following:

Theorem 7. *l is the only prime number ramified in* $\mathbb{Q}(\zeta_m)/\mathbb{Q}$. □

Now let m be an arbitrary natural number and let $m = \prod_{i=1}^{s} l_i^{h_i}$ be the prime decomposition of m. We set $m_i = l_i^{h_i}$. Then

$$\mathbb{Q}(\zeta_m) = \prod_{i=1}^{s} \mathbb{Q}(\zeta_{m_i}), \quad \mathbb{Q}(\zeta_{m_u}) \cap \prod_{i=1}^{u-1} \mathbb{Q}(\zeta_{m_i}) = \mathbb{Q} \quad \text{for } u = 2,3,\ldots,s$$

It follows from the last equation and Chapter 22, (3), that $\mathbb{Q}(\zeta_{m_i})$ is fully ramified for l_i and is unramified for all other prime numbers. By Section 17.5 we have

$$[\mathbb{Q}(\zeta_m) : \mathbb{Q}] = \prod_{i=1}^{s} [\mathbb{Q}(\zeta_{m_i}) : \mathbb{Q}].$$

We have thereby proved the following theorem:

Theorem 8. $[\mathbb{Q}(\zeta_m) : \mathbb{Q}] = \phi(m) = \prod_{i=1}^{s} l_i^{h_i-1}(l_i-1)$. *A prime number* p *with* $p \nmid m$ *is unramified in* $\mathbb{Q}(\zeta_m)$. □

We now come to the Galois theory of the field $\mathbb{Q}(\zeta_m)$.

Theorem 9. *The Galois group* G_m *of* $\mathbb{Q}(\zeta_m)/\mathbb{Q}$ *is canonically isomorphic to* $(\mathbb{Z}/m\mathbb{Z}$

Proof. By Theorem 8 the degree of $\mathbb{Q}(\zeta_m)/\mathbb{Q}$ equals the order of $(\mathbb{Z}/m\mathbb{Z})^\times$. Let $f_m(x)$ be the minimal polynomial associated with ζ_m; $f_m(x)$ is a divisor of $x^m - 1$. The zeros of $f_m(x)$ are therefore of the form ζ_m^r, where r must be prime to m, otherwise ζ_m^r would satisfy an equation of lower degree. In this way we obtain $\phi(m)$ roots of unity. Since the degree of $f_m(x)$ is also $\phi(m)$, we thus obtain all the zeros of $f_m(x)$. It follows that

$$g_r\zeta_m = \zeta_m^r,$$

where r is prime to m, determines an automorphism of G_m. The mapping $\bar{r} \mapsto g_r$ defines a bijection of $(\mathbb{Z}/m\mathbb{Z})^\times$ onto G_m which does not depend on the choice of r in the associated residue class mod m, nor on the choice of the primitive root of unity ζ_m. As one easily sees, this bijection is an isomorphism. □

Theorem 10. *Let* p *be a prime number relatively prime to* m *and let* $\mathfrak{P}$ *be a prime divisor of* p *in* O_m. *Then the decomposition group* Z_m *of* $\mathfrak{P}$ *is generated by* g_p.

Proof. By Theorem 3, (ii),

$$g\alpha \equiv \alpha^p (\mathrm{mod}\ \mathfrak{P}) \quad \text{for all} \quad \alpha \in O_m$$

determines a generator g of $Z_{\mathfrak{P}}$; g is of the form g_r. In particular,

$$g_r\zeta_m = \zeta_m^r \equiv \zeta_m^p (\mathrm{mod}\ \mathfrak{P}). \tag{6}$$

By differentiation of x^{m-1}-1, followed by substitution of ζ_m^r, one obtains

$$m\zeta_m^{r(m-1)} = \Pi(\zeta_m^r - \zeta_m^s), \tag{7}$$

where the product is taken over a system of representatives of the residue classes mod m that are different from r. It follows from (7) that $\zeta_m^r - \zeta_m^p$ is either 0 or a divisor of m. The second possibility is excluded by (6). □

As a consequence of Theorem 10 one obtains:

Theorem 11. *The inertia degree of* $\mathfrak{P}$ *equals the order of* $\bar{p}$ *in* $(\mathbb{Z}/m\mathbb{Z})^\times$. □

With this we have found the decomposition theorem for prime numbers in the field $\mathbb{Q}(\zeta_m)$. In what follows we want to use it to give a new proof of the law of quadratic reciprocity (Theorem 10, Chapter 1).

In Section 18.9 we have shown that the Legendre symbol $(\frac{q}{p})$ for the odd primes p, q is characterised by

$$\left(\frac{q^*}{p}\right) = \begin{cases} 1 & \text{when } p \text{ splits in } \mathbb{Q}(\sqrt{q^*}), \\ -1 & \text{when } p \text{ is inert in } \mathbb{Q}(\sqrt{q^*}), \end{cases}$$

where $q^* = (-1)^{(q-1)/2}q$.

$\mathbb{Q}(\sqrt{q^*})$ is the only quadratic field in which only q is ramified. Since $\mathbb{Q}(\zeta_q)$ contains a subfield with this property, $\mathbb{Q}(\sqrt{q^*})$ is contained in $\mathbb{Q}(\zeta_q)$. $\mathbb{Q}(\sqrt{q^*})$ is the fixed field of the subgroup $\{g^2 \mid g \in G_q\}$ of G_q. Hence p splits in $\mathbb{Q}(\sqrt{q^*})$ if and only if g_p is a square in G_q, i.e. when p is a quadratic residue mod q. Consequently, by definition of the Legendre symbol,

$$\left(\frac{p}{q}\right) = \left(\frac{q^*}{p}\right) = \left(\frac{(-1)^{(q-1)/2}}{p}\right)\left(\frac{q}{p}\right) = (-1)^{(p-1)(q-1)/4}\left(\frac{q}{p}\right).$$

Here we have used the first extension law (Theorem 7, Chapter 1).

In contrast to the proof given in Section 3.7, the one just given illuminates the background of the law of quadratic reciprocity. It is a consequence of general laws in cyclotomic fields. Already in the last quarter of the 19th century, Kronecker, Weber and Hilbert began to extend the laws to abelian extensions of arbitrary number fields, thereby paving the way for the general reciprocity theorem, which was stated by Artin in 1924 and proved by him in 1927. We shall go into this in the second volume of this book.

25.4 The ζ-function of a cyclotomic field

In this section we relate Chapter 6 to Chapter 20. We connect the Dirichlet L-series for characters of $(\mathbb{Z}/m\mathbb{Z})^\times$ with the ζ-function of the field $\mathbb{Q}(\zeta_m)$, and make good the proof of Theorem 4, Chapter 6. In this section σ denotes a real variable.

Theorem 12. *Let* M *be an arbitrary natural number and let*

$$G_m(\sigma) := \prod_{\mathfrak{P}|m}\left(1 - \frac{1}{N(\mathfrak{P})^\sigma}\right)^{-1} \quad \textit{for} \quad \sigma > 0,$$

where the product is taken over all prime divisors $\mathfrak{P}$ *of* m *in* $\mathbb{Q}(\zeta_m)$. *Then*

$$\zeta_{\mathbb{Q}(\zeta_m)}(\sigma) = G_m(\sigma)\prod_\chi L(\sigma,\chi) \quad \textit{for} \quad \sigma > 1, \tag{8}$$

where the product is taken over all characters χ *of* $(\mathbb{Z}/m\mathbb{Z})^\times$.

Proof. Since the series

$$L(\sigma,\chi) = \prod_p \left[1 - \frac{\chi(p)}{p^\sigma}\right]^{-1}$$

are absolutely convergent for $\sigma > 1$ (Chapter 6), we can order the factors in

$$\prod_\chi \prod_p \left[1 - \frac{\chi(p)}{p^\sigma}\right]^{-1}$$

arbitrarily. Therefore

$$\prod_\chi L(\sigma,\chi) = \prod_p \prod_\chi \left[1 - \frac{\chi(p)}{p^\sigma}\right]^{-1}.$$

Hence to prove Theorem 12 it suffices to show that, for $p \nmid m$,

$$\left[1 - \frac{1}{N(\mathfrak{P})^\sigma}\right]^{\phi(m)/f_p} = \prod_\chi \left[1 - \frac{\chi(p)}{p^\sigma}\right], \tag{9}$$

where f_p is the inertia degree of the prime divisor $\mathfrak{P}$ of p in $\mathbb{Q}(\zeta_m)$ (Chapter 20, (25)). Since $N(\mathfrak{P}) = p^{f_p}$, (9) results from the following

Lemma 1. *Let* x *be an indeterminate. Then*

$$(x^{f_p}-1)^{\phi(m)/f_p} = \prod_\chi (x-\chi(p))$$

where the product is taken over all characters χ *of* $(\mathbb{Z}/m\mathbb{Z})^\times$.

Proof. By Theorem 11, the class $\bar{p}$ of p in $(\mathbb{Z}/m\mathbb{Z})^\times$ generates a cyclic group C of order f_p. A character χ' of C admits continuation to $[(\mathbb{Z}/m\mathbb{Z})^\times/C] = \phi(m)/f_p$ different characters of $(\mathbb{Z}/m\mathbb{Z})^\times$. Thus χ' is determined by its value $\chi'(\bar{p})$, which can be an arbitrary f_p^{th} root of unity. □

Theorem 13. *(Kummer 1847). In the notation of Section 6.3 and 20.1, with* $L = \mathbb{Q}(\zeta_m)$, *the following holds for* $m > 2$:

$$G_m(1) \prod_{p|m} (1-\frac{1}{p}) \prod_{\chi\neq\chi_0} L(1,\chi) = \frac{(2\pi)^{\phi(m)/2} Rh}{w\sqrt{|D|}}. \tag{10}$$

Proof. The Riemann ζ-function satisfies

$$\zeta(\sigma) = \prod_{p|m}\left(1 - \frac{1}{p^{\sigma}}\right)^{-1} L(\sigma,\chi_0) \quad \text{for } \sigma > 1. \tag{11}$$

Theorem 13 then follows by dividing (8) by (11) and letting $\sigma \to 1$. □

Theorem 4, Chapter 6, is now an immediate consequence of Theorem 13, since the numbers on the right-hand side of (10) are all different from 0.

Kummer proved (10) for the case where m is a prime number, and made it his point of departure for deep results on the class number h of $\mathbb{Q}(\zeta_m)$, with application to the solution of Fermat's problem in special cases. (See Borevich-Shafarevich [1], Chapter 5 or Edwards [1].)

25.5 The theorem of Kronecker and Weber

Theorem 9 shows that the Galois group of $\mathbb{Q}_m(\zeta)/\mathbb{Q}$ is abelian. This result has the following converse:

Theorem 14. (*Theorem of Kronecker and Weber*). *Let* $L/\mathbb{Q}$ *be a normal extension with abelian Galois group. Then there is an* m *with* $L \subset \mathbb{Q}(\zeta_m)$.

Theorem 14 was formulated by Kronecker in 1853 in his work "*Über die algebraisch auflösbaren Gleichungen*" (*Monatsber. Preuss. Akad. Wiss.* (1853)) and first proved, though with gaps, by Weber in 1886 (*Theorie der Abelschen Zahlkörper*, Acta. Math. 8 (1886)).[1)] The first complete proof is due to Hilbert and is presented in his Zahlbericht. It makes essential use of the Minkowski discriminant theorem (Theorem 6, Chapter 24) and the theory of ramification groups (Section 25.1). Here we essentially follow Hilbert's proof.

We prove Theorem 14 with the help of a series of reductions and propositions. Let G be the Galois group of $L/\mathbb{Q}$. Then G is a direct product of cyclic groups of prime power order (Theorem 7', A1). It therefore suffices to prove the theorem for such groups. Hence we assume that G is a cyclic group of order l^h, where l is a prime number. Let M be the unique subfield of L of degree l over $\mathbb{Q}$.

[1)] See O. Neumann, *Two proofs of the Kronecker-Weber Theorem 'according to Kronecker and Weber'*, *J. reine angew. math.* 323 (1981).

Let $p \neq l$ be a prime that is ramified in M, and let $\mathfrak{P}$ be a prime divisor of p in L. Then $V_0(\mathfrak{P}) = G$, and $[V_0] = l^h$ is a divisor of $[O_L:\mathfrak{P}] - 1$ (Theorem 3). Since $\mathfrak{P}$ is fully ramified we have

$$p \equiv 1 \pmod{l^h}.$$

Let ζ_p be a primitive p^{th} root of unity and let K_p be the subfield of $\mathbb{Q}(\zeta_p)$ of degree l^h. Let the composite $L' = LK_p$ have degree $l^{h+h'}$ over $\mathbb{Q}$. Also let $\mathfrak{P}'$ be a prime ideal of L' with $\mathfrak{P}'|\mathfrak{P}$. The inertia group $V_0(\mathfrak{P}')$ is cyclic (Theorem 3, (v)). Since $G' = G(L'/\mathbb{Q})$ is a homomorphic image of $G \times G(K_p/\mathbb{Q})$ (Theorem 8, Chapter 17), $[V_0(\mathfrak{P}')] = l^h$. Let L_1 be the fixed field of $V_0(\mathfrak{P}')$. Then $\mathfrak{P}' \cap L_1$ is unramified in $L_1/\mathbb{Q}$ and hence $L_1 \cap K_p = \mathbb{Q}$. Thus, by Theorem 8, Chapter 17,

$$[L_1K_p:\mathbb{Q}] = [L_1:\mathbb{Q}][K_p:\mathbb{Q}] = l^{h+h'} = [LK_p:\mathbb{Q}]$$

and therefore $L_1K_p = LK_p$. Hence it suffices to show that L_1 is a cyclotomic field.

L_1 contains one fewer ramified prime p than L. By repetition of the process one finally arrives at an extension whose single subfield of degree l over $\mathbb{Q}$ is ramified only for the prime l. We can therefore assume that L has this property.

We now concern ourselves with the extension $M/\mathbb{Q}$, which is cyclic of degree l and ramified at most for the prime l. By Dedekind's and Minkowski's discriminant theorems (Theorem 8, Chapter 22, Theorem 6, Chapter 24) l is ramified, since there are no extensions unramified over $\mathbb{Q}$. We want to show that M is contained in $\mathbb{Q}(\zeta_{l^2})$ for $l \neq 2$ and in $\mathbb{Q}(\zeta_8)$ for $l = 2$. To do this we need a few propositions, which are also interesting in their own right.

Proposition 1. *Let* N/F *be a normal extension of prime degree* l. *Suppose that the field* F *contains the* l^{th} *roots of unity. Also suppose that* α *is a generator of* N/F *with* $\alpha^l = a \in F$. *Then each element* $\beta \in N$ *such that* $\beta^l \in F$ *has the form*

$$\beta = \alpha^t b,$$

where $t \in \mathbb{Z}$ *and* $b \in F$.

Proof. β has the form $\beta = b_0 + b_1\alpha + \ldots + b_{l-1}\alpha^{l-1}$. Let τ be the automorphism of N/F that carries α to $\zeta_l\alpha$ (Section 8.7). Then one has $\tau\beta = \zeta_l^t\beta$ for a certain $t \in \mathbb{Z}$ and

$$b_0 + b_1\zeta_l\alpha + \ldots + b_{l-1}(\zeta_l\alpha)^{l-1} = \zeta_l^t b_0 + \zeta_l^t b_1\alpha + \ldots + \zeta_l^t b_{l-1}\alpha^{l-1}.$$

This implies

$$b_i = b_i\zeta_l^{i-t} \quad \text{for} \quad i = 0,1,\ldots,l-1$$

and therefore $b_i = 0$ for $i \neq t$. □

Proposition 2. *Let* α *be a generator of* $M(\zeta_l)/\mathbb{Q}(\zeta_l)$ *which satisfies an equation* $\alpha^l = a$ *with* $a \in \mathbb{Q}(\zeta_l)$. *Let* g *be a primitive root* mod 1 *and let* $\sigma : \zeta_l \mapsto \zeta_l^g$ *be the associated automorphism of* $\mathbb{Q}(\zeta_l)$. *Then*

$$\sigma a = a^g b^l$$

for some $b \in \mathbb{Q}(\zeta_l)$.

Proof. By Theorem 3, Chapter 17 there is an extension σ' of σ to $M(\zeta_l)$. For the latter we have

$$(\sigma'\alpha)^l = \sigma a.$$

By Proposition 1, $\sigma'\alpha$ is of the form

$$\sigma'\alpha = \alpha^t b' \quad \text{with} \quad t \in \mathbb{Z},\ b' \in \mathbb{Q}(\zeta_l).$$

Let τ be the automorphism of $M(\zeta_l)/\mathbb{Q}(\zeta_l)$ with $\tau\alpha = \zeta_l\alpha$. Then

$$\sigma'\tau\alpha = \sigma'\zeta_l\alpha = \zeta_l^g\alpha^t b' = \tau\sigma'\alpha = \tau\alpha^t b' = \zeta_l^t\alpha^t b'$$

and therefore $t \equiv g \pmod{l}$. It follows that

$$\sigma a = a^t b'^l = a^g(a^{(t-g)/l}b')^l. \quad □$$

We now want to normalise the generator α of $M(\zeta_l)/\mathbb{Q}(\zeta_l)$. For this purpose we assume that $l \neq 2$.

Let $\mathfrak{l}$ be the prime ideal of $\mathbb{Q}(\zeta_l)$ that divides l. Then $\mathfrak{l} = \lambda O_l$ with $\lambda = 1 - \zeta_l$ and $\mathfrak{l}^{l-1} = lO_l$ (Theorem 5). As one easily sees

$$\sigma\lambda = 1 - \zeta_l^g \equiv g\lambda \pmod{\mathfrak{l}^2}. \tag{12}$$

Proposition 3. *There is a generator* α *of* $M(\zeta_l)/\mathbb{Q}(\zeta_l)$ *with* $\alpha^l \in O_l$ *and* $\alpha^l \equiv 1$ $\pmod{\mathfrak{l}}$.

Proof. Let α_1 be a generator of $M(\zeta_l)/\mathbb{Q}(\zeta_l)$ with $\alpha_1^l \in \mathbb{Q}(\zeta_l)$. We set $a_1 := \alpha_1^l$. Since $\sigma\mathfrak{l} = \mathfrak{l}$, $a_1^{-1}\sigma a_1$ is an element of $\mathbb{Q}(\zeta_l)$, the prime ideal decomposition of which contains no power of $\mathfrak{l}$. Consequently there is a natural number c prime to $\mathfrak{l}$ such that $c^l a_1^{-1}\sigma a_1$ is an integer. By the Fermat theorem (Chapter 18, (14))

$$(c^l a_1^{-1}\sigma a_1)^{l-1} \equiv 1 \pmod{\mathfrak{l}}.$$

We set $a := (c^l a_1^{-1}\sigma a_1)^{l-1}$. By Proposition 2 there is a $b \in \mathbb{Q}(\zeta_l)$ with

$$a = c^{l(l-1)} a_1^{(g-1)(l-1)} b^{l(l-1)}.$$

From the hypothesis $l \neq 2$ it follows that $g - 1 \not\equiv 0 \pmod{l}$. Therefore $\alpha := (c\alpha_1^{g-1}b)^{l-1}$ does what is required. □

Proposition 4. *Let* α *be as in Proposition* 3, *let* $a := \alpha^l$ *and let* i *be a natural number with*

$$a \equiv 1 + i\lambda \pmod{\mathfrak{l}^2}.$$

Then

$$a\zeta_l^i \equiv 1 \pmod{\mathfrak{l}^l}.$$

Proof. We set $d := a\zeta_l^i$. Let j be the greatest natural number with

$$a \equiv 1 + k\lambda^j \pmod{\mathfrak{l}^{j+1}} \text{ for some } k \in \mathbb{Z},\ k \not\equiv 0 \pmod{l}.$$

By Proposition 2, $d^{-g}\sigma d = b^l$ with $b \in \mathbb{Q}(\zeta_l)$. Since no power of $\mathfrak{l}$ appears in the prime ideal decomposition of b, there are elements b_1, b_2 of O_l with $b = b_1/b_2$. It follows from

$$b_2^l \sigma d = b_1^l d^g$$

and (12) that

$$b_2^l(1+kg^j\lambda^j) \equiv b_1^l(1+gk\lambda^j) \pmod{l^{j+1}}. \tag{13}$$

By (13) and the Fermat theorem we have $b_1 \equiv b_2 \pmod{l}$, whence

$$b_1^l \equiv b_2^l \pmod{l^l} \tag{14}$$

by the binomial theorem. When $j \geq l$, this proves Proposition 4. When $j < l$ it follows from (13) and (14) that

$$1 + kg^j\lambda^j \equiv 1 + gk\lambda^j \pmod{l^{j+1}}$$

and hence $g^{j-1} \equiv 1 \pmod{l}$. Since $j \geq 2$ this is impossible. □

We now choose an l^2-th root of unity ζ_{l^2} with $\zeta_{l^2}^l = \zeta_l$. Then we get $d = (\alpha\zeta_{l^2}^i)^l$. We also set $\beta := \alpha\zeta_{l^2}^i$. By construction, β is a generator of $M(\zeta_{l^2})/\mathbb{Q}(\zeta_{l^2})$ with $\beta^l \in O_l$ and $\beta^l \equiv 1 \pmod{l^l}$. We want to show that β lies in $\mathbb{Q}(\zeta_{l^2})$.

The number $\xi = (\beta-1)/\lambda$ satisfies the equation

$$\sum_{\nu=1}^{l} \binom{l}{\nu} \xi^\nu \lambda^\nu = d.$$

It follows that

$$f(\xi) := \xi^l + \sum_{\nu-1}^{1-l} \binom{l}{\nu} \lambda^{\nu-1}\xi^\nu + (1-d)\lambda^{-l} = 0, \tag{15}$$

whence it is clear that ξ is an integer. We now suppose that ξ does not lie in $\mathbb{Q}(\zeta_l)$. Then f is an irreducible polynomial over $\mathbb{Q}(\zeta_l)$, and the different $D(\xi)$ satisfies

$$D(\xi) = f'(\xi) \equiv l\lambda^{1-l} \pmod{l}. \tag{16}$$

It also follows from (16) and the Dedekind different theory (Chapter 22) that $\mathbb{Q}(\xi,\xi_l)/\mathbb{Q}(\zeta_l)$ is unramified.

Along with $\mathbb{Q}(\xi,\zeta_l)/\mathbb{Q}(\zeta_l)$, the extension of degree l over $\mathbb{Q}$ contained in $\mathbb{Q}(\xi,\zeta_l)$ is also unramified, contrary to the Minkowski discriminant theorem. Thus we have proved that M is contained in $\mathbb{Q}(\zeta_{l^2})$.

There are exactly three quadratic fields in which only 2 is ramified (Section 18.9): $\mathbb{Q}(\sqrt{-1})$, $\mathbb{Q}(\sqrt{2})$, $\mathbb{Q}(\sqrt{-2})$. $\mathbb{Q}(\zeta_8)$ is the composite of these three fields.

$\mathbb{Q}(\zeta_{l^2})$ contains exactly one field of degree l over $\mathbb{Q}$, which we denote by M_l. Now let $L/\mathbb{Q}$ be cyclic of order l^h and let the subfield of L of degree l be M_l for $l \neq 2$ and one of the three quadratic fields $\mathbb{Q}(\sqrt{-1})$, $\mathbb{Q}(\sqrt{2})$, $\mathbb{Q}(\sqrt{-2})$ for $l = 2$. When $h = 1$ we are finished. Suppose $h > 1$ and suppose first that $l \neq 2$. Also let K_l be the subfield of degree l^h contained in $\mathbb{Q}(\zeta_{l^{h+1}})$. Then $K_l \cap L \supset M_l$, whence it follows that K_lL has degree less than l^{2h}. $G(K_lL/\mathbb{Q})$ is therefore of the form

$$G(K_lL/\mathbb{Q}) = G(K_l/\mathbb{Q}) \times H$$

with a cyclic group H whose order is less than l^h (Section 17.5). To prove Theorem 14 it suffices to show that the fixed field L' of H is abelian. Since L' has smaller degree than L we can apply induction and obtain the desired result.

Now suppose $l = 2$ and L is not real. Then $L(\sqrt{-1})$ contains a real subfield L' of degree l^h. We have $L'(\sqrt{-1}) = L(\sqrt{-1})$. It therefore suffices to show Theorem 14 for real fields L. Let K_2 be the real subfield of degree 2^h contained in $\mathbb{Q}(\zeta_{2^{h+1}})$. The same reasoning as in the case $l \neq 2$ now leads to the proof of Theorem 14. □

Exercises

25.1 Let K/F be a normal extension of an algebraic number field of degree n. Suppose F contains a prime ideal $\mathfrak{p}$ with $(N(\mathfrak{p}),n) = 1$ and the decomposition $\mathfrak{p}O_K$ in L. Show that the Galois group G(K/F) is cyclic.

25.2 Let $F = \mathbb{Q}(\sqrt{-23})$ and let $K = F(\alpha)$ with $\alpha^3 - \alpha - 1$. Show that all prime ideals in K/F are ramified.

25.3 Let K/F be a normal extension of algebraic number fields and let $\mathfrak{P}$ be a prime ideal of K. Show that the decomposition group of $\mathfrak{P}$ is solvable.

25.4 Let N/F and K/F be normal extensions of algebraic number fields with $N \supset K$

a) Show that the decomposition group and the inertia group of a prime ideal $\mathfrak{P}$ of N are mapped by the projection $G(N/F) \to G(K/F)$ onto the corresponding groups of the prime ideal $\mathfrak{P} \cap O_K$ of K.

b) Give an example in which $n > 1$ and the ramification group $V_n(N/F)$ is not mapped onto $V_n(K/F)$ by the projection $G(N/F) \to G(K/F)$.

25.5 Let K be an abelian extension of $\mathbb{Q}$. The conductor K is the smallest natural number f with $K \subset \mathbb{Q}(\zeta_f)$.

a) Show that a prime p in $K/\mathbb{Q}$ is ramified if and only if p is a divisor of f.

b) Show that the splitting behaviour of a prime p in $K/\mathbb{Q}$ depends only on the residue class of p modulo f.

c) Let K be a quadratic number field with discriminant D. Show that the conductor of K equals $|D|$.

d) For an arbitrary abelian extension $K/\mathbb{Q}$ show that the conductor of K is a divisor of the discriminant of K.

25.6 Compute the ramification group of the cyclotomic field $\mathbb{Q}(\zeta_m)$ and determine the different and discriminant of this field.

25.7 Let $K = \mathbb{Q}(\zeta_7+\zeta_7^{-1})$. Compute the coefficients of the cubic form

$$f(x_1,x_2,x_3) := N_{K/\mathbb{Q}}(x_1+x_2(\zeta_7+\zeta_7^{-1}) + x_3(\zeta_7+\zeta_7^{-1})^2)$$

and show that, for certain integer values of the indeterminates x_1,x_2,x_3, $f(x_1,x_2,x_3)$ represents a prime p if and only if $p \equiv \pm 1 \pmod 7$ (cf. Section 24.2).

26. Entire functions with growth of finite order

26.1 Formulation of the problem

The Weierstrass product representation characterises an entire function by its zeros, up to a factor of the form $e^{g(z)}$, where $g(z)$ is also an entire function (Section 9.2). In general it is not possible to say anything more precise about $g(z)$, but in the previous examples of $\sin \pi z$, $1/\Gamma(z)$ (Chapter 9) and $\xi_1(z)$ (Section 15.3), the function $g(z)$ has an extremely simple form.

Picard, Poincaré and Hadamard succeeded in delimiting a class of entire functions $f(z)$, by a growth condition as $|z| \to \infty$, for which $g(z)$ is a polynomial. In his work "*Étude sur les propriétés des fonctions entières et en particulier d'une fonction considéré par Riemann*" (J. math. pures appl., sér. 4, 9 (1893)) Hadamard was able to prove the product representation of $\xi(t)$ asserted by Riemann. In the present chapter we give a presentation of the main results of Hadamard.

26.2 Entire functions of finite order

Let $f(z)$ be an entire function, i.e. a function that is defined and regular over the whole complex plane (Chapter 9). Then $f(z)$ is called a *function of finite order* when there is an $a > 0$ such that

$$M_f(r) := \max_{|z|=r} |f(z)| \le \exp r^a \tag{1}$$

for sufficiently large r. Since a regular function takes its maximum on the boundary (Section 8.7), $M_f(r)$ is a monotonically increasing function of r. The lower limit $\alpha(f)$ of all $a > 0$ which satisfy the inequality (1) is called the *order of* f. When no $a > 0$ satisfying (1) exists, f has infinite order by definition. As one easily sees, a polynomial has order 0, $e^{g(z)}$ has order m when $g(z)$ is a polynomial of degree m, and $\exp(\exp z)$ has infinite order.

Entire functions f_1, f_2 have the properties

$$\alpha(f_1f_2) \le \max\{\alpha(f_1),\alpha(f_2)\}, \quad \alpha(f_1+f_2) \le \max\{\alpha(f_1),\alpha(f_2)\}. \tag{2}$$

In what follows we assume for simplicity that $f(0) \neq 0$.

We want to connect the order of f with the zeros of f, and for that purpose we define the convergence exponent of a sequence $z_1,z_2,\ldots$ of complex numbers. We assume that the latter numbers are all different from 0 and that their absolute values are ordered as follows:

$$0 < |z_1| \le |z_2| \le \ldots \le |z_n| \le \ldots \tag{3}$$

The sequence $z_1,z_2,\ldots$ is called a *sequence with finite convergence exponent* when there is a $b > 0$ such that

$$\sum_{n=1}^{\infty} |z_n|^{-b} < \infty. \tag{4}$$

The lower limit β of the $b > 0$ satisfying (4) is called the *convergence exponent* of the sequence $z_1,z_2,\ldots$. When there is no $b > 0$ satisfying (4), the sequence has convergence exponent ∞ by definition.

When $z_1,z_2,\ldots$ is the sequence of zeros of f, satisfying condition (3) and with each zero taken according to its multiplicity, then we call $\beta = \beta(f)$ the *convergence exponent of the zeros* of f.

Theorem 1. *Let* f *be an entire function with* $f(0) \neq 0$. *Then*

$$\beta(f) \le \alpha(f).$$

We first prove

Proposition 1. *Suppose* $0 < r < R$ *and let* $m(r)$ *be the number of zeros* z *of* f *with* $|z| \le r$. *Then*

$$|f(0)| \left(\frac{R}{r}\right)^{m(r)} \le M_f(R). \tag{5}$$

Proof. We consider the function

$$F(z) := f(z)\prod_{n=1}^{m(r)} \frac{R^2 - z\bar{z}_n}{R(z-z_n)}.$$

$F(z)$ is an entire function with $|F(z)| = |f(z)|$ for $|z| = R$.

By the maximum modulus theorem (Section 8.7) we therefore have

$$|F(0)| = |f(0)|\prod_{n=1}^{m(r)} \frac{R}{|z_n|} \le M_f(R). \quad \square$$

We now come to the proof of Theorem 1. We have to show that the series (4) converges for each $b < \alpha$.

We take logarithms of both sides of (5) and set $R = 2r$. Then for each $\varepsilon > 0$ and sufficiently large r we get

$$\log|f(0)| + m(r)\log 2 \le \log M_f(2r) \le (2r)^{\alpha+\varepsilon}.$$

It follows that there is a $c > 0$ such that, for all $r > 0$,

$$m(r) \le cr^{\alpha+\varepsilon}.$$

If one sets $r = |z_n|$ then one gets $n \le c|z_n|^{\alpha+\varepsilon}$, hence for $\varepsilon < b-\alpha$

$$\sum_{n=1}^{\infty} |z_n|^{-b} \le \sum_{n=1}^{\infty} c^{b/(\alpha+\varepsilon)} < \infty \quad \text{(Chapter 6, (6))}. \quad \square$$

The investigations which follow will show when $\beta(f) = \alpha(f)$.

When $\alpha(f)$ is finite, Theorem 1 gives an integer $p \ge 0$ with

$$\sum_{n=1}^{\infty} |z_n|^{-p-1} < \infty. \tag{6}$$

In what follows let $p = p(f)$ be the smallest such number. Then in particular $p+1 \ge \beta \ge p$. The Weierstrass product representation of f has the form

$$f(z) = e^{g(z)}\prod_{n=1}^{\infty}\left(1 - \frac{z}{z_n}\right)\exp\left(\frac{z}{z_n} + \frac{1}{2}\left(\frac{z}{z_n}\right)^2 + \dots + \frac{1}{p}\left(\frac{z}{z_n}\right)^p\right). \tag{7}$$

Theorem 2. *Let* $z_1, z_2, \ldots$ *be a sequence of complex numbers satisfying* (3) *and with convergence exponent* $\beta < \infty$. *Also let* $p \geq 0$ *be the smallest integer satisfying* (6). *Then*

$$h(z) := \prod_{n=1}^{\infty} \left(1 - \frac{z}{z_n}\right) \exp\left(\frac{z}{z_n} + \frac{1}{2}\left(\frac{z}{z_n}\right)^2 + \ldots + \frac{1}{p}\left(\frac{z}{z_n}\right)^p\right) \tag{8}$$

has order $\alpha(h) = \beta$.

If also

$$\sum_{n=1}^{\infty} |z_n|^{-\beta} < \infty, \tag{9}$$

then there is a constant $c > 0$ *with*

$$M_h(r) \leq \exp(cr^{\beta}). \tag{10}$$

Proof. By Theorem 1 we have $\beta \leq \alpha(h)$. Hence, to prove $\alpha(h) = \beta$, it suffices to establish the estimate

$$\log M_h(r) \leq c(\varepsilon) r^{\beta+\varepsilon} \tag{11}$$

for all $\varepsilon > 0$. Assume without loss of generality that $z \neq z_n$ for $n = 1,2,\ldots$. We introduce the following notation:

$$V_1 := \left\{\frac{z}{z_n} \,\middle|\, n = 1,2,\ldots, \left|\frac{z}{z_n}\right| \leq \frac{1}{2}\right\},\ V_2 := \left\{\frac{z}{z_n} \,\middle|\, n = 1,2,\ldots, \left|\frac{z}{z_n}\right| > \frac{1}{2}\right\}.$$

$$u(v) := (1-v)\exp(v + \frac{1}{2}v^2 + \ldots + \frac{1}{p}v^p),$$

$$\Sigma_1 := \sum_{v \in V_1} \log|u(v)|,\ \Sigma_2 := \sum_{v \in V_2} \log|u(v)|.$$

We have to show

$$\Sigma_1 + \Sigma_2 \leq c(\varepsilon)|z|^{\beta+\varepsilon}.$$

For $v \leq \frac{1}{2}$ we have

$$\log|u(v)| \leq \frac{1}{p+1}|v|^{p+1} + \frac{1}{p+2}|v|^{p+2} + \ldots \leq 2|v|^{p+1}, \tag{12}$$

and for $|v| > \frac{1}{2}$,

$$\log|u(v)| \leq \log(1+|v|) + |v| + \ldots + \frac{1}{p}|v|^p \leq c_1(\varepsilon)|v|^{p+\varepsilon}. \tag{13}$$

First we bound Σ_1 with the help of (12). For $\beta = p+1$,

$$\Sigma_1 \leq 2|z|^b \sum_{|\frac{z}{z_n}|\leq\frac{1}{2}} \frac{1}{|z_n|^{p+1}} = c_2|z|^\beta.$$

For $\beta < p+1$ we can assume $\beta + \varepsilon < p+1$. Then $|v|^{p+1} \leq |v|^{\beta+\varepsilon}$ and therefore

$$\Sigma_1 \leq 2|z|^{\beta+\varepsilon} \sum_{|\frac{z}{z_n}|\leq\frac{1}{2}} \frac{1}{|z_n|^{\beta+\varepsilon}} = c_3(\varepsilon)|z|^{\beta+\varepsilon}$$

For Σ_2 one finds, with the help of (13),

$$\Sigma_2 \leq c_1(\varepsilon) \sum_{v\in V_2} |v|^{p+\varepsilon} = c_1(\varepsilon)|z|^{\beta+\varepsilon} \sum_{|\frac{z}{z_n}|>\frac{1}{2}} \frac{1}{|z_n|^{\beta+\varepsilon}}\left|\frac{z}{z_n}\right|^{p-\beta} \leq c_4(\varepsilon)|z|^{\beta+\varepsilon}.$$

When (9) holds one can carry out the above estimation with β in place of $\beta + \varepsilon$ to obtain the desired result. □

Theorem 3. *Let* f *be an entire function of finite order* α *with* $f(0) \neq 0$. *Then* f *has a representation in the form* (7) *with a polynomial* g *of degree* $\leq \alpha$.

When $r^{-\alpha} \log M_f(r)$ *is unbounded with increasing* r *then* $\alpha = \beta(f)$ *and the series*

$$\sum_{n=1}^{\infty}|z_n|^{-\alpha} \tag{14}$$

diverges.

We first prove the following proposition, which is a variant of Cauchy's coefficient estimate (Section 8.6).

Proposition 2. *Suppose* $R > r > 0$ *and that the function*

$$f(z) = \sum_{n=0}^{\infty} a_n z^n$$

is regular for $|z| \le R$. *Suppose* $\text{Re } f(z) \le M$ *on* $|z| = R$. *If* $B := 2(2M\text{-Re } a_0)$ *the*

$$a_n \le \frac{B}{R^n} \quad \textit{for} \quad n = 1,2,... \tag{15}$$

$$|f(z) - a_0| \le \frac{Br}{R-r} \quad \textit{for} \quad |z| \le r \tag{16}$$

and

$$|f^{(m)}(z)| \le \frac{m!BR}{(R-r)^{m+1}} \quad \textit{for} \quad |z| \le r,\ m = 1,2,... \,. \tag{17}$$

Proof. Let $a_n = |a_n|e^{i\phi_n}$, $z = re^{i\phi}$. Then

$$\text{Re } f(Re^{i\phi}) = \sum_{n=0}^{\infty} |a_n|\cos(n\phi+\phi_n)R^n. \tag{18}$$

This series converges uniformly with respect to ϕ and therefore can be integrated term by term. This yields

$$\int_0^{2\pi} \text{Re } f(Re^{i\phi})d\phi = 2\pi \text{ Re } a_0$$

and

$$\int_0^{2\pi} \text{Re } f(Re^{i\phi})\cos(n\phi+\phi_n)d\phi = \pi|a_n|R^n \quad \text{for} \quad n = 1,2,... \,.$$

Since $0 \le 1 + \cos(n\phi+\phi_n) \le 2$ we have then

$$\pi|a_n|R^n = \int_0^{2\pi} \text{Re } f(Re^{i\phi})(1+\cos(n\phi+\phi_n))d\phi - 2\pi \text{ Re } a_0 \le 4\pi M - 2\pi \text{ Re } a_0.$$

This proves (15); (16) and (17) follow easily:

$$|f(z)-a_0| \le \sum_{n=1}^{\infty} |a_n|r^n \le B\sum_{n=1}^{\infty}\left[\frac{r}{R}\right]^n = \frac{Br}{R-r} \quad \text{for} \quad |z| \le r,$$

$$|f^{(m)}(z)| \le \sum_{n=m}^{\infty} |a_n|n(n-1)...(n-m+1)r^{n-m}$$

$$\leq B \sum_{n=m}^{\infty} n(n-1)\ldots(n-m+1)\frac{r^{n-m}}{R^m}$$

$$= B\frac{d^m}{dr^m}\left(\sum_{n=0}^{\infty}\left(\frac{r}{R}\right)^n\right)$$

$$= \frac{Bm!R}{(R-r)^{m+1}}$$

for $|z| \leq r$, $m = 1,2,\ldots$. □

We now come to the proof of Theorem 3. Let f be an entire function of finite order α with $f(0) \neq 0$. By Theorem 1, f has a representation (7) and the convergence exponent of its zeros $\beta(f) \leq \alpha$. By Theorem 2 the order of

$$h(z) = \sum_{n=1}^{\infty}\left(1-\frac{z}{z_n}\right)\exp\left(\frac{z}{z_n}+\frac{1}{2}\left(\frac{z}{z_n}\right)^2+\ldots+\frac{1}{p}\left(\frac{z}{z_n}\right)^p\right)$$

equals $\beta(f)$. Let $k := [\alpha]$. We show that $g^{(k+1)}(z)$ vanishes:

By taking logarithms and $(k+1)$-fold differentiation of (7) one finds

$$g^{(k+1)}(z) = \frac{d^k}{dz^k}\left(\frac{f'(z)}{f(z)}\right) + k!\sum_{n=1}^{\infty}(z_n-z)^{-k-1}. \tag{19}$$

For any $R > 0$ we split the right-hand sum into two partial sums $\sum_{|z_n|\leq R}(z_n-z)^{-k-1}$ and $\sum_{|z_n|>R}(z_n-z)^{-k-1}$. For all $|z| \leq R/2$ we have

$$\left|\sum_{|z_n|>R}(z_n-z)^{-k-1}\right| \leq \sum_{|z_n|>R}(z_n-z)^{-k-1} \leq 2^{k+1}\sum_{|z_n|>R}|z_n|^{-k-1}.$$

By Theorem 1, $\sum_{n=1}^{\infty}|z_n|^{-k-1}$ converges. Hence, for all $z \in \mathbb{C}$,

$$\lim_{R\to\infty}\sum_{|z_n|>R}(z_n-z)^{-k-1} = 0. \tag{20}$$

Now we consider the function

$$g^{(k+1)}(z) - k! \sum_{|z_n|>R} (z_n - z)^{-k-1} = \frac{d^k}{dz^k}\left[\frac{f'(z)}{f(z)}\right] + \sum_{|z_n|>R} (z_n - z)^{-1}. \tag{21}$$

We set

$$f_R(z) := \frac{f(z)}{f(0)} \prod_{|z_n| \le R} \left[1 - \frac{z}{z_n}\right]^{-1}.$$

The function $f_R(z)$ is entire and without zeros for $|z| \le R$. Hence there is a function $g_R(z)$, regular for $|z| \le R$, with

$$f_R(z) = \exp g_R(z),$$

and the function (21) is the $(k+1)$th derivative of $g_R(z)$.

On the circle $|z| = 2R$ we have $|f_R(z)| < c(\varepsilon)\exp(2R)^{\alpha+\varepsilon}$ for any $\varepsilon > 0$. Then, by the maximum modulus theorem, this bound also holds for all $|z| \le 2R$. Thus for $|z| \le R$ we have $\operatorname{Re} g_R(z) = \log|f_R(z)| \le c_1(\varepsilon)R^{\alpha+\varepsilon}$. We also have $\operatorname{Re} g_R(0) = 0$. Now we apply (17) to $g_R(z)$ with $r := R/2$ and obtain

$$|g_R^{(k+1)}(z)| \le c_2(\varepsilon)R^{\alpha+\varepsilon-k-1}.$$

Since $\varepsilon > 0$ is arbitrarily small and $\alpha < k+1$, it follows that

$$\lim_{R\to\infty} g_R^{(k+1)}(z) = 0 \tag{22}$$

for all $z \in \mathbb{C}$.

It follows from (20) and (22) that $g^{(k+1)}(z) = 0$.

Now suppose $r^{-\alpha}\log M_f(r)$ is unbounded with increasing r. Since $\log M_{e^g}(r) = O(r^{\alpha})$, $r^{-\alpha}\log M_h(r)$ is also unbounded. The second part of Theorem 3 now follows from Theorems 1 and 2. □

26.3 Application to the Riemann ζ-function

We are now in a position to prove some of Riemann's assertions about the ζ-function (Section 15.3).

Theorem 4. *Let* $\rho_1, \rho_2, \ldots$ *be the non-trivial zeros of* $\zeta(s)$, *ordered according to increasing absolute value. The function*

$$\xi(s) := \frac{1}{2}s(s-1)\pi^{-s/2}\Gamma(\frac{s}{2})\zeta(s)$$

is an entire function of order 1. *The series* $\sum_{n=1}^{\infty} |\rho_n|^{-1}$ *is divergent. In particular,* $\zeta(s)$ *has infinitely many non-trivial zeros.*

Proof. At $s = 1$, $\zeta(s)$ has the residue 1 (Section 15.2). Since $\Gamma(\frac{1}{2}) = \sqrt{\pi}$ (Chapter 9, (13)), we therefore have $\xi(1) = \frac{1}{2}$. It then follows from the functional equation $\xi(1-s) = \xi(s)$ (Section 15.2) that $\xi(0) = \frac{1}{2}$. In particular, we can directly apply the considerations of the previous section to $\xi(s)$.

Since $\xi(1-s) = \xi(s)$ we can confine ourselves to $\text{Re } s := \sigma \geq \frac{1}{2}$ in estimating $\xi(s)$. Then $(s-1)\zeta(s) = O(|s|^2)$. This follows from

Proposition 3. *For* $\sigma > 0$

$$\zeta(s) = 1 + \frac{1}{s-1} - s\int_1^{\infty} \frac{x-[x]}{x^{s+1}}dx.$$

Proof. For $\sigma > 0$

$$\begin{aligned}
\zeta(s) &= \sum_{n=1}^{\infty} \frac{1}{n^s} \\
&= \sum_{n=1}^{\infty} n\left(\frac{1}{n^s} - \frac{1}{(n+1)^s}\right) \\
&= \sum_{n=1}^{\infty} sn\int_n^{n+1} \frac{dx}{x^{s+1}} \\
&= \sum_{n=1}^{\infty} s\int_n^{n+1} \frac{[x]dx}{x^{s+1}} \\
&= -s\int_1^{\infty} \frac{x-[x]}{x^{s+1}}dx + s\int_1^{\infty} \frac{dx}{x^s} \\
&= 1 + \frac{1}{s-1} - s\int_1^{\infty} \frac{x-[x]}{x^{s+1}}dx.
\end{aligned}$$

Since the latter expression represents a regular function for $\sigma > 0$, the assertion follows. □

$\pi^{-s/2}$ has order 1. Finally, $\log|\Gamma(s)| \le c|s|\log|s|$ (Chapter 9, Theorem 5). From this it is clear that $\xi(s)$ has order less than or equal to 1.

On the other hand, $\lim\limits_{x\to\infty} \dfrac{\log \Gamma(x)}{x \log x} = 1$ by Stirling's formula (Chapter 9, Theorem 5). Hence there is no constant c with $\log M_\xi(r) \le cr$. □

Theorem 5. *There are complex numbers* A, B *with*

$$\xi(s) = \exp(A+Bs) \sum_{n=1}^{\infty} (1-\frac{s}{\rho_n})\exp \frac{s}{\rho_n}. \tag{23}$$

Proof. This follows from Theorem 4 and the preceding section. □

Exercises

26.1 Show that the Weierstrass σ-function (Section 13.2) has order 2.

26.2 Let f be an entire function of finite order. Show, without using the elliptic modular function, that f omits at most one complex value. When the order of f is not an integer, then f takes all complex values.

26.3 Derive the product representation of the sine function (Chapter 9. (4)) with the help of Theorem 3.

26.4 Determine the constants A, B in Theorem 5.

a) Show $e^A = \frac{1}{2}$

b) Show $B = \frac{1}{2}\gamma - 1 + \frac{1}{2}\log 4\pi - \lim\limits_{s\to 1}\left[\frac{\zeta'(s)}{\zeta(s)} + \frac{1}{s-1}\right]$,

where γ denotes Euler's constant.

c) Show $\lim\limits_{s\to 1}\left[\frac{\zeta'(s)}{\zeta(s)} + \frac{1}{s-1}\right] = 1 - \int_1^\infty \frac{(x-[x])}{x^2}dx = \gamma$.

26.5 Show $B = -\frac{1}{2}\gamma \sum_{n=1}^{\infty}\left(\frac{1}{\rho_n} + \frac{1}{\bar{\rho}_n}\right)$, where the sum is taken over the non-trivial zeros ρ_n of the ζ-function.

26.6 Show $\zeta(0) = -\frac{1}{2}$.

26.7 Show $\zeta(\sigma) \neq 0$ for $0 \leq \sigma < 1$.

27. Proof of the prime number theorem

27.1 Hadamard and de la Vallée Poussin

Hadamard's proof of some assertions of Riemann about the ζ-function (Section 26.3) paved the way for the proof of the prime number theorem (Section 15.1). Here we give a proof of this theorem in the sharper form proved by de la Vallée Poussin.

Theorem 1. *There is a constant* $c > 0$ *with*

$$\pi(x) = \int_2^x \frac{dt}{\log t} + O(xe^{-c\sqrt{\log x}}).$$

Proof. We first remark that Theorem 1 implies the prime number theorem in Hadamard's form:

$$\lim_{x\to\infty} \pi(x)/\frac{x}{\log x} = 1. \tag{1}$$

To prove this it suffices to show

$$\int_2^x \frac{dt}{\log x} = \frac{x}{\log x} + O\left(\frac{x}{(\log x)^2}\right). \tag{2}$$

Integrating by parts, one finds

$$\int_2^x \frac{dt}{\log t} = \frac{x}{\log x} - \frac{2}{\log 2} + \int_2^x \frac{dt}{(\log t)^2}. \tag{3}$$

Also, for $x \geq 4$

$$\int_2^x \frac{dt}{(\log t)^2} = \int_2^{\sqrt{x}} \frac{dt}{(\log t)^2} + \int_{\sqrt{x}}^x \frac{dt}{(\log t)^2} \leq \frac{\sqrt{x}}{(\log 2)^2} + \frac{x}{(\log \sqrt{x})^2} = O\left(\frac{x}{(\log x)^2}\right).$$

This yields the assertion.

27.2 The Chebyshev function

The relationship between $\zeta(s)$ and the prime numbers that we are going to use is stated in the following:

Theorem 2. *Let* $\Lambda(n) = \log p$ *when* n *is a power of a prime* p, *and let* $\Lambda(n) = 0$ *for all other natural numbers* n. *Then for* $\operatorname{Re} s =: \sigma > 0$ *the meromorphic function* $-\zeta'(s)/\zeta(s)$ *has the representation*

$$-\frac{\zeta'(s)}{\zeta(s)} = \sum_{n=1}^{\infty} \frac{\Lambda(n)}{n^s}. \tag{4}$$

Proof. For $\sigma \geq \sigma_0 > 1$ the infinite product

$$\zeta(s) = \prod_p \frac{1}{1-p^{-s}}$$

is absolutely and uniformly convergent. Therefore

$$-\frac{\zeta'(s)}{\zeta(s)} = \sum_p p^{-s}\frac{\log p}{1-p^{-s}} = \sum_p \sum_{k=1}^{\infty} \frac{\log p}{p^{ks}} = \sum_{n=1}^{\infty} \frac{\Lambda(n)}{n^s}. \qquad \square$$

The function

$$\psi(x) := \sum_{n\leq x} \Lambda(n)$$

is called the *Chebyshev function.* It satisfies

Theorem 3. *There is a constant* $c > 0$ *with*

$$\psi(x) = x + O(xe^{-c\sqrt{\log x}}).$$

In this section we show that Theorem 1 follows from Theorem 3. To do this we consider

$$S(x) := \sum_{n\leq x} \frac{\Lambda(n)}{\log n} = \pi(x) + \sum_k \frac{\Lambda(h)}{\log h},$$

where the sum on the right is taken over all $h = p^k \leq x$ with $k \geq 2$. This sum is harmless, because $k \leq \log x$ and for fixed $k \geq 2$ the sum contains less than $\sqrt{x}$ non-zero summands. Therefore

$$\sum_h \frac{\Lambda(h)}{\log h} = O(\sqrt{x}\,\log x).$$

Thus it suffices to show

$$S(x) = \int_2^x \frac{dt}{\log t} + O(xe^{-c\sqrt{\log x}}).$$

To do this we use the following lemma, which goes back to Abel (Section 6.3).

Lemma 1. *Suppose the function* f(x) *is continuously differentiable in the interval* [a,b], *let* $c_1, c_2, \ldots$ *be arbitrary complex numbers, and let*

$$c(x) = \sum_{a<n\leq x} c_n.$$

Then

$$\sum_{a<n\leq b} c_n f(n) = -\int_a^b c(x)f'(x)dx + c(b)f(b). \quad \square$$

We set $c_n = \Lambda(n)$ and $f(x) = \frac{1}{\log x}$. Then by Theorem 3

$$c(x) = \sum_{3/2<n\leq x} c_n = \psi(x) = x + O(xe^{-c\sqrt{\log x}}),$$

and

$$S(x) = \int_2^x \frac{\psi(t)}{t(\log t)^2}dt + \frac{\psi(x)}{\log x}$$

$$= \int_2^x \frac{dt}{(\log t)^2} + \frac{x}{\log x} + O\left(\int_2^x \frac{e^{-c\sqrt{\log t}}}{(\log t)^2}dt\right) + O\left(\frac{xe^{-c\sqrt{\log x}}}{\log x}\right).$$

By (3) we have, in turn,

$$S(x) = \int_2^x \frac{dt}{\log t} + O\left(\int_2^{\sqrt{x}} dt\right) + O\left(\int_{\sqrt{x}}^{x^2} e^{-c\sqrt{\log \sqrt{x}}}dt\right) + O\left(xe^{-c\sqrt{\log x}}\right),$$

$$S(x) = \int_2^x \frac{dt}{\log t} + O\left(xe^{-(c/2)\sqrt{\log x}}\right).$$

27.3 The method of complex integration

To prove Theorem 3 we use a method that goes back to Riemann, the *method of complex integration*, which makes it possible to estimate the sum of the first N coefficients $a_1, \dots, a_N$ of the Dirichlet series

$$f(s) := \sum_{n=1}^{\infty} \frac{a_n}{n^s}. \tag{5}$$

We set $s = \sigma + it$ and

$$\phi_f(x) = \sum_{n<x} a_n \quad \text{for } x > 0,\ x \notin \mathbb{Z},$$

$$\phi_f(x) = \sum_{n<x} a_n + \frac{1}{2} a_x \quad \text{for } x > 0,\ x \in \mathbb{Z}.$$

Also let $\langle x \rangle$, for $x \in \mathbb{R} - \mathbb{Z}$, denote the distance from x to the nearest integer, and let $\langle x \rangle = 1$ for $x \in \mathbb{Z}$.

Lemma 2. *Suppose the series* (5) *is absolutely convergent for* $\sigma > 1$, *and suppose*

$$\sum_{n=1}^{\infty} |a_n| n^{-\sigma} = O((\sigma-1)^{-\alpha})$$

as $\sigma \to 1$, *for some* $\alpha > 0$. *Also suppose* $|a_n| \le A(n)$, *where* $A(x)$ *is a monotonically increasing function of* x. *We set* $b = b(x) = 1 + 1/\log x$.

Then for arbitrary $T \ge 1$, $x \ge 2$ *we have the estimate*

$$\phi_f(x) = \frac{1}{2\pi i} \int_{b-iT}^{b+iT} f(s) \frac{x^s}{s} ds + O\left(\frac{x \log^{\alpha} x}{T}\right) + O\left(\frac{xA(2x)(\log x + \frac{1}{\langle x \rangle})}{T}\right). \tag{6}$$

Remark 1. When $T \to \infty$ one obtains the Riemann equation

$$\phi_f(x) = \frac{1}{2\pi i} \int_{b-i\infty}^{b+i\infty} f(s) \frac{x^s}{s} ds \text{ (Section 15.1).} \tag{7}$$

At points of discontinuity $\phi_f(x)$ is defined as a function for which the Dirichlet theorem of Fourier analysis holds (Chapter 6). Riemann was content to prove (6) with a reference to "the Fourier theorem".

Remark 2. To estimate the order of growth of $\phi_f(x)$ in the case $a_n = \Lambda(n)$ which is of interest to us it suffices to set $x = N + \frac{1}{2}$ with N a natural number. Then the inconvenient term $1/\langle x \rangle$ drops out of the estimate (6). For arbitrary x, (6) is of interest in deriving an "exact formula" for $\psi_0(x)$ (see Exercise 27.6).

To prove Lemma 2 we need the following

Proposition 1. *If* $b > 0,\ T > 0$ *then*

$$\left|\frac{1}{2\pi i}\int_{b-iT}^{b+iT}\frac{a^s}{s}ds - \delta(a)\right| \le \begin{cases} \dfrac{2b}{T} & \textit{for } a = 1 \\[2mm] \dfrac{2a^b}{T\,|\log a|} & \textit{for } a \ne 1, a > 0, \end{cases}$$

where $\delta(a) = 1$ *for* $a > 1$, $\delta(a) = 0$ *for* $0 < a < 1$ *and* $\delta(a) = \frac{1}{2}$ *for* $a = 1$.

Proof. For $a > 1$ (resp. $0 < a < 1$) we integrate a^s/s over the rectangle with the sides

$$\begin{aligned} C_1 &= \{b+it \mid -T \le t \le T\} \\ C_2 &= \{\sigma+iT \mid b \ge \sigma \ge -h \text{ (resp. } b \le \sigma \le h) \text{ with an } h > b\}, \\ C_3 &= \{-h + it \text{ (resp. } h + it) \mid T \ge t \ge -T\}, \\ C_4 &= \{\sigma - iT \mid -h \le \sigma \le b \text{ (resp. } h \ge \sigma \ge b)\} \text{ (Fig. 16).} \end{aligned}$$

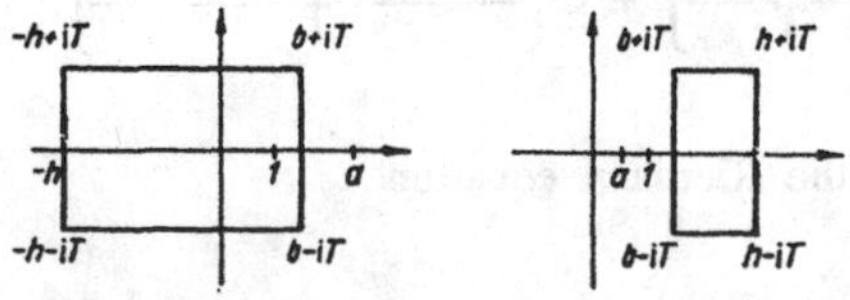

Fig. 16

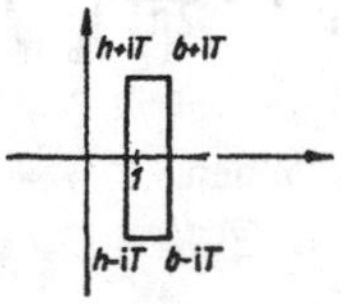

Fig. 17

By the residue theorem (Chapter 8, Theorem 12),

$$\frac{1}{2\pi i}\int_{C_1+C_2+C_3+C_4}\frac{a^s}{s}ds = \begin{cases}1 \text{ for } a > 1\\ 0 \text{ for } 0 < a < 1.\end{cases}$$

We now estimate the integrals over C_2, C_3, C_4. For $a > 1$

$$\left|\int_{C_2}\frac{a^s}{s}ds\right| = \left|\int_{C_4}\frac{a^s}{s}ds\right| \le \int_{-h}^{b}\frac{a^{\sigma}d\sigma}{\sqrt{T^2+\sigma^2}} \le \frac{a^b}{T\log a},$$

$$\left|\int_{C_3}\frac{a^s}{s}ds\right| \le \int_{-T}^{T}\frac{a^{-h}dt}{\sqrt{h^2+t^2}} \le \frac{2Ta^{-h}}{h} \to 0 \text{ as } h \to \infty.$$

For $0 < a < 1$

$$\left|\int_{C_2}\frac{a^s}{s}ds\right| = \left|\int_{C_4}\frac{a^s}{s}ds\right| \le \int_{b}^{h}\frac{a^{\sigma}d\sigma}{\sqrt{T^2+\sigma^2}} \le \frac{a^b}{T|\log a|},$$

$$\left|\int_{C_3}\frac{a^s}{s}ds\right| \le \int_{-T}^{T}\frac{a^{h}dt}{\sqrt{h^2+t^2}} \le \frac{2Ta^{h}}{h} \quad 0 \text{ as } h \to \infty.$$

For $a = 1$

$$\int_{b-iT}^{b+iT}\frac{ds}{s} = \log(b+iT) - \log(b-iT)$$

$$= \pi i + \log(1-i\tfrac{b}{T}) - \log(1+i\tfrac{b}{T})$$

$$= \pi i + R, \ |R| \le \frac{2b}{T},$$

where log denotes the principal value of the logarithm. □

We now come to the proof of Lemma 2. Since (5) is absolutely and uniformly convergent for $s = b + it$, we can integrate it term by term, and by Proposition 1 we obtain

$$\frac{1}{2\pi i}\int_{b-iT}^{b+iT}f(s)\frac{x^s}{s}ds = \frac{1}{2\pi i}\sum_{n=1}^{\infty}a_n\int_{b-iT}^{b+iT}\left(\frac{x}{n}\right)^s\frac{ds}{s} = \phi_f(x) + R$$

with

$$|R| \le 2\,\frac{A(x)b}{T} + 2\sum_{\substack{n=1\\ n\neq x}}^{\infty} a_n\left(\frac{x}{n}\right)^b T^{-1}\left|\log\frac{x}{n}\right|^{-1}. \tag{8}$$

We partition the sum on the right into two partial sums R_1, R_2 with

$$\left|\log\frac{x}{n}\right| \ge \log 2 \quad \text{resp.} \quad \left|\log\frac{x}{n}\right| < \log 2.$$

For R_1 one finds

$$R_1 \le (\log^{-1}2)x^bT^{-1}\sum_{n=1}^{\infty}|a_n|n^{-b}$$

$$= O(x^bT^{-1}(b-1)^{-\alpha})$$

$$= O(x\log^{\alpha}x/T).$$

R_2 has the form

$$R_2 = \sum_{\substack{x/2<n<2x\\ n\neq x}} |a_n|\left(\frac{x}{n}\right)^b T^{-1}\left|\log\frac{x}{n}\right|^{-1}$$

$$\le A(2x)2^bT^{-1}\sum_{x/2<n<2x}\left|\log\frac{x}{n}\right|^{-1}.$$

It remains to show

$$\sum_{\substack{x/2<n<2x\\ n\neq x}}\left|\log\frac{x}{n}\right|^{-1} = O\left(x(\log x + \frac{1}{<x>}\right). \tag{9}$$

To do this we note that, for $x/2 < n < x$,

$$\left|\log\frac{x}{n}\right| = \log\frac{x}{n} = \log\left(1+\frac{x-n}{n}\right) \ge \frac{1}{2}\left(\frac{x-n}{n}\right) \ge \frac{1}{2}\left(\frac{x-n}{n}\right),$$

$$\sum_{x/2<n<x}\left|\log\frac{x}{n}\right|^{-1} \le 2x\sum_{x/2<n<x}\frac{1}{x-n} \le 2x\left(\frac{1}{<x>} + \int_{<x>}^{x}\frac{dt}{t}\right) = O\left[x(\log x + \frac{1}{<x>})\right],$$

and for $x < n < 2x$,

$$\left|\log \frac{x}{n}\right| = \log \frac{n}{x} = \log\left(1 + \frac{n-x}{x}\right) \geq \frac{n-x}{2x},$$

$$\sum_{x<n<2x} \left|\log \frac{n}{x}\right|^{-1} \leq 2x \sum_{x<n<2x} \frac{1}{n-x} \leq 2x\left(\frac{1}{<x>} + \int_{<x>}^{x} \frac{dt}{t}\right) = O\left(x(\log x + \frac{1}{<x>})\right). \quad \square$$

Remark. It is easy to see from the proof that $<x>$ in (6) can be replaced by the distance from x to the next natural number n such that $a_n \neq 0$.

27.4 Commencement of the proof of Theorem 3

We apply Lemma 2 to the function $f(s) := -\zeta'(s)/\zeta(s)$. For brevity we set $\psi_0(x) := \phi_{-\zeta'(s)/\zeta(s)}(x)$. Since $\psi(x) = \psi_0(x) + O(\log x)$, it suffices to prove that Theorem 3 holds for $\psi_0(x)$ instead of $\psi(x)$. By (4), and the fact that $\zeta(s)$ has a simple pole for $s = 1$, we can set $A(x) = \log x$ and $\alpha = 1$. With $x = N + \frac{1}{2}$, where N is a natural number > 2, we get

$$\psi_0(x) = \frac{1}{2\pi i}\int_{b-iT}^{b+iT} (-\zeta'(s)/\zeta(s))\frac{x^s}{s}ds + O\left(\frac{x \log^2 x}{T}\right). \tag{10}$$

We now consider the integral over the rectangle $R(T)$ with boundary $C = C_1 + C_2 + C_3 + C_4$, where

$$C_1 = \{b+it\,|\,-T \leq t \leq T\},\ C_2 = \{\sigma+iT\,|\,b \geq \sigma \geq h\},$$

$$C_3 = \{h+it\,|\,T \geq t \geq -T\},\ C_4 = \{-iT\,|\,h \leq \sigma \leq b\}$$

(Fig. 17). We make the following assumptions, which we shall later show to be satisfied.

$R(T)$ *can be chosen so that* $\zeta(s)$ *is non-zero for* $s \in R(T)$; $h < 1$ *is given as a function of* T *so that*

$$-\zeta'(s)/\zeta(s) = O(\log^2 T) \quad \textit{for} \quad s \in R(T).$$

By the residue theorem

$$\frac{1}{2\pi i}\int_C (-\zeta'(s)/\zeta(s))\frac{x^s}{s}ds = x. \tag{11}$$

The proof of Theorem 3 therefore comes down to estimating the subintegrals of (11) over C_2, C_3, C_4. Bearing in mind that $b = 1 + 1/\log x$ we get

$$\left|\int_{C_2}\right| = \left|\int_{C_4}\right| = O\left(\int_h^b \log^2 T \frac{x^\sigma}{T} d\sigma\right) = O\left(\frac{x \log^2 T}{T}\right)$$

and

$$\left|\int_{C_3}\right| = O\left(\log^2 T \int_0^T \frac{x^h}{|h+it|}\right) = O\left(x^h \log^2 T\left(\int_0^1 \frac{dt}{h} + \int_1^T \frac{dt}{h}\right)\right) = O(x^h \log^3 T).$$

Altogether we have

$$\psi_0(x) = x + O\left(\frac{x \log^2 x}{T}\right) + O\left(\frac{x \log^2 T}{T}\right) + O(x^h \log^3 T). \qquad (12)$$

The decisive term in this estimate is $O(x^h \log^3 T)$. The smaller h can be chosen, the better the bound on $\psi_0(x)$.

27.5 On the zeros of the ζ-function

We shall now determine a rectangle $R(T)$ with the desired properties. To do this we must acquire some knowledge of the non-trivial zeros of the ζ-function.

Theorem 4. *Let* $\rho_n = \beta_n + i\gamma_n$, $n = 1,2,\ldots$, *be the zeros of* $\zeta(s)$ *with* $0 \le \beta_n \le 1$, *and suppose* $T \ge 2$. *Then*

$$\sum_{n=1}^{\infty} \frac{1}{1+(T-\gamma_n)^2} = O(\log T).$$

First we prove the following simple

Proposition 2. *Let* $s = \sigma + iT$ *with* $T \ge 2$, $-1 \le \sigma \le 2$. *Then*

$$\left|\sum_{n=1}^{\infty}\left(\frac{1}{s+2n} - \frac{1}{2n}\right)\right| = O(\log T).$$

Proof. Since $\left|\frac{1}{s+2n}\right| = \frac{1}{\sqrt{(\sigma+2n)^2+T^2}} \le \frac{1}{n}$,

$$\left|\sum_{n=1}^{\infty}\left(\frac{1}{s+2n} - \frac{1}{2n}\right)\right| \le \sum_{n \le T}\left(\frac{1}{n} + \frac{1}{2n}\right) + \sum_{n>T} \frac{|s|}{2n^2} = O(\log T).\square$$

Proof of Theorem 4. We consider the logarithmic derivative of $\xi(s)$ (Chapter 26, (19)). By consideration of the product representation of $\Gamma(s)$ (Chapter 9, (5)) one finds

$$\frac{\zeta'(s)}{\zeta(s)} = -\frac{1}{s-1} + \sum_{n=1}^{\infty}\left(\frac{1}{s-\rho_n} + \frac{1}{\rho_n}\right) + \sum_{n=1}^{\infty}\left(\frac{1}{s+2n} - \frac{1}{2n}\right) + d \tag{13}$$

with a constant $d = \frac{\zeta'(0)}{\zeta(0)} - 1$. Hence, by Proposition 2, for $\sigma > 1$, $T \geq 2$ and

$$|\zeta'(s)/\zeta(s)| = \left|\sum_{n=1}^{\infty} \Lambda(n)n^{-\sigma-iT}\right| \leq \sum_{n=1}^{\infty} \Lambda(n)n^{-\sigma}$$

we have the estimate

$$\sum_{n=1}^{\infty}\left(\frac{1}{s-\rho_n} + \frac{1}{\rho_n}\right) = O(\log T). \tag{14}$$

We set $\sigma = 2$. The assertion now follows from

$$\begin{aligned}
\operatorname{Re}\frac{1}{s-\rho_n} &= \operatorname{Re}\frac{1}{(2-\beta_n)+i(T-\gamma_n)} \\
&= \frac{2-\beta_n}{(2-\beta_n)^2+(T-\gamma_n)^2} \\
&\geq \frac{1}{4+(T-\gamma_n)^2} \\
&\geq \frac{1}{4}\cdot\frac{1}{1+(T-\gamma_n)^2},
\end{aligned}$$

$$\operatorname{Re}\frac{1}{\rho_n} = \frac{\beta_n}{\beta_n^2+\gamma_n^2} \geq 0. \quad \square$$

Theorem 4 easily yields

Theorem 5. *The number* $r(T)$ *of* ρ_n *with* $T \leq |\gamma_n| \leq T+1$ *satisfies*

$$r(T) = O(\log T). \quad \square$$

Theorem 6. *For* $T \geq 2$,

$$\sum_{|T-\gamma_n|>1} \frac{1}{|T-\gamma_n|^2} = O(\log T). \qquad \square$$

Theorem 7. *For* $-1 \le \sigma \le 2$, $s = \sigma + iT$, $T \ge 2$,

$$\frac{\zeta'(s)}{\zeta(s)} = \sum_{|T-\gamma_n|\le 1} \frac{1}{s-\rho_n} + O(\log T).$$

Proof. It follows from (13) and Proposition 2 that, for $-1 \le \sigma \le 2$, $T \ge 2$,

$$\frac{\zeta'(s)}{\zeta(s)} = \sum_{n=1}^{\infty} \left(\frac{1}{s-\rho_n} + \frac{1}{\rho_n}\right) + O(\log T).$$

By (14) we have

$$\sum_{n=1}^{\infty} \left(\frac{1}{2+iT-\rho_n} + \frac{1}{\rho_n}\right) = O(\log T)$$

and therefore $$\frac{\zeta'(s)}{\zeta(s)} = \sum_{n=1}^{\infty} \left(\frac{1}{s-\rho_n} - \frac{1}{2+iT-\rho_n}\right) + O(\log T). \tag{15}$$

By Theorem 6,

$$\begin{aligned}\sum_{|T-\gamma_n|>1} \left|\frac{1}{\sigma+iT-\rho_n} - \frac{1}{2+iT-\rho_n}\right| &\le \sum_{|T-\gamma_n|>1} \frac{2-\sigma}{(T-\gamma_n)^2} \\ &\le \sum_{|T-\gamma_n|>1} \frac{3}{(T-\gamma_n)^2} \\ &= O(\log T). \end{aligned} \tag{16}$$

By Theorem 5,

$$\sum_{|T-\gamma_n|\le 1} \left|\frac{1}{2+iT-\rho_n}\right| = O(\log T). \tag{17}$$

Theorem 7 follows from (15) to (17). $\square$

Theorem 8. *There is a constant* $c > 0$ *such that* $\zeta(\sigma+it) \ne 0$ *for*

$$\sigma \ge 1 - \frac{c}{\log(|t|+2)}.$$

Proof. Since $\zeta(s)$ has a pole at 1, there is a $c_0 > 0$ with $\zeta(s) \neq 0$ for $|s-1| < c_0$. Hence it suffices to show that there is a constant $c > 0$ such that each zero $\rho = \beta + i\gamma$ of $\zeta(s)$ with $|\rho-1| \geq c_0$ satisfies the inequality

$$\beta \leq 1 - \frac{c}{\log(|\gamma|+2)}. \tag{18}$$

We use the inequality

$$3 + 4 \cos \phi + \cos 2\phi = 2(1+\cos \phi)^2 \geq 0. \tag{19}$$

For $\sigma > 1$ we have

$$-\frac{\zeta'(s)}{\zeta(s)} = \sum_{n=1}^{\infty} \frac{\Lambda(n)}{n^s} = \sum_{n=1}^{\infty} \Lambda(n)n^{-\sigma}e^{-it\log n}$$

and therefore

$$-\mathrm{Re}\,\frac{\zeta'(s)}{\zeta(s)} = \sum_{n=1}^{\infty} \Lambda(n)n^{-\sigma}\cos(t \log n).$$

Thus, by (19),

$$3\left[-\frac{\zeta'(s)}{\zeta(s)}\right] + 4\left[-\mathrm{Re}\,\frac{\zeta'(\sigma+it)}{\zeta(\sigma+it)}\right] + \left[-\mathrm{Re}\,\frac{\zeta'(\sigma+2it)}{\zeta(\sigma+2it)}\right] \geq 0. \tag{20}$$

The function $-\frac{\zeta'(s)}{\zeta(s)} - \frac{1}{\sigma-1}$ is regular in the interval $1 \leq \sigma \leq 2$. Hence there is a constant $c_1 > 0$ with

$$-\frac{\zeta'(s)}{\zeta(s)} < \frac{1}{\sigma-1} + c_1 \quad \text{for } 1 < \sigma \leq 2. \tag{21}$$

By (13) and Proposition 2, for $|s-1| \geq c_0$ and $1 < \sigma \leq 2$ there is a constant $c_2 > 0$ with

$$-\mathrm{Re}\,\frac{\zeta'(s)}{\zeta(s)} < c_2 \log(|t|+2) - \sum_{n=1}^{\infty} \mathrm{Re}\left[\frac{1}{s-\rho_n} + \frac{1}{\rho_n}\right]. \tag{22}$$

Also, for $\sigma > 1$,

$$\mathrm{Re}\,\frac{1}{s-\rho_n} = \frac{\sigma-\beta_n}{(\sigma-\beta_n)^2+(t-\gamma_n)^2} > 0,\ \mathrm{Re}\,\frac{1}{\rho_n} \geq 0.$$

Hence it follows from (22) that

$$- \operatorname{Re} \frac{\zeta'(\sigma+i\gamma)}{\zeta(\sigma+i\gamma)} < c_2 \log(|\gamma|+2) - \frac{1}{\sigma-\beta} \tag{23}$$

and

$$- \operatorname{Re} \frac{\zeta'(\sigma+2i\gamma)}{\zeta(\sigma+2i\gamma)} < c_3 \log(|\gamma|+2). \tag{24}$$

The inequalities (20) to (24) now give

$$\frac{3}{\sigma-1} - \frac{4}{\sigma-\beta} + c_4 \log(|\gamma|+2) \geq 0, \tag{25}$$

where the constant c_4, like c_2 and c_3, depends on c_0 but not on σ. We now set

$$\sigma = 1 + \frac{1}{2c_4 \log(|\gamma|+2)}.$$

Solving (25) for β yields

$$\beta \leq 1 - \frac{1}{14c_4 \log(|\gamma|+2)}. \qquad \square$$

Theorem 9. *Let* $T \geq 2$ *and* $c > 0$ *be the constants of Theorem* 8. *Then for all* $s = \sigma + it$ *with*

$$\sigma \geq 1 - \frac{c}{2 \log(T+2)}, \quad |t| \leq T$$

we have the estimate

$$\frac{\zeta'(s)}{\zeta(s)} = O(\log^2 T).$$

Proof. By Theorem 7

$$\left|\frac{\zeta'(s)}{\zeta(s)}\right| \leq \sum_{|t-\gamma_n| \leq 1} \frac{1}{|\sigma-\beta_n+i(t-\gamma_n)|} + O(\log T).$$

Since $\beta_n \leq 1-c/\log(T+2)$, $\sigma \geq 1-c/2 \log(T+2)$, Theorem 5 gives

$$\left|\frac{\zeta'(s)}{\zeta(s)}\right| \leq \frac{2}{c} \log(T+2) \sum_{|t-\gamma_n| \leq 1} 1 + O(\log T) = O(\log^2 T). \qquad \square$$

We now come to the proof of Theorem 3. By Theorem 9 we can set

$$h = 1 - c/\log(T+2)$$

in (12). With $T = e^{\sqrt{\log x}}$ we get

$$\psi_0(x) = x + O(xe^{\sqrt{\log x}}\log^2 x) + O(xe^{-c_5\sqrt{\log x}}\log^{3/2} x)$$

$$= x + O(xe^{-c_6\sqrt{\log x}})$$

for certain constants $c_5, c_6 > 0$. □

If the Riemann hypothesis holds (Section 15.3), one can set $h = \frac{1}{2} + \varepsilon$ and $T = \sqrt{x}$ in (12), where $\varepsilon > 0$ is arbitrary. Then one obtains the essentially more precise estimates

$$\psi(x) = x + O(x^{1/2+\varepsilon}),\ \pi(x) = \int_2^x \frac{dt}{\log t} + O(x^{1/2+\varepsilon}),$$

where the constants which belong to the O depend on ε.

Exercises

27.1 Let $H(x,s) = -\frac{\zeta'(s)x^s}{\zeta(s)s}$. Show that

$$\psi_0(x) = \frac{1}{2\pi i}\int_{b-iT}^{b+iT} H(x,s)ds + O\left(\frac{x\log x(\log x + \frac{1}{<x>})}{T}\right)$$

for $T \geq 1$, $x \geq 2$, where $b := 1 + 1/\log x$.

27.2 Show that there is a sequence of positive numbers $T_1, T_2, \ldots$ with $\lim_{n\to\infty} T_n = \infty$ such that

a) $\frac{1}{\log T_n} = O(|\gamma - T_n|)$ for all non-trivial zeros $\beta + i\gamma$ of $\zeta(s)$.

b) $\frac{\zeta'(s)}{\zeta(s)} = O(\log^2 T_n)$ for $-1 \leq \operatorname{Re} s \leq 2$.

27.3 Let s be a complex number with $\operatorname{Re} s \leq -1$ and $|s+2m| \geq \frac{1}{2}$ for $m = 1,2,\ldots$. Show that $\zeta'(s)/\zeta(s) = O(\log(2|s|))$. (Hint: Use the functional equation for the ζ-function (Chapter 15).)

27.4 Let $U \geq 1$ be an odd integer. Use exercises 27.2 and 27.3 to show

$$\int_{-U-iT_n}^{-U+iT_n} H(x,s)ds = O\left(\frac{x \log^2 T_n}{T_n \log x}\right).$$

27.5 Show

$$\int_{-U-iT_n}^{-U+iT_n} H(x,s)ds = O\left(\frac{T_n \log U}{U x^U}\right).$$

27.6 Use exercises 27.1 to 27.5 to show

$$\psi_0(x) = x - \sum_{|\gamma|<T} \frac{x^\rho}{\rho} - \frac{\zeta'(0)}{\zeta(0)} - \frac{1}{2}\log\left(1 - \frac{1}{x^2}\right) + O\left(\frac{x \log x(\log x + \frac{1}{<x>})}{T}\right)$$

$$+ O\left(\frac{x \log^2 T}{T \log x}\right)$$

for all $T \geq 1$ and $x \geq 2$, where $\rho = \beta + i\gamma$ runs through the non-trivial zeros of $\zeta(s)$.

27.7 Compute $\zeta'(0)/\zeta(0)$ with the help of exercise 26.4.

27.8 Let $\mu(n)$ be the Möbius function (Section 15.3). Show that

$$\sum_{n\leq x} \mu(n) = O(x \exp(-c\sqrt{\log x}))$$

for a positive constant c.

The following exercises 27.9 – 27.15 yield a proof of the Riemann prime number formula (16), Chapter 15. (Here we follow H. Cramér, *Arkiv für Mat., Astron. och Fysik* 13, Nr. 24 (1919).)

27.9 Suppose $x > 1$ is not a prime power. Show that the series

$$\sum_\rho \frac{x^\rho}{\rho} := \lim_{T\to\infty} \sum_{|\gamma|<T} \frac{x^\rho}{\rho}$$

converges uniformly in a neighbourhood of x.

27.10 Let p be a prime number. Show that

a) $$\sum_{|\rho|<T} \frac{x^\rho}{\rho} = -\frac{\log p}{2\pi i}\int_{b-iT}^{b+iT}\left(\frac{x}{p^m}\right)^s\frac{ds}{s} + \phi_1(T,x),$$

b) $$\int_{b-iT}^{b+iT}\left(\frac{x}{p^m}\right)^s\frac{ds}{s} = 2i\left(\frac{x}{p^m}\right)^b\int_0^{T\log\frac{x}{p^m}}\frac{\sin t}{t}dt + \phi_2(T,x)$$

for $p^m - \frac{1}{2} \le x \le p^m + \frac{1}{2}$, where $\phi_1(T,x)$ and $\phi_2(T,x)$ are continuous functions which converge uniformly as $T \to \infty$.

27.11 Show that $\sum_{|\gamma|<T} \frac{x^\rho}{\rho}$ is uniformly bounded in the interval $p^m - \frac{1}{2} \le x \le p^m + \frac{1}{2}$ as $T \to \infty$ (Chapter 5, Proposition 2).

27.12 Suppose $\rho = \beta + i\gamma$, $\gamma \ne 0$, $x > 1$. Show

a) $$\mathrm{Li}(x^\rho) = \mathrm{Li}(e^{\rho\log x}) = \int_{-\infty}^{\beta}\frac{x^{\sigma+i\gamma}}{(\sigma+i\gamma)}d\sigma + \begin{cases}\pi i & \text{if } \gamma > 0\\ -\pi i & \text{if } \gamma < 0\end{cases}$$

b) $$\sum_{|\gamma|<T}\mathrm{Li}(x^\rho) = \frac{1}{\log x}\sum_{|\gamma|<T}\frac{x^\rho}{\rho} - \sum_{|\gamma|<T}\frac{x^{i\gamma}}{\rho}\int_{-\infty}^{\beta}\frac{(\sigma-\beta)x^\sigma}{\sigma+i\gamma}d\sigma$$

c) $$\lim_{T\to\infty}\sum_{|\gamma|<T}\frac{x^{i\gamma}}{\rho}\int_{-\infty}^{\beta}\frac{(\sigma-\beta)x^\sigma}{\sigma+i\gamma}d\sigma \text{ converges absolutely.}$$

27.13 Let $f(x)$ be the function defined in Chapter 15, (7). In what follows, c_1, c_2, c_3 denote constants. Show

a) $$f(x) = \frac{\psi_0(x)}{\log x} + \int_2^x\frac{\psi_0(u)du}{u(\log u)^2},$$

b) $$f(x) = \int_2^x \frac{du}{\log u} - \left(\frac{1}{\log x} \sum_\rho \frac{x^\rho}{\rho} + \int_2^x \frac{du}{u(\log u)^2} \left[\sum_\rho \frac{u^\rho}{\rho} \right] \right)$$

$$+ \int_x^\infty \frac{du}{u(u^2-1)\log u} + c_1,$$

$$= \int_2^x \frac{du}{\log u} - \sum_\rho \int_2^x \frac{u^{\rho-1}}{\log u} du + \int_x^\infty \frac{du}{u(u^2-1)\log u} + c_2,$$

c) $$f(x) = \mathrm{Li}((x) - \sum_\rho \mathrm{Li}(x^\rho) + \int_x^\infty \frac{du}{u(u^2-1)\log u} + c_3.$$

27.14* Show $c_3 = -\log 2$.

27.15 Suppose $\varepsilon > 0$. Show that there is a number $n_0(\varepsilon)$ such that for all $n \geq n_0(\varepsilon)$ there is a prime number between n and $n(1+\varepsilon)$.

Hint: Use Theorem 1.

Remark. In 1845 J. Bertrand conjectured the assertion of exercise 27.15 for $\varepsilon = 1$ (*J. École Polytechnique* 17). Chebyshev was able to prove it in 1852 for $\varepsilon = 1/5$ (*J. Math. pures et appl.* 17).

28. Combinatorial topology

Riemann recognised that the topology of surfaces is of fundamental importance for complex function theory in one variable (Chapter 10). Betti generalised Riemann's ideas to manifolds M of arbitrary dimension n. He defined numbers $b_0(M)$, $b_1(M)$,..., $b_n(M)$ that are now known as the *Betti numbers* of M. However, the real creator of topology as an independent mathematical discipline was Poincaré. In his works on topology in the years 1892 to 1904 he developed the basic ideas, which were elaborated, but scarcely extended, up to the end of the 1920s. In the year 1901 Poincaré wrote about his entrance to this discipline which he, following Riemann, called "*analysis situs*":

"*A method which makes known to us the qualitative relations holding in space of more than three dimensions can, to a certain degree, perform for us the same service as figures. This method can be nothing but analysis situs in more than three dimensions. However, this branch of science has been very little cultivated until now. After Riemann there was Betti, who introduced a few fundamental concepts, but no-one followed Betti. As far as I am concerned, it seems that all the paths I have followed have led to analysis situs. I have needed the results of this science in order to pursue my study of curves defined by differential equations, and to extend it to differential equations of higher order, particularly those for the three body problem*" (*Acta Math.* 38 (1921)).

28.1 Polyhedra in $\mathbb{R}^n$

The basic idea of Poincaré for the study of topological properties of figures in n-dimensional euclidean space $\mathbb{R}^n$ is to divide the figures into simple pieces and to consider the positions of these pieces relative to each other. We shall first define these simple pieces, which are called *simplexes*.

Points $P_0,...,P_k$ of $\mathbb{R}^n$ are called *linearly independent* when the hyperplane spanned by them, consisting of all points P of the form

$$P = x_0P_0 + x_1P_1 + ... + x_kP_k$$

for $x_0,x_1,...,x_k \in \mathbb{R}$ with $x_0 + x_1 + ... + x_k = 1$, has dimension k. In this case the point P determines the numbers $x_0,x_1,...,x_k$ uniquely. They are called the *barycentric coordinates* of P relative to the system $P_0,P_1,...,P_k$.

Suppose now that $P_0, P_1, ..., P_k$ are linearly independent points of $\mathbb{R}^n$. The smallest convex subset S (Section 24.1) of $\mathbb{R}^n$ which contains $P_0, P_1, ..., P_k$ consists of all the points

$$P = x_0 P_0 + x_1 P_1 + ... + x_k P_k$$

with $x_0 + x_1 + ... + x_k = 1$ and $x_i \geq 0$ for $i = 0,1,...,k$. One easily proves this assertion by performing a translation by $-P_0$ so as to come back to the case $P_0 = 0$. We write $S = (P_0; P_1; ...; P_k)$ and call S a *k-dimensional linear simplex*.

In what follows we need *oriented simplexes*. The oriented simplex $(P_0, P_1, ..., P_k)$ is the set S with the orientation given by the ordering $P_0, P_1, ..., P_k$ of vertices. Let π be a permutation of $0,1,...,k$. Then $(P_0, P_1, ..., P_k)$ and $(P_{\pi(0)}, P_{\pi(1)}, ..., P_{\pi(k)})$ are considered to be the same when π is an even permutation. A zero-dimensional simplex is a point, a one-dimensional simplex is a directed line segment, and a two-dimensional simplex is an oriented triangle. Let $(P_0, ..., P_k)$ be a simplex and suppose $i_0, ..., i_r$ belong to $\{0, ..., k\}$. Then the simplex $(P_{i_0}, ..., P_{i_r})$ is called an *r-dimensional face* of $(P_0, ..., P_k)$. Thus the faces of a simplex include the simplex itself. This should be borne in mind in the following definition.

A set $\mathfrak{K}$ of simplexes in $\mathbb{R}^n$ is called a *simplicial complex* when the following conditions are satisfied:

(i) *If a simplex* A *belongs to* $\mathfrak{K}$ *so do all faces of* A.

(ii) *The intersection of two simplexes of* $\mathfrak{K}$ *is empty or a face.*

(iii) *Let* P *be a point of* $\cup\mathfrak{K}$. *Then there is a neighbourhood of* P *in* $\mathbb{R}^n$ *that has points in common with only finitely many simplexes of* $\mathfrak{K}$.

The zero-dimensional simplexes of $\mathfrak{K}$ are called *vertices* of $\mathfrak{K}$. The *dimension* of $\mathfrak{K}$ is the maximal dimension of simplexes in $\mathfrak{K}$. The collection of k-dimensional simplexes of $\mathfrak{K}$ is denoted by $\mathfrak{K}_k$. Finally, a *polyhedron* is the union of all simplexes of a complex.

28.2 Topological polyhedra

In combinatorial topology we are concerned only with the schema of a complex $\mathfrak{K}$, i.e. the set E of vertices of $\mathfrak{K}$ and the subsets of E that generate simplexes of $\mathfrak{K}$. Hence it is natural to consider polyhedra not only in $\mathbb{R}^n$, but in an arbitrary topological space T (A2.1).

A subset S of T is called a k-dimensional *simplex of* T when there is a homeomorphism ϕ from a k-dimensional linear simplex onto S. More precisely, the simplex is given by the pair (S,ϕ). Two pairs (S,ϕ_1), (S,ϕ_2), where ϕ_1 resp. ϕ_2 maps the k-dimensional linear simplex S_1 resp. S_2 onto S, are considered *equal* when the mapping $\phi_2^{-1}\phi_1$ is a linear mapping of S_1 onto S_2.

A decomposition of T into a complex $\mathfrak{K}$ is given by a set of simplexes of T whose union is T and which satisfies conditions (i), (ii) and (iii) of Section 28.1.

A topological space which admits decomposition into a complex is called a *topological polyhedron.*

A two-dimensional complex is called a *triangulation* of the associated polyhedron.

The real objects of investigation in combinatorial topology are the polyhedra. The concept of complex enters as a technical tool. Two polyhedra are regarded as *not essentially different* when they are homeomorphic.

Example 1. *The n-dimensional ball* B_n consists of all points $(x_1,\dots,x_n)$ of $\mathbb{R}^n$ with $x_1^2 + \dots + x_n^2 \le 1$. It is homeomorphic to an n-dimensional linear simplex.

Example 2. The *n-dimensional sphere* S_n consists of the boundary points of B_{n+1}, i.e. all points $(x_0,\dots,x_n)$ of $\mathbb{R}^{n+1}$ with $x_0^2 + \dots + x_n^2 = 1$. The n-dimensional sphere is realised by the complex consisting of all faces of $(P_0,\dots,P_{n+1})$ with dimension $\le n$.

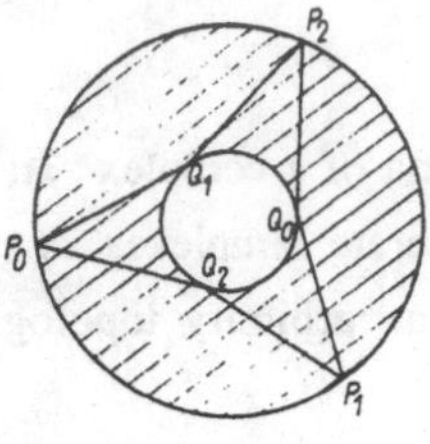

Fig. 18

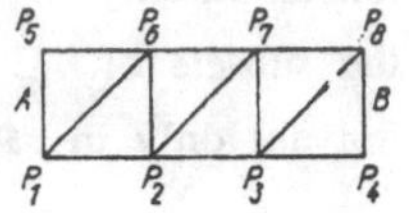

Fig. 19

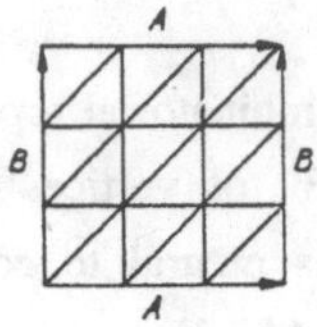

Fig. 20

Example 3. An annulus can be triangulated as shown in Figure 18.

Example 4. The annulus can also be represented topologically by joining the sides A, B of the band (Fig. 19) so that P_1 is identified with P_4 and P_5 with P_8. A second possibility is to twist the band and join A to B so that P_1 is identified with P_8 and P_4 with P_5. The resulting polyhedron is called the *Möbius band* (Section 10.7).

Example 5. A two-dimensional simplicial complex with finitely many vertices is a polygon complex (Section 10.7). Conversely, an arbitrary polygon complex can be converted into a simplicial complex by elementary transformations (Section 10.8). For example, a triangulation of the torus is obtained as shown in Figure 20.

28.3 The homology groups of a polyhedron

Associated with a simplicial complex $\mathfrak{K}$ are certain abelian groups called its *homology groups*. Their most important property is that they depend only on the polyhedron corresponding to $\mathfrak{K}$. Two decompositions of the polyhedron into simplicial complexes yield isomorphic homology groups. These groups were first defined in 1928 by Hopf (*Nachr. Königl. Gesellsch. Wiss. Göttingen, Math. Phys. Kl.*), following a suggestion of E. Noether. Poincaré had considered the ranks (Betti numbers) and torsion coefficients of these groups (A1.4). Their independence from the simplicial decomposition was first proved completely by Alexander (*Trans. Amer. Math. Soc.* 16 (1915)).

Let $\mathfrak{K}$ be a simplicial complex. We choose an orientation for each simplex S and denote the oriented simplex by S′. A *k-dimensional chain* of $\mathfrak{K}$ is a linear combination

$$\sum_{S \in \mathfrak{K}_k} a_S S'$$

with coefficients a_S from $\mathbf{Z}$, almost all of them zero. S' is identified with the chain $1.S'$. The simplex S with the orientation opposite to that of S' is denoted by $-S'$. Addition of chains is performed coefficientwise. The collection $K_k(\mathfrak{K})$ of all chains of $\mathfrak{K}$ then forms an abelian group.

Now for $k = 1,2,\dots$ we define a homomorphism δ_k of $K_k(\mathfrak{K})$ into $K_{k-1}(\mathfrak{K})$ called the *boundary operator.* It suffices to define δ_k for oriented simplexes $S' = (P_0,\dots,P_k)$. Then δ_k is defined for a chain by

$$\delta_k\Big(\sum_{S\in\mathfrak{K}} a_S S'\Big) = \sum_{S\in\mathfrak{K}} a_S \delta_k S'.$$

We set

$$\delta_k S' = \sum_{i=0}^{k} (-1)^i (P_0,P_1,\dots,P_{i-1},P_{i+1},\dots,P_k).$$

We abbreviate $(P_0,P_1,\dots,P_{i-1},P_{i+1},\dots,P_k)$ by $(P_0,\dots,\hat{P}_i,\dots,P_k)$. We also set $\delta_0 = 0$.

The boundary operators satisfy the following fundamental relation:

$$\delta_k\delta_{k+1} = 0 \quad \text{for } k = 0,1,\dots\,. \tag{1}$$

In fact

$$\begin{aligned}
\delta_k\delta_{k+1}(P_0,\dots,P_{k+1}) &= \sum_{i=0}^{k+1} (-1)^i \delta_k(P_0,\dots,\hat{P}_i,\dots,P_{k+1}) \\
&= \sum_{i=0}^{k+1} (-1)^i \Big[\sum_{j=0}^{i-1} (-1)^j (P_0,\dots,\hat{P}_j,\dots,\hat{P}_i,\dots,P_{k+1}) \\
&\qquad + \sum_{j=i+1}^{k+1} (-1)^{j-1} (P_0,\dots,\hat{P}_i,\dots,\hat{P}_j,\dots,P_{k+1})\Big] \\
&= \sum_{j<i} (-1)^{i+j} (P_0,\dots,\hat{P}_j,\dots,\hat{P}_i,\dots,P_{k+1}) \\
&\quad - \sum_{j>i} (-1)^{i+j} (P_0,\dots,\hat{P}_i,\dots,\hat{P}_j,\dots,P_{k+1}) \\
&= 0.
\end{aligned}$$

We define the subgroups $Z_k(\mathfrak{K})$ of *cycles* and $R_k(\mathfrak{K})$ of *boundaries* by

$$Z_k(\mathfrak{K}) := \text{Ker}\,\delta_k = \{x \in K_k(\mathfrak{K}) \,|\, \delta_k x = 0\}, \quad R_k(\mathfrak{K}) := \delta_{k+1} K_{k+1}(\mathfrak{K})$$

for $k = 0,1,2,\ldots$. Because of (1) we have

$$R_k(\mathfrak{K}) \subset Z_k(\mathfrak{K}).$$

The quotient group $H_k(\mathfrak{K}) := Z_k(\mathfrak{K})/R_k(\mathfrak{K})$ is called the kth *homology group* of $\mathfrak{K}$.

The definition of $H_k(\mathfrak{K})$ depends on the choice of orientation of the simplex. However, one easily convinces oneself that $H_k(\mathfrak{K})$ is independent of this choice, up to isomorphism.

The following *fundamental theorem on homology groups* will be proved in the second volume of this book.

Fundamental theorem. *The homology groups* $H_k(\mathfrak{K})$ *of a complex* K *depend only on the associated polyhedron. They are invariant under homeomorphisms.*

Let $T := M(\mathfrak{K})$ be the polyhedron associated with the complex $\mathfrak{K}$. In what follows we also write $H_k(T)$ in place of $H_k(\mathfrak{K})$, in order to emphasise the independence of the homology groups from the decomposition of T into a complex.

28.4 Computation of homology groups in simple cases

A polyhedron M is called *connected* when any two points P, Q of M can be connected by a path which runs entirely in M. A path from P to Q is a continuous mapping ϕ of the interval [0,1] into M with $\phi(0) = P$, $\phi(1) = Q$.

M decomposes uniquely into connected subpolyhedra M_i, where i runs through a set I of indices. The M_i are called the *connected components* of M. They are pairwise disconnected from each other. As one easily sees

$$H_k(M) = \sum_{i \in I} H_k(M_i) \quad \text{for } k = 0,1,\ldots$$

and $H_0(M_i) = \mathbb{Z}$. In particular, the rank of $H_0(M)$ as a $\mathbb{Z}$-module is the number of connected components of M.

We now consider the examples from Section 28.2.

Example 1. The n-dimensional ball B_n is homeomorphic to the simplex $(P_0,\dots,P_n)$ whose subsimplexes form a complex $\mathfrak{K}$ associated with B_n.

Let $(P_{i_0},\dots,P_{i_k})$ be a k-simplex of $\mathfrak{K}$ with $0 < i_0 < \dots < i_k$. Then

$$(P_{i_0},\dots,P_{i_k}) = \sum_{s=0}^{k} (-1)^s (P_0,P_{i_0},\dots,\hat{P}_{i_s},\dots,P_{i_k}) \pmod{R_k(\mathfrak{K})}.$$

Hence each k-cycle Z of $\mathfrak{K}$ may be written in the form

$$Z \equiv \sum_{0<i_1<\dots<i_k} a_{i_1\dots i_k}(P_0,P_{i_1},\dots,P_{i_k}) \pmod{R_k(\mathfrak{K})}.$$

Then, for $k \geq 1$

$$\delta_k Z = \sum_{0<i_1<\dots<i_k} a_{i_1\dots i_k}(P_{i_1},\dots,P_{i_k}) + W = 0,$$

where W is a linear combination of simplexes involving the vertex P_0. Therefore $a_{i_1\dots i_k} = 0$, i.e. $Z \in R_k(\mathfrak{K})$. Thus we obtain the result that $H_k(B_n) = \{0\}$ for $k \geq 1$.

Example 2. The n-dimensional sphere S_n is realised by the complex $\mathfrak{K}$ consisting of all proper subsimplexes of $(P_0,\dots,P_{n+1})$. It then follows immediately from Example 1 that $H_k(S_n) = \{0\}$ for $0 < k < n$. Also $Z_n(\mathfrak{K}) = Z_n(\mathfrak{K}_1) = R_n(\mathfrak{K}_1)$, where $\mathfrak{K}_1$ is the complex consisting of all subsimplexes of $(P_0,\dots,P_{n+1})$. Thus $H_n(S_n) = Z_n(\mathfrak{K}) = \mathbb{Z}$.

Example 3. The annulus C. Since each 2-simplex has an edge which appears in no other 2-simplex, there is only a trivial 2-cycle, i.e. $H_2(C) = \{0\}$. For the same reason, each 1-chain is uniquely expressible in the form

$$Z \equiv \sum_{i\neq j} a_{ij}(P_i,Q_j) \pmod{R_1(\mathfrak{K})}$$

(see Figure 18 of Section 28.2). Z is a cycle if and only if $a_{01} = -a_{02} = -a_{21} = a_{12} = a_{20} = -a_{10}$, i.e. $H_1(C) \cong \mathbb{Z}$.

Example 4. The Möbius band C'. As in Example 3, $H_2(C') = \{0\}$ and $H_1(C') \cong \mathbb{Z}$

Example 5. To compute the first homology group of the torus T we allow ourselves to be guided by Riemann's ideas on the decomposition of a surface (Section 10.7). We orient all 2-simplexes (Figure 20, Section 28.2) in the clockwise sense. Then each 1-simplex appears exactly once in each direction. One easily sees from this that multiples of the sum of all 2-simplexes, and only these, are cycles. I.e., $H_2(T) \cong \mathbb{Z}$. In Section 10.8 we have constructed the torus from a square. To do this we can use the oriented paths A, B, which are obviously 1-cycles. We shall show that A and B constitute a basis of $H_1(T)$. First one convinces oneself easily that each 1-cycle can be converted, by addition of boundaries, into a chain in which only edges from A or B appear. But a chain of this kind can only be a cycle when it is of the form $\alpha A + \beta B$ with $\alpha, \beta \in \mathbb{Z}$. Finally, it follows from $\alpha A + \beta B \in R_1(\mathfrak{K})$ that $\alpha = \beta = 0$, because $R_1(\mathfrak{K})$ contains no boundary, except 0, consisting entirely of edges in A or B. Thus we have $H_1(T) \cong \mathbb{Z} + \mathbb{Z}$.

One shows similarly that the first homology group of a closed Riemann surface of genus g is a free abelian group of rank 2g.

In all the previous examples the homology groups are torsion-free. The following example, of the projective plane $\mathbb{P}^2(\mathbb{R})$, shows that this is not always the case.

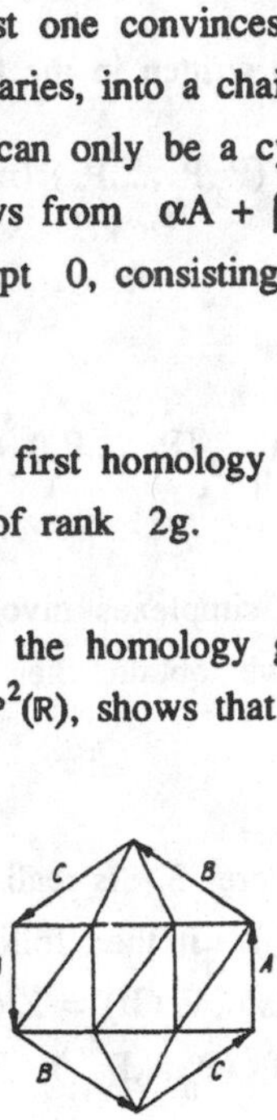

Fig. 21

In Section 10.7 we realised the projective plane topologically as a closed disc with identification of diametrically opposite boundary points. The projective plane can therefore be triangulated as shown in Figure 21. One sees from this that $\mathbb{P}^2(\mathbb{R})$ also results from extension of a Möbius band by a disc to form a closed surface.

We call the 1-simplexes of our triangulation $\mathfrak{K}$ other than A, B, C the *inner edges*. As in Example 5 we orient the 2-simplexes in the clockwise sense. Then each of the inner edges appears exactly once in each direction, but A, B, C each appear twice in the same direction. From this it is clear that 0 is the only 2-cycle. Each 1-cycle is congruent, modulo the boundaries, to a 1-cycle in which only A, B, C appear. From these 1-simplexes we can form the cycles $\alpha(A+B+C)$ for $\alpha \in \mathbb{Z}$. A + B + C cannot lie in $R_1(\mathfrak{K})$. On the other hand, 2(A+B+C) equals $\delta_2\Sigma$, where Σ is the sum of all 2-simplexes of $\mathfrak{K}$. I.e., $H_1(\mathbb{P}^2(\mathbb{R})) = \mathbb{Z}/2\mathbb{Z}$.

28.5 Betti numbers and Euler characteristic

The kth homology group of an n-dimensional complex $\mathfrak{K}$, consisting of finitely many simplexes, is a finitely generated abelian group. The rank of this group (Section A1.4) is called the kth *Betti number* b_k, and the alternating sum $\sum_{k=0}^{n} (-1)^k b_k$ is called the *Euler characteristic.* The use of the latter term from Chapter 10, Theorem 5, is justified by the following theorem.

Theorem 1. *Let* $\mathfrak{K}$ *be an* n*-dimensional complex, consisting of finitely many simplexes, and let* s_k *be the number of* k*-simplexes of* $\mathfrak{K}$*. Then*

$$\sum_{k=0}^{n} (-1)^k s_k = \sum_{k=0}^{n} (-1)^k b_k .$$

Proof. It follows from the exact sequences

$$0 \to R_k(\mathfrak{K}) \to Z_k(\mathfrak{K}) \to H_k(\mathfrak{K}) \to 0 \quad \text{for } k = 0,\dots,n$$

$$0 \to Z_k(\mathfrak{K}) \to K_k(\mathfrak{K}) \xrightarrow[\delta_k]{} R_{k-1}(\mathfrak{K}) \to 0 \quad \text{for } k = 1,\dots,n$$

that

$$r(Z_k(\mathfrak{K})) = r(R_k(\mathfrak{K})) + b_k,$$

$$s_k = r(Z_k(\mathfrak{K})) + r(R_{k-1}(\mathfrak{K})),$$

where r(A) denotes the rank of the abelian group A (Section A1.4). □

28.6 The fundamental group

Among Poincaré's most important discoveries in topology is the *fundamental group of a polyhedron,* which was anticipated by the investigations of Jordan on surfaces (*Des contours tracées sur les surfaces,* J. math. pures appl. 11 (1866)).

In contrast to the homology groups we define the fundamental group in an invariant way. We also define an *edge path group* of a complex analogous to the first homology group. The proof of the isomorphism between the fundamental group and the edge path group is closely related to the invariance of the homology groups.

Let M be a topological space. An *oriented path* from point $P \in M$ to point $Q \in M$ is a continuous mapping W of the closed unit interval $[0,1]$ into M with $W(0) = P$ and $W(1) = Q$. If W' is a path from Q to a point $R \in M$ then the product WW' is equal to the concatenation of the paths W and W':

$$WW'(s) = W(2s) \quad \text{for} \quad 0 \le s \le 1/2,$$
$$WW'(s) = W'(-1+2s) \quad \text{for} \quad 1/2 \le s \le 1.$$

W^{-1}, the path inverse to W, goes from Q to P and is defined by

$$W^{-1}(s) = W(1-s) \quad \text{for} \quad 0 \le s \le 1.$$

As one easily sees, $(WW')^{-1} = W'^{-1}W^{-1}$.

Two paths W_1, W_2 from P to Q are called *homotopic*, $W_1 \sim W_2$, when they are continuously deformable into each other, i.e. when there is a continuous mapping ϕ of $[0,1] \times [0,1]$ into M such that

$$\phi(s,0) = W_1(s),\ \phi(s,1) = W_2(s) \quad \text{for} \quad 0 \le s \le 1.$$

As one easily sees, the homotopy relation $\sim$ is an equivalence relation, compatible with the formation of products and inverses, i.e. $W_1 \sim W_2$, $W_1' \sim W_2'$ implies $W_1W_1' \sim W_2W_2'$, $W_1^{-1} \sim W_2^{-1}$.

The trivial path 1_Q from Q (to Q) is the mapping $1_Q(s) = Q$ for $0 \le s \le 1$. For any path W from P to Q we have

$$WW^{-1} \sim 1_P,\ W^{-1}W \sim 1_Q.$$

The homotopies are effected by the functions $\phi_P(s,t) = WW^{-1}(st)$ and $\phi_Q(s,t) = W^{-1}W(st)$ respectively.

Now suppose M is *arcwise connected*, i.e. any two points of M are connected by a path. We fix a point P of M. The *fundamental group* F(M) of M is defined to be the set of all homotopy classes of paths from P to P under the multiplication induced by the product of paths. It is easy to see that F(M) is a group and that, up to isomorphism, it does not depend on the choice of P.

In Chapter 8 we have called a region simply connected when each closed path in it is contractible to a point. By the definition of fundamental group, an equivalent statement is that this group consists only of the identity element. (Strictly speaking, in Chapter 8 we considered only piecewise smooth paths, for the sake of simplicity. However, one can show that the fundamental group of piecewise smooth paths is isomorphic to the fundamental group defined above.)

28.7 The edge path group of a polygon complex

In order to be able to compute the fundamental group of a polyhedron M, we define the *edge path* group of a simplicial complex $\mathfrak{K}$ with which the polyhedron is associated. In what follows we confine ourselves to finite complexes $\mathfrak{K}$. Since the fundamental group depends only on the two-dimensional subcomplex of $\mathfrak{K}$, we proceed more generally from a polygon complex $\mathfrak{P}$ (Section 10.7), which in any case is convertible to a simplicial complex $\mathfrak{P}'$ by elementary transformations (Section 10.8). The homology groups (Exercise 28.2) and the edge path groups of $\mathfrak{P}$ and $\mathfrak{P}'$ are isomorphic. However, the polygon complex can often be chosen so as to make the computation of the homology groups and the edge path groups clearer. The reader will already have noticed how awkward the computation of homology groups can be from the examples of the torus and the projective plane.

An edge path W of $\mathfrak{P}$ is a path in M which runs along edges of $\mathfrak{P}$. We denote it by the product of the edges traversed, with each factor provided with an exponent + 1 or - 1 according as the edge is traversed with or against the chosen orientation (Section 10.7). Two edge paths are called *combinatorially homotopic* when one is convertible to the other by finitely many elementary homotopies of the following two kinds:

a) Insertion or deletion of AA^{-1} or $A^{-1}A$, where A is an edge, e.g. $W_1AA^{-1}W_2 \sim W_1W_2$;

b) Insertion or deletion of the boundary path of a polygon.

A *boundary path* is the sequence of images of the edges of a preimage polygon. For example, we have the representation of the projective plane by a polygon with one edge A, which is the image of a polygon with two edges (see Figure 22 and compare with Figure 21 in 28.5) and boundary path AA.

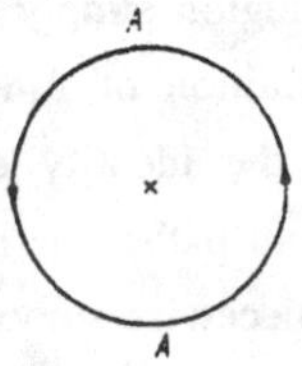

Fig. 22

The collection of combinatorial homotopy classes of edge paths from a fixed vertex P of $\mathfrak{P}$ to P is the edge path group $F'(\mathfrak{P})$, where the group operation is again induced by the product of paths.

We associate each class of edge paths in $F'(\mathfrak{P})$ with the corresponding class in $F(\cup\mathfrak{P})$, obtaining a homomorphism ϕ of the group $F'(\mathfrak{P})$ into the group $F(\cup\mathfrak{P})$.

Fundamental theorem on edge path groups. *The homomorphism* ϕ *is an isomorphism of* $F'(\mathfrak{P})$ *onto* $F(\cup\mathfrak{P})$.

The proof of this theorem will be carried out in the second volume of this book.

One easily convinces oneself that two complexes related by elementary transformations have isomorphic edge path groups. Hence in the proof of the main theorem on edge path groups one can confine oneself to simplicial complexes.

If one compares the definition of the edge path group with the definition of the first homology group of a simplicial complex $\mathfrak{K}$, then one sees that the only difference between the two is that in the edge path group the order of edges matters, while in the homology group it does not. The homology group $H_1(\mathfrak{K})$ is therefore the "largest" abelian quotient group of $F'(\mathfrak{K})$. Underlying this is the following group theoretic process: let G be an arbitrary group and suppose $x,y \in G$. If, for a normal subgroup N of G, the quotient group G/N is to be abelian, then obviously the commutator $x^{-1}y^{-1}xy =: [x,y]$ must lie in N. Thus N contains the smallest normal subgroup G^c of G generated by the elements $[x,y]$ for all $x,y \in G$. On the other hand, G/G^c is obviously abelian. G^c is called the *commutator subgroup* of G and G/G^c the *commutator quotient* group of G. We have

$$H_1(\mathfrak{K}) = F'(\mathfrak{K})/F'(\mathfrak{K})^c. \tag{2}$$

28.8 Presentation of the edge path group by generators and relations

We want to present the group $F'(\mathfrak{K})$ of a polygon complex by *generators* and *relations*. Such a presentation is based on the following group theoretic construction: let $s_1,\dots,s_m$ be arbitrary objects; we call them *generators*, and the set of them an *alphabet*. An expression $w := s_{i_1}^{\varepsilon_1} \dots s_{i_h}^{\varepsilon_h}$, where $i_1,\dots,i_h$ are any numbers between 1 and m and each ε_k is $+1$ or -1, for $k = 1,\dots,h$, is called a *word* in $s_1,\dots,s_m$. Let W be the set of words in $s_1,\dots,s_m$. Among them we include the "empty word" 1. A multiplication is defined in W by concatenation of words. Two words are called *equivalent* when one is convertible to the other by finitely many *elementary transformations*. An elementary transformation is the insertion or deletion of $s_k s_k^{-1}$ or $s_k^{-1} s_k$ for $k = 1,\dots,$ or m. In this way we obtain an equivalence relation compatible with the multiplication in W. The set of equivalence classes is a group F_m, called the free group with generators $s_1,\dots,s_m$. The equivalence classes are also denoted by words in $s_1,\dots,s_m$. They have a unique shortest representation, in which s_k and s_k^{-1} do not appear next to each other. The proof of the latter assertion is left to the reader.

Let $r_1,\dots,r_l$ be words in F_m and let R be the normal subgroup of F_m generated by $r_1,\dots,r_l$. Then F_m/R is called the group with *generators* $s_1,\dots,s_m$ and *relations* $r_1,\dots,r_l$.

We now come back to the presentation of the group $F'(\mathfrak{K})$ by generators and relations. Let the edges of $\mathfrak{K}$ be denoted by $A_1,\dots,A_m$.

We connect each vertex Q of the complex, assumed to be connected, to P by an edge path W_Q. In particular, we set $W_P = 1_P$. The edge A_k with initial point Q and final point Q' is associated with the path

$$s_k = W_Q A_k W_{Q'}^{-1}, \ k = 1,\dots,m, \tag{3}$$

from P to P. Thanks to the elementary homotopies of the first kind, (3) induces a homomorphism ϕ of F_m into $F(\mathfrak{K})$. One easily sees that ϕ is onto $F(\mathfrak{K})$.

The kernel of ϕ is generated by *relations* of the *first* and *second kind*: the relations of the first kind result from the fact that the generators s_k are themselves words in $A_1,\dots,A_m$. When $s_k = A_{i_1}^{\varepsilon_1} \dots A_{i_h}^{\varepsilon_h}$ one obtains a relation $r_k := s_k^{-1} s_{i_1}^{\varepsilon_1} \dots s_{i_h}^{\varepsilon_h}$. The relations of the second kind result from the elementary homotopies of the second kind: corresponding to the polygon G with boundary $A_{i_1}^{\varepsilon_1} \dots A_{i_h}^{\varepsilon_h}$ we have the relation $r_G = s_{i_1}^{\varepsilon_1} \dots s_{i_h}^{\varepsilon_h}$.

In this way we have presented $F(\mathfrak{K})$ as a group with generators $s_1,\dots,s_m$ and relations r_k, r_G, where k runs through the numbers 1 to m and G runs through all polygons of $\mathfrak{K}$.

We now consider a few examples. A closed Riemann surface of genus g is represented by a polygon with one vertex P and edges $A_1,B_1,\dots,A_g,B_g$. The boundary path of the polygon is

$$A_1^{-1}B_1^{-1}A_1B_1 \dots A_g^{-1}B_g^{-1}A_gB_g \tag{4}$$

(Section 10.8). Since we have only one vertex, $F(\mathfrak{K})$ is the group with generators $A_1,B_1,\dots,A_g,B_g$ and the single relation (4).

The fundamental group of the two-dimensional sphere resp. projective plane is isomorphic to $\{1\}$ resp. $\mathbb{Z}/2\mathbb{Z}$.

Let $S(m)$ be the two-dimensional sphere with m points removed. Since $S(m)$ is not compact, the method of the edge path group is not at first applicable. However, it follows from the definition of the fundamental group that the latter does not change when one removes m disjoint open discs instead of m points (i.e. if the m holes are enlarged somewhat). The resulting surface may be represented by a polygon with 2m-1 edges and m vertices. The corresponding edge path group is isomorphic to the free group with m-1 generators (see also Exercise 28.5).

28.9 Covering spaces

The starting point of Riemann's topological reflections was the consideration of surfaces covering the plane (Chapter 10). This concept can easily be formulated for arbitrary topological spaces:

Let M_1 and M be topological spaces and let ϕ be a continuous mapping of M_1 onto M. Then $\{M_1,\phi\}$ is called a *covering* of M when, for each point $P \in M_1$ and each neighbourhood U(P), $\phi(U)$ is also a neighbourhood of $\phi(P)$. $\{M_1,\phi\}$ is called *unramified (unbranched)* at a point $P \in M_1$ when there is a neighbourhood U of P such that U is mapped homeomorphically onto $\phi(U)$.

The covering surfaces considered by Riemann are ramified at isolated points. If these points and their images are omitted, then one obtains an unramified covering. In what follows we consider only unramified coverings. More precisely, we make the following assumptions:

For each point of M there is a neighbourhood U which is arcwise and simply connected, and which has the property that $\phi^{-1}(U)$ consists of disjoint subsets which are mapped homeomorphically onto U by ϕ. We also assume that M_1, and hence M, is arcwise connected.

Two coverings $\{M_1,\phi\}$, $\{M_1',\phi'\}$ of M are called *homeomorphic* when there is a homeomorphism ψ of M_1 onto M_1' which is compatible with ϕ and ϕ', i.e. $\phi = \phi'\psi$.

Let $\{M_1,\phi\}$ be a covering of P, let $P \in M_1$, and suppose w(s), with $0 \le s \le 1$, is a path in M originating from $\phi(P) = w(0)$. We consider paths $\tilde{w}$ in M_1, originating from P, with $\phi(\tilde{w}(s)) = w(s)$ for $0 \le s < s_0$ for a certain s_0 with $0 < s_0 \le 1$. For a fixed s_0, $\tilde{w}$ is uniquely determined by the non-ramification condition. Also, one easily sees that $\tilde{w}$ always can be continued to a path that covers all of w. In addition to that, one has the following theorem, called the *monodromy theorem*:

Theorem 2. *Let* $\{M_1,\phi\}$ *be a covering of* M, *let* $P \in M_1$, *and let* w_1, w_2 *be homotopic paths which originate from* $\phi(P)$. *Then the paths* $\tilde{w}_1, \tilde{w}_2$ *originating from* P *are also homotopic.* □

In what follows our goal is to clarify the connection between coverings and the fundamental group.

Let $\{M_1,\phi\}$ be a covering of M. The induced mapping of closed paths originating from a fixed point yields a homomorphism ϕ_* of $F(M_1)$ into F(M). By Theorem 2, ϕ_* is a monomorphism. The image of ϕ_* in F(M) depends on the choice of initial point P in M_1. If one starts, not from P, but from a point P′ with the same image in M, and if w is a path from P to P′, then $\phi(w)$ is a closed path out of $\phi(P)$ which corresponds to the element $\overline{\phi(w)} \in F(M)$. Thus we obtain a homomorphism with image $\overline{\phi(w)}^{-1}\phi_*(F(M_1))\overline{\phi(w)}$. Hence $\{M_1,\phi\}$ determines a class of conjugate subgroups in F(M).

We now want to show that, conversely, each class of conjugate subgroups of F(M) determines a covering $\{M_1,\phi\}$. Let H be a subgroup of F(M). Then we consider all paths w_Q which go from a fixed point 0 of M to a point Q. Two such paths w_Q, w'_Q are called *equivalent* when the homotopy class $w_Q w'^{-1}_Q$ lies in H. M_1 is defined to be the set of classes of equivalent paths as Q varies. Then ϕ is the mapping that associates Q with the class of w_Q. The topology of M_1 is the topology induced by ϕ. One easily convinces oneself that $\{M_1,\phi\}$ is a covering of M. A subgroup conjugate to H leads to a homeomorphic covering.

In this way we have obtained a mapping Φ_1 from the set $\mathfrak{M}_1$ of homeomorphism classes of coverings of M into the set $\mathfrak{M}$ of classes of conjugate subgroups of F(M), as well as a mapping Φ_2 of $\mathfrak{M}$ into $\mathfrak{M}_1$. As one easily sees, the mappings are inverse to each other. Thus we have proved the following theorem:

Theorem 3. *There is a one-to-one correspondence between the homeomorphism classes of coverings of a topological space* M *and the classes of conjugate subgroups of* F(M). □

When $\{M_1,\phi\}$ is a covering of M the same number of points of M_1 lie over each point of M. The number of these points is called the *sheet number* of the covering. The index of the subgroup H of F(M) corresponding to M_1 is equal to the sheet number of M_1.

The covering corresponding to the subgroup $\{1\}$ of $F(M)$ is called the *universal covering*. It follows from the above considerations that each covering of M is homeomorphic to a covering which is itself covered by the universal covering. In particular, the latter has no proper covering and hence it is simply connected.

28.10 Covering transformations

A *covering transformation* ψ of the covering $\{M_1,\phi\}$ of M is a homeomorphism of $\{M_1,\phi\}$ onto itself. Thus ψ permutes the points of M_1 lying over a point of M.

Theorem 4. *Let* O *be a point of* M *and let* O_1, O_2 *be two points of* M_1 *lying over* O. *Then there is at most one covering transformation* ψ *which carries* O_1 *to* O_2.

Proof. Let P be an arbitrary point of M, let P_1 be a point lying over P, let $\tilde{w}$ be a path from O_1 to P_1 and let w be the corresponding base path from O to P. Then there is exactly one path over w in M_1 from O_2 to a point P_2 over P, and it follows by continuity that $\psi P_1 = P_2$. □

A covering of M is called *regular* when, for any two points O_1, O_2 lying over the same point O of M, there is a covering transformation which carries O_1 to O_2. For example, the universal covering is regular. As one easily sees, a covering of M is regular if and only if the conjugacy class of subgroups of $F(M)$ in Theorem 3 consists of only one group, which is therefore a normal subgroup. In this case the number of covering transformations equals the number of sheets of the covering.

Theorem 5. *The fundamental group of a topological space* M *is isomorphic to the group of covering transformations of the universal covering.*

Proof. Let O be a point of M and let $F(M)$ be realised as the group of homotopy classes of paths from O to O. Also let O_1 be a point of the universal covering over O, let w be a path from O to O, and let $\tilde{w}$ be the corresponding path originating from O_1. By definition of the universal covering, the final point O_2 of $\tilde{w}$ does not depend on the choice of w within its homotopy class. We associate the class of w with the covering transformation that maps O_1 onto O_2. This gives the desired isomorphism. □

There is a far-reaching analogy between the Galois theory of a normal field extension (Chapter 17) and the theory of subcoverings of a regular covering. The Galois group corresponds to the group of covering transformations. It is left to the reader to work out the details of this theory.

28.11 Quotient spaces

Let M be a topological space and let G be a group of homeomorphisms of M onto itself. Two points P_1, P_2 of M are called *equivalent* under G when there is a $g \in G$ with $P_2 = gP_1$. The relation defined in this way is an equivalence. The classes of equivalent points of M are called *orbits* under G. The collection of orbits of M is called the *quotient space* G\M of M by G. The projection $\pi : M \to G \setminus M$ sends a point $P \in M$ to the associated orbit $GP := \{gP \mid g \in G\}$.

We call G admissible when each point of M has a neighbourhood U such that the sets U and gU are disjoint for all $g \in G, g \neq 1$. When G is admissible we introduce a topology on $G \setminus M$ whose open sets are precisely the images of open sets in M. Then π is a continuous mapping.

Theorem 6. *Let* M *be arcwise connected and let* G *be an admissible group of homeomorphisms of* M *onto itself. Then* $\{M,\pi\}$ *is a regular covering of* $G \setminus M$ *and* G *is the group of covering transformations of* $\{M,\pi\}$.

Conversely, if $\{M,\phi\}$ *is a regular covering of a topological space* M′*, then the group* G *of covering transformations of* $\{M,\phi\}$ *is an admissible group of homeomorphisms of* M *and* M′ *is homeomorphic to* $G \setminus M$.

The easy proof of Theorem 6 is left to the reader.

Exercises

28.1 Show that the polygon complex (Section 10.7) can be triangulated by elementary transformations (Section 10.8).

28.2 Define the homology groups of a polygon complex analogously to those of a simplicial complex and show that these groups are invariant, up to isomorphism, under elementary transformations.

28.3 Compute the homology groups of the closed Riemann surfaces.

28.4 Compute the homology groups of the non-orientable closed surfaces (Exercise 10.9).

28.5 Let $\mathcal{F}$ be a closed surface and let $\mathcal{F}(m)$ be the surface that results from $\mathcal{F}$ by removal of m points. Compute the fundamental group of $\mathcal{F}(m)$.

28.6 Let G be an arbitrary finite group with m generators, let z be an indeterminate and let $z_1,\dots,z_{m+1}$ be arbitrary points of the extended complex plane $\bar{\mathbb{C}}$. Show that there is a normal extension K of the field $\mathbb{C}(z)$, ramified (Chapter 23) at $z_1,\dots,z_{m+1}$ at most, whose Galois group $G(K/\mathbb{C}(z))$ is isomorphic to G. (Hint: Represent G as a quotient of the fundamental group of $\bar{\mathbb{C}} - \{z_1,\dots,z_{m+1}\}$ and use Section 28.9 as well as Chapter 11, Theorem 13.)

29. The idea of a Riemann surface

Riemann defined what were later called "Riemann surfaces" as coverings of the plane (Chapter 10). The function theory constructed on this basis played a prominent rôle in the mathematics of the second half of the 19th century. However, it was by no means as rigorous as the other branches of mathematics had become in the meantime. Despite this, it lasted until the year 1913, when the first rigorous definition of a Riemann surface was given by Weyl. His book *"Die Idee der Riemannschen Fläche"* showed the way to an exact definition of the concept of manifold, and therefore represented an important step in the development of mathematics. Prerequisite was topology in the form of an abstract theory, which at that time occurred in conjunction with the set theory of Cantor and Dedekind. Weyl defined the concept of a two-dimensional manifold along with the concept of a topological space. However, we assume here that the reader is familiar with the basic concepts of set-theoretic topology (A.2) and we define, as is usual today, an n-dimensional manifold by beginning with a given topological space.

29.1 The concept of a real n-dimensional manifold

An *n-dimensional real manifold* is a Hausdorff topological space M which is covered by open sets U homeomorphic to open subsets of $\mathbb{R}^n$. Let ϕ_U be the corresponding mapping of U into $\mathbb{R}^n$. The open subset U is called a *chart* of M, and the collection $\mathfrak{A}$ of sets covering M is called an *atlas* of M. If two charts U_1, U_2 have a non-empty intersection U_{12}, then they give rise to a homeomorphism $\phi_{12} = \phi_{U_2}\phi_{U_1}^{-1}$ of $\phi_{U_1}(U_{12})$ onto $\phi_{U_2}(U_{12})$. Since these sets are open subsets of $\mathbb{R}^n$ it is meaningful to speak of ϕ_{12} being differentiable. We call ϕ_{12} the *transition function* from U_1 to U_2. When all the transition functions for $\mathfrak{A}$ are differentiable, M is called a *differentiable manifold.*

Example 1. $\mathbb{R}^n$ is a differentiable manifold. The system of covering subsets consists of the single set $U := \mathbb{R}^n$, and ϕ_U is the identity mapping.

Example 2. Let $P^n_{\mathbb{R}}$ be the n-dimensional real projective space (A.5). The mappings ϕ_i given in A.5 define an atlas of $P^n_{\mathbb{R}}$, with respect to which $P^n_{\mathbb{R}}$ is a differentiable manifold.

29.2 Definition of a Riemann surface

In what follows we confine ourselves to two-dimensional real manifolds which are triangulable in the sense of 28.2. We view the images of the associated charts as subsets of $\mathbb{C}$. Such a manifold $\mathcal{F}$ is called a *Riemann surface* when the transition functions are holomorphic.

The basic concepts of function theory may be carried over to a Riemann surface in the same way that they were carried over to coverings of the extended plane in Section 10.4.

A complex-valued function f, defined on an open subset V of $\mathcal{F}$, is called *holomorphic* at a point $P \in V$ when there is a U in the atlas $\mathfrak{A}_{\mathcal{F}}$ of $\mathcal{F}$ such that $P \in U$ and the mapping $f\phi_U^{-1}$ of a neighbourhood of $\phi_U(P)$ into $\phi_U(U \cap V)$ is *holomorphic*. Since the transition functions are holomorphic, this definition does not depend on the choice of $U \in \mathfrak{A}_{\mathcal{F}}$.

A *uniformising variable* for P is a holomorphic function t, defined on a neighbourhood V of P, such that t maps V homeomorphically onto an open set in $\mathbb{C}$ and $t(P) = 0$. If u is another uniformising variable for P, then $u = c_1t + c_2t^2 + \ldots$ is a power series which converges in a neighbourhood of 0. Since t must also be expressible in terms of u in this way, it must be that $c_1 \neq 0$. For each $P \in \mathcal{F}$ and $U \in \mathfrak{A}_{\mathcal{F}}$ with $P \in U$, one uniformising variable is $\phi_U - \phi_U(P)$.

Let $\mathcal{F}_1$ and $\mathcal{F}_2$ be Riemann surfaces. A mapping ϕ of $\mathcal{F}_1$ into $\mathcal{F}_2$ is called *holomorphic* at the point $P \in F_1$ if, for any two uniformising variables t_1, t_2 of P and $\phi(P)$, the function $t_2\phi t_1^{-1}$ is holomorphic at the point 0. A mapping holomorphic on the whole of $\mathcal{F}_1$ is called *conformal*.

Local properties of a Riemann surface $\mathcal{F}$ are those that concern a sufficiently small neighbourhood of a point of $\mathcal{F}$. It follows from the definition of $\mathcal{F}$ that local properties and concepts are equivalent to local function-theoretic properties and concepts on $\mathbb{C}$.

A triangulation (Section 28.2) of a Riemann surface $\mathcal{F}$ is called *analytic* when the associated simplex mappings are analytic. It can be proved that every Riemann surface is analytically triangulable. We do not go into the proof of this theorem here, but make analytic triangulability part of the definition of a Riemann surface. In all the examples we consider this property is easily proved.

In order to better compare the triangulation of a Riemann surface with its atlas one *refines* the triangulation by decomposing each triangle into six subtriangles by halving the sides. Obviously this gives another analytic triangulation. We now give examples of Riemann surfaces.

Example 1. The plane $\mathbb{C}$. We can take the single chart to be the whole of $\mathbb{C}$ and take $\phi_{\mathbb{C}}$ to be the identity mapping.

Example 2. The extended plane $\bar{\mathbb{C}} := \mathbb{C} \cup \{\infty\}$. The open neighbourhoods of ∞ can be taken to have the form $(\mathbb{C}\text{-}X) \cup \{\infty\}$, where X is a bounded closed subset of $\mathbb{C}$. We cover $\bar{\mathbb{C}}$ by two charts $U_1 := \mathbb{C}$ and $U_2 := (\mathbb{C} - \{0\}) \cup \{\infty\}$ and set $\phi_{U_1}(z) = z$, $\phi_{U_2}(z) = 1/z$ for $z \neq \infty$, $\phi_{U_2}(\infty) = 0$. The transition function ϕ_{12} on $U_1 \cap U_2 = \mathbb{C} - \{0\}$ is $\phi_{12}(z) = 1/z$. One can also interpret $\bar{\mathbb{C}}$ as the complex projective line $P^1_{\mathbb{C}}$ (cf. Section 29.1, Example 2).

Example 3. Let $\Gamma := \omega_1\mathbb{Z} + \omega_2\mathbb{Z}$ with $\omega_1, \omega_2 \in \mathbb{C}^\times$ be a lattice in $\mathbb{C}$, i.e. let ω_1 and ω_2 be linearly independent over $\mathbb{R}$. We interpret $\mathbb{C}/\Gamma$ as a topological space by taking the open sets in $\mathbb{C}/\Gamma$ to be all images of open sets V in $\mathbb{C}$ under the projection $\pi : \mathbb{C} \to \mathbb{C}/\Gamma$. When V contains no numbers z_1, z_2 which are mapped by π into the same class in $\mathbb{C}/\Gamma$, then the mapping $\phi_V := \pi|_V$ is a homeomorphism of V onto $\pi(V)$. We define $\mathbb{C}/\Gamma$ as a Riemann surface by the atlas with charts $\pi(V)$ and mappings ϕ_V^{-1}. A meromorphic function on $\mathbb{C}/\Gamma$ is none other than an elliptic function on $\mathbb{C}$ with period lattice Γ (Section 13.1).

Example 4. The Riemann surfaces defined in Chapter 10 are Riemann surfaces in the sense of our present definition.

29.3 Orientability of Riemann surfaces

In Section 10.7 we have defined what it means for a triangulated surface to be orientable. To do this one has to provide each triangle of the triangulation with an orientation. If this is possible in such a way that the orientations induced in an edge by the two neighbouring triangles are always opposite, then the surface is *orientable*.

Theorem 1. *Each Riemann surface is orientable.*

Proof. We choose the triangulation so fine that each triangle Δ lies in *one chart* U. Then Δ is mapped by ϕ_U into a topological triangle in $\mathbb{C}$, which we provide with the positive orientation. This orientation is carried over to Δ by ϕ_U^{-1}. Since a holomorphic and homeomorphic mapping of a subregion of $\mathbb{C}$ preserves orientation (Section 14.12), this definition does not depend on the chart chosen. By passing to a finer triangulation one can always arrange that neighbouring triangles of Δ also lie in the chart U, whence it follows that the edges of Δ are oppositely oriented in neighbouring triangles. □

When we are concerned in particular with a closed Riemann surface, then the latter is determined, as a topological manifold, by its genus alone (Section 10.8).

29.4 Meromorphic differentials

A meromorphic differential ω on the Riemann surface $\mathcal{F}$ is defined, as in Section 11.1, to be a mapping which associates with each point P of $\mathcal{F}$ and each uniformising variable t of P a function ω/dt meromorphic in a neighbourhood of P, and such that

$$\frac{\omega}{dt} = \frac{dt'}{dt}\,\frac{\omega}{dt'}$$

holds at all points Q of $U_P \cap U_{P'}$ when t,t' are uniformising variables of P,P'.

The residue of ω at the point P is defined to be a_{-1} when ω/dt has the form

$$\frac{\omega}{dt} = \sum_{i=i_0}^{\infty} a_i t^i .$$

This definition is independent of the choice of uniformising variable.

Let f be a meromorphic function on $\mathcal{F}$. Then the differential df is defined as the mapping that associates with the point P of $\mathcal{F}$ and the uniformising variable t of P the function df/dt.

Let C be a (piecewise smooth) curve in $\mathcal{F}$ on which ω is regular. We then want to define the integral $\int_C \omega$.

When C lies in the neighbourhood U_P of a uniformising variable t we set

$$\int_C \omega = \int_C \frac{\omega}{dt} dt.$$

In general we decompose C into finitely many subcurves $C_1, C_2, \ldots, C_s$, each of which lies in the domain of definition of a uniformising variable $t_1, \ldots, t_s$ (this is always possible because C is compact by definition). Then we set

$$\int_C \omega = \sum_{i=1}^{s} \int_{C_i} \frac{\omega}{dt_i} dt_i.$$

This definition is independent of the choice of decomposition of C and of the choice of uniformising variables.

Theorem 2. *If* f *is a meromorphic function on a Riemann surface* $\mathcal{F}$, *regular on a curve* C *from* P_1 *to* P_2, *then*

$$\int_C df = f(P_2) - f(P_1). \quad \square$$

Theorem 3. (*Cauchy integral theorem*). *Let* G *be a simply connected region on the Riemann surface* $\mathcal{F}$ *and let* C *be a closed curve in* G. *Also let* ω *be a differential which is regular in* G. *Then*

$$\int_C \omega = 0.$$

Theorem 4. (*Residue theorem*). *Let the hypotheses about* G *and* C *be as in Theorem 3. Also let* ω *be a differential which is meromorphic in* G *and regular on* C. *If* C *is traversed in the anticlockwise sense then*

$$\int_C \omega = 2\pi i \sum_P \mathrm{Res}_P \omega,$$

where the sum is taken over the finitely many non-zero residues of ω *in the region enclosed by* 0.

The proof of Theorems 3 and 4 is carried out by replacing the curve C by an edge path in a sufficiently fine triangulation of $\mathcal{F}$, as is possible by the Cauchy integral theorem in the plane. In this way Theorems 3 and 4 are reduced to the case where C is the boundary of a triangle in the plane (Section 8.3). $\square$

29.5 The Dirichlet integral on a Riemann surface

Suppose that G is a region in $\mathbb{R}^2$ bounded by a closed curve C, and let H be a region which contains the closure $\overline{G}$ of G. Then if f is a continuous function on H the integral

$$\iint_G f(x,y)dx\,dy$$

is defined. Let ϕ be a smooth homeomorphism of H onto a region H_1 in $\mathbb{R}^2$, given by the functions $x_1(x,y)$, $y_1(x,y)$. Then

$$\iint_G f(x,y)dx\,dy = \iint_{\phi G} (f/d)(x,y)dx_1dy_1,$$

where d is the determinant of the matrix

$$\begin{bmatrix} \partial x_1/\partial x & \partial x_1/\partial y \\ \partial y_1/\partial x & \partial y_1/\partial y \end{bmatrix}.$$

We let S_H denote the vector space of continuously differentiable functions on H and let

$$D_G(v_1,v_2) := \iint_G \left(\frac{\partial v_1}{\partial x} \cdot \frac{\partial v_2}{\partial x} + \frac{\partial v_1}{\partial y} \cdot \frac{\partial v_2}{\partial y} \right) dx\,dy \quad \text{for} \quad v_1,v_2 \in S_H.$$

$D_G(v_1,v_2)$ is a bilinear form on S_H. The quadratic form $D_G(v) := D_G(v,v)$ for $v \in S_H$ is called the *Dirichlet integral*. It is invariant under a conformal mapping ϕ of H onto H_1 : because of the Cauchy-Riemann differential equations

$$\frac{\partial x_1}{\partial x} = \frac{\partial y_1}{\partial y}, \quad \frac{\partial x_1}{\partial y} = -\frac{\partial y_1}{\partial x}$$

we have, in this case,

$$\left(\frac{\partial v}{\partial x}\right)^2 + \left(\frac{\partial v}{\partial y}\right)^2 = \left(\left(\frac{\partial v}{\partial x_1}\right)^2 + \left(\frac{\partial v}{\partial y_1}\right)^2\right)\left(\left(\frac{\partial x_1}{\partial x}\right)^2 + \left(\frac{\partial x_1}{\partial y}\right)^2\right),$$

$$d = \left(\frac{\partial x_1}{\partial x}\right)^2 + \left(\frac{\partial x_1}{\partial y}\right)^2,$$

hence

$$\iint_G \left(\left(\frac{\partial v}{\partial x}\right)^2 + \left(\frac{\partial v}{\partial y}\right)^2\right) dx\, dy = \iint_{\phi G} \left(\left(\frac{\partial v}{\partial x_1}\right)^2 + \left(\frac{\partial v}{\partial y_1}\right)^2\right) dx_1 dy_1.$$

This shows that one can define the Dirichlet integral on an arbitrary Riemann surface.

For this purpose, let a triangulation of $\mathcal{F}$ be given, so fine that each triangle Δ lies inside a chart of $\mathcal{F}$. Then $D_\Delta(v)$ is defined. We define $D_F(v)$ to be the sum of the $D_\Delta(v)$ over all triangles of the triangulation. $D_F(v)$ can also be equal to ∞. $D_F(v)$ is independent of the triangulation chosen, as one sees by superimposing two triangulations.

A function $v \in S_F$ is called *harmonic* at a point $P \in \mathcal{F}$ when v is twice continuously differentiable and satisfies the equation $\Delta v := \dfrac{\partial^2 v}{\partial x^2} + \dfrac{\partial^2 v}{\partial y^2} = 0$ for the uniformising variable $t = x + iy$. Because of Cauchy-Riemann differential equations another uniformising variable $t_1 = x_1 + iy_1$ satisfies

$$\frac{\partial^2 v}{\partial x_1^2} + \frac{\partial^2 v}{\partial y_1^2} = \left(\frac{\partial^2 v}{\partial x^2} + \frac{\partial^2 v}{\partial y^2}\right)\left(\left(\frac{\partial x}{\partial x_1}\right)^2 + \left(\frac{\partial x}{\partial y_1}\right)^2\right).$$

Thus v is harmonic at P independently of the choice of uniformising variable.

When v is harmonic at all points P of $\mathcal{F}$, v is called a *potential function* on $\mathcal{F}$. For such a function, $\omega := \left(\dfrac{\partial v}{\partial x} - i\dfrac{\partial v}{\partial y}\right) dt$ is a regular differential on $\mathcal{F}$. In a simply connected region G of $\mathcal{F}$, integration of ω gives a regular function f with $\operatorname{Re} f = v$. By application of the maximum modulus principle (Theorem 9, Chapter 8) to e^f and e^{-f} one obtains

Theorem 5. *The maximum (minimum) of a potential function in a region* G *is taken on the boundary of* G.

The differentials sought in the Riemann existence theorem (Theorem 3, Chapter 11) will be constructed with the help of potential functions.

29.6 Dirichlet's principle

Let G be a simply connected region of the plane, bounded by a closed curve C. We call a function *admissible* if it is continuous on $\bar{G} = G \cup C$ and differentiable in G.

Let $Z(v_0)$ be the set of admissible functions on G which agree with a prescribed admissible function v_0 on C. The first boundary value problem of potential theory seeks a function u in $Z(v_0)$ which is a potential function in G.

Dirichlet "solved" this problem in his lectures as follows: let u be a function in $Z(v_0)$ for which $D_G(u)$ is finite and equal to the infimum of the $D_G(v)$ for $v \in Z(v_0)$. We show that u is a potential function in G. To do this, let w be an admissible function which equals 0 on C and let λ be a real parameter. Then

$$D_G(u+\lambda w) = D_G(u) + 2\lambda D_G(u,w) + \lambda^2 D_G(w) \geq D_G(u).$$

It follows that $D_G(u,w) = 0$, otherwise one could choose the parameter λ so that $2\lambda D_G(u,w) + \lambda^2 D_G(w) < 0$. Also by Green's theorem (A.3)

$$0 = \int_C \left(\frac{\partial u}{\partial x} w\, dy - \frac{\partial u}{\partial y} w\, dx\right)$$

$$= \iint_G (\Delta u) w\, dx\, dy + D_G(u,w)$$

$$= \iint_G (\Delta u) w\, dx\, dy.$$

Since this equation holds for all admissible functions w which vanish on the boundary, it follows that $\Delta u = 0$.

Conversely, suppose that u is a function in $Z(v_0)$ which is a potential function in G. Then $D_G(u) \leq D_G(v)$ for all $v \in Z(v_0)$. Namely, with $w = v - u$ we have

$$D_G(v) = D_G(u) + 2D_G(u,w) + D_G(w),$$

and since

$$D_G(u,w) = \int_C \left(\frac{\partial u}{\partial x} w\, dy - \frac{\partial u}{\partial y} w\, dx\right) - \iint_G (\Delta u) w\, dx\, dy = 0,$$

it follows that $D_G(v) \geq D(u)$.

This implies, as does Theorem 5, that the potential function u is uniquely determined.

The above idea was called *Dirichlet's principle* by Riemann, and it was used by him in a form to suit his purpose, assuming as obvious the existence of a function u which realises the infimum of $D_G(v)$ for $v \in Z(v_0)$, as well as the applicability of the divergence theorem. Weierstrass showed the incorrectness of this assumption in 1870 in his lecture "*Über das sogenannte Dirichletsche Princip*" (*Preussische Akademie der Wissenschaften*, 14.7.1870). However, Schwarz (*Berliner Berichte*) and Neumann (*Sächsische Berichte*) also showed in 1870 that the existence theorem for differentials (Chapter 11, Theorem 3) is correct nevertheless. Finally, in 1901 Hilbert gave a rigorous proof of Dirichlet's principle (see Section 11.2).

29.7 The Poisson integral

The Poisson integral solves the first boundary value problem of potential theory for the circle.

For $R > 0$, let $K := \{z \in \mathbf{C} \,\big|\, |z| \leq R\}$ and let ∂K be the boundary of K, traversed in the anticlockwise direction.

Theorem 6. *Let* $u(\phi)$ *be a continuous function on* $u(\phi)$, *where* ϕ *is the parameter for the point* $\zeta = Re^{i\phi}$ *on* ∂K. *Then*

$$f(z) := \frac{1}{2\pi}\int_0^{2\pi} \frac{\zeta+z}{\zeta-z}u(\phi)d\phi \tag{1}$$

is a regular function for $|z| < R$, *whose real part extends continuously to* K *and equals* $u(\phi)$ *on* ∂K.

(1) is called the *Poisson integral.*

Proof of Theorem 6. By definition, $f(z)$ is regular for $|z| < R$. Also

$$\operatorname{Re}\frac{\zeta+z}{\zeta-z} = \frac{|\zeta|^2-|z|^2}{|\zeta-z|^2} \quad \text{for} \quad \zeta \neq z$$

and therefore

$$2\pi \mathrm{Re}\ f(z) = \int_0^{2\pi} \frac{|\zeta|^2-|z|^2}{|\zeta-z|^2} u(\phi)d\phi \quad \text{for} \quad |z| < R. \tag{2}$$

By the residue theorem,

$$\int_0^{2\pi}\left[\frac{\zeta+z}{\zeta-z}\right]d\phi = -i\int_{\partial K}\frac{(\zeta-z)d\zeta}{(\zeta-z)\zeta} = 2\pi.$$

Hence, for a fixed $\zeta_0 = Re^{i\phi_0}$

$$2\pi(\mathrm{Re}\ f(z)-u(\phi_0)) = \int_0^{2\pi}\frac{\zeta+z}{\zeta-z}(u(\phi)-u(\phi_0))d\phi. \tag{3}$$

Now suppose $\varepsilon > 0$ is given. Since $u(\phi)$ is continuous, there is a $\delta_0 > 0$ with $|u(\phi)-u(\phi_0)| \le \varepsilon/2$ for $|\zeta-\zeta_0| \le \delta_0$ and a constant $M > 0$ with $|u(\phi)| \le M$ for all ϕ. We now divide the interval $[0,2\pi]$ into two subsets:

$$A := \{\phi \in [0,2\pi] \,\big|\, |Re^{i\phi}-\zeta_0| \le \delta_0\}, \quad B := [0,2\pi] - A.$$

Then by (2) and (3)

$$2\pi|\mathrm{Re}\ f(z)-u(\phi_0)| \le \int_A \frac{(|\zeta|^2-|z|^2)}{|\zeta-z|^2}\frac{\varepsilon}{2}d\phi + \int_B \frac{(|\zeta|^2-|z|^2)}{|\zeta-z|^2}2Md\phi$$

$$\le \pi\varepsilon + 2M\int_B \frac{(|\zeta|^2-|z|^2)}{|\zeta-z|^2}d\phi.$$

Suppose $|z-\zeta_0| \le \delta_0/2$. Then for $\phi \in B$

$$|\zeta-z| \ge |\zeta-\zeta_0| - |z-\zeta_0| \ge \delta_0/2,$$

$$\frac{|\zeta|^2-|z|^2}{|\zeta-z|^2} = \frac{(R+|z|)(R-|z|)}{|\zeta-z|^2} \le \frac{2R\delta}{(\delta_0/2)^2} = \frac{8\delta}{\delta_0^2}$$

and hence

$$|\mathrm{Re}\ f(z)-u(\phi_0)| \le \frac{\varepsilon}{2} + \frac{16M\delta}{\delta_0^2} < \varepsilon,$$

when $|z-\zeta_0| = \delta$ is sufficiently small. □

We first use the Poisson integral to prove a convergence theorem of Harnack which is needed in what follows.

Theorem 7. *Let* $u_1, u_2, \ldots$ *be a sequence of potential functions which are harmonic for* $|z| < R$ *and which converge uniformly, for* $|z| \le q$, *to a limit function* u, *where* q *is an arbitrary positive number* $< R$. *Then* u *is likewise a potential function for* $|z| < R$. *The derivatives* $u_i^{(n)}$ *of order* n *converge uniformly to* $u^{(n)}$ *for* $|z| \le q$.

Proof. One uses the Poisson integral for the circle of radius q. The uniform convergence allows the limit and the integral to be interchanged. The second assertion follows from the representation

$$2u_i^{(n)}(z) = \mathrm{Re}\int_0^{2\pi} \left[\frac{\zeta+z}{\zeta-z}\right]^{(n)} u_i(\zeta)d\phi. \quad \square$$

29.8 Dirichlet's principle for the circle

In this section we show that Dirichlet's principle is valid for the circle. The proof given here follows the approach of Hadamard (*Sur le principle de Dirichlet, Bull. Soc. math. France* 34 (1906)) and Zaremba (*Sur le principle du minimum, Bull. Acad. sci. Cracovie, Juli* 1909).

We set $K_r := \{z \,\big|\, |z| \le r\}$ and $D_r := D_{K_r}$.

Theorem 8. *Suppose* $0 < q < R$ *and let* v *be an admissible function for the circle* K_R, *such that* $D_R(v) = \lim_{q\to R} D_q(v)$ *is finite. Also let* u *be the unique potential function which coincides with* v *on* ∂K_R.

Then $D_R(u) \le D_R(v)$.

Proof. We introduce polar coordinates $z = re^{i\theta}$, $\zeta = Re^{i\phi}$, and write $v(\phi)$ instead of $v(R\,e^{i\phi})$. Then for $r < R$

$$\mathrm{Re}\frac{\zeta+z}{\zeta-z} = \mathrm{Re}(1+2\sum_{\nu=1}^{\infty}(z/\zeta)^{\nu}) \tag{4}$$

$$= 1 + 2\sum_{\nu=1}^{\infty}(r/R)^{\nu}(\cos\nu\phi\,\cos\nu\theta + \sin\nu\phi\,\sin\nu\theta)),$$

and because of the uniform convergence of the series on the right-hand side, the Fourier expansion (cf. Chapter 5) holds for $r < R$:

$$u(r,\theta) := u(re^{i\theta}) = \frac{a_0}{2} + \sum_{\nu=1}^{\infty} (r/R)^{\nu}(a_{\nu}\cos \nu\theta + b_{\nu}\sin\nu\theta) \tag{5}$$

with

$$a_{\nu} = \frac{1}{\pi}\int_0^{2\pi} v(\phi)\cos \nu\phi d\phi,\ b_{\nu} = \frac{1}{\pi}\int_0^{2\pi} v(\phi)\sin \nu\phi d\phi.$$

We set $g_{\nu}(r,\theta) := r^{\nu}\cos \nu\theta$, $h_{\nu}(r,\theta) := r^{\nu}\sin \nu\phi$ for $\nu = 1,2,\ldots$. By Green's theorem (A.3),

$$D_q(v_1,v_2) = -\iint_{K_q} v_1(\Delta v_2)rdrd\theta + q\int_0^{2\pi}\left[v_1\frac{\partial v_2}{\partial r}\right]_{r=q} d\theta, \tag{6}$$

for admissible functions v_1,v_2. (Here we assume that v_2 is twice continuously differentiable.) Hence one easily calculates that the potential functions g_{ν},h_{ν} are orthogonal relative to D_q, and that

$$D_q(g_{\nu}) = D_q(h_{\nu}) = \pi\nu q^{2\nu}. \tag{7}$$

One also sees that

$$\lim_{q\to R} D_q(v,g_{\nu}) = \lim_{q\to R} \nu q^{\nu}\int_0^{2\pi} v \cos \nu\theta d\theta = \pi\nu R^{\nu}a_{\nu}. \tag{8}$$

Exchange of limit and integral is allowed here, since $D_q(v,g_{\nu})$ is bounded. Similarly we have

$$\lim_{q\to R} D_q(v,h_{\nu}) = \pi\nu R^{\nu}b_{\nu}. \tag{9}$$

The series (5) converges for $|z| \le q < R$, uniformly and absolutely in all derivatives. Hence one may compute $D_q(u)$ by termwise squaring and integration. One obtains

$$D_q(u) = \sum_{\nu=1}^{\infty} \pi\nu q^{2\nu}(a_{\nu}^2+b_{\nu}^2)/R^{2\nu}. \tag{10}$$

On the other hand

$$0 \le D_q(v - \sum_{\nu=1}^{n} (a_\nu g_\nu + b_\nu h_\nu)/R^\nu)$$

$$= D_q(v) - 2\sum_{\nu=1}^{n} (a_\nu D_q(v,g_\nu) + b_\nu D_q(v,h_\nu))/R^\nu$$

$$+ \sum_{\nu=1}^{n} \pi\nu q^{2\nu}(a_\nu^2 + b_\nu^2)/R^{2\nu}. \tag{11}$$

As $q \to R$ one obtains, by (8) and (9),

$$0 \le D_R(v) - \sum_{\nu=1}^{n} \pi\nu(a_\nu^2 + b_\nu^2).$$

It follows that

$$D_R(u) = \lim_{q \to R} D_q(u) = \sum_{\nu=1}^{\infty} \pi\nu(a_\nu^2 + b_\nu^2) \le D_R(v). \quad \Box$$

Theorem 9. *(Dirichlet's principle for the circle). Let* Z *be the set of continuous functions on* K_R *which are continuously differentiable in the interior of* K_R *and which take given values on the boundary of* K_R*. We suppose that there is a* $v \in Z$ *for which* $D_R(v)$ *is finite. Then the infimum of* $D_R(v)$ *for* $v \in Z$ *is taken for a unique* $u \in Z$*, and* u *is a potential function in the interior of* K_R*.*

Proof. By Theorem 8 it suffices to prove the uniqueness of u. Let $v \in Z$ be such that $D_R(v) = D_R(u)$. Then $D_R(u-v) \le 2D_R(u) + 2D_R(v)$ is also finite. Let λ be a real parameter and let $w := u-v$. Then $D_R(u) \le D_R(u+\lambda w)$ and hence $2\lambda D_R(u,w) + \lambda^2 D_R(w) \ge 0$. It follows that $D_R(u,w) = 0$ (Section 29.6), and hence $D_R(w) = D_R(v) - D_R(u) = 0$. Since w equals 0 on ∂K_R, w is 0 on the whole of K_R. $\Box$

29.9 The smoothing process

Let K be a circle with the same midpoint as K_R, but with greater radius, and let v be a function continuously differentiable in K. We want to apply Theorem 9 in the following way. When v is not a potential function in K_R, then by Theorem 9 a regular potential function u in the interior of K_R, which agrees with v on ∂K_R, satisfies

$$D_R(u) < D_R(v).$$

We need a function $\tilde{v}$, continuously differentiable in K, with

$$D_K(\tilde{v}) < D_K(v).$$

We want to construct this function by setting $\tilde{v}(z) := v(z)$ for $z \in K - K_R$ and then smoothing u in the interior of K_R in such a way that the result is a function $\tilde{v}$ for which $D_R(\tilde{v})$ departs arbitrarily little from $D_R(u)$. For this purpose we choose a q with $R/2 < q < R$ and set

$$\tilde{v}(z) = \begin{cases} u(z) & \text{for } |z| < q, \\ u(z)+\chi(r)(v(z)-u(z)) & \text{for } q \le |z| = r \le R, \end{cases}$$

where $\chi(r) = (r-q)^2/(R-q)^2$. Then $\tilde{v}$ is continuously differentiable in K. In order to see this we first prove

Proposition 1. *If* $r = |z|$ *with* $R/2 < r < R$ *then*

$$\left(\frac{\partial u}{\partial x}\right)^2 + \left(\frac{\partial u}{\partial y}\right)^2 \le \frac{1}{\pi(R-r)^2}(D_R(u)-D_{2r-R}(u)).$$

Proof. By (5), $\left(\frac{\partial u}{\partial x}\right)_{z=0} = a_1/R$, $\left(\frac{\partial u}{\partial y}\right)_{z=0} = b_1/R$, and hence by (12)

$$\left(\left(\frac{\partial u}{\partial x}\right)^2 + \left(\frac{\partial u}{\partial y}\right)^2\right)_{z=0} \le \frac{1}{\pi R^2}D_R(u).$$

If we apply this inequality, not to K_R but to the circle $K_{R-r}(z)$ with centre z and radius R-r (Figure 23), then it follows that

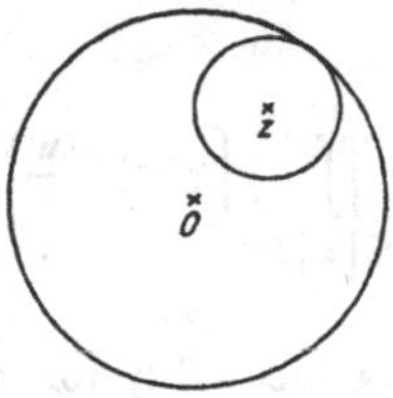

Fig. 23

$$\left(\left(\frac{\partial u}{\partial x}\right)^2 + \left(\frac{\partial u}{\partial y}\right)^2\right) \le \frac{1}{\pi(R-r)^2}\iint_{K_{R-r}(z)}\left(\left(\frac{\partial u}{\partial x}\right)^2 + \left(\frac{\partial u}{\partial y}\right)^2\right)dx\,dy \tag{12}$$

$$\le \frac{1}{\pi(R-r)^2}(D_R(u)-D_{2r-R}(u)),$$

since $K_{R-r}(z)$ lies in the annulus $\{t \in \mathbb{C} \mid R \ge |t| \ge 2r - R\}$. □

From now on we use the abbreviation

$$D_s^r := D_r - D_s \quad \text{for} \quad 0 < s \le r.$$

We consider $\frac{\partial \tilde{v}}{\partial x}$ for $q \le |z| < R$:

$$\frac{\partial \tilde{v}}{\partial x} = (1-\chi(r))\frac{\partial u}{\partial x} + \chi(r)\frac{\partial v}{\partial x} + \frac{\partial \chi(r)}{\partial x}(v-u).$$

By Proposition 1 the following holds uniformly in θ

$$\lim_{r\to R}\left|(1-\chi(r))\frac{\partial u}{\partial x}\right| \le \lim_{r\to R}\left|(1-\chi(r))\frac{1}{\sqrt{\pi}(R-r)}\sqrt{D_{2r-R}^R(u)}\right| = 0$$

and hence $\lim\limits_{r\to R} \frac{\partial \tilde{v}}{\partial x} = \frac{\partial v}{\partial x}$. Similarly one obtains $\lim\limits_{r\to R} \frac{\partial \tilde{v}}{\partial y} = \frac{\partial v}{\partial y}$. It follows that v is continuously differentiable in K.

We now estimate $D_R(\tilde{v}-u) = D_q^R(\chi(r)(v-u))$ by setting $w := v-u$. Since

$$\frac{\partial}{\partial x}(\chi(r)w) = \chi(r)\frac{\partial w}{\partial x} + w\frac{2(r-q)x}{(R-q)^2 r}$$

we get

$$\frac{1}{2}\iint_{q\le|z|\le R}\left[\frac{\partial}{\partial x}(\chi(r)w)\right]^2 dx\,dt \le \iint_{q\le|z|\le R}\left[\chi^2(r)\left(\frac{\partial w}{\partial x}\right)^2 + \frac{4}{(R-q)^4}w^2(r-q)^2\left(\frac{x}{r}\right)^2\right]dx\,dy,$$

which, together with the corresponding inequality for the partial y derivative, implies

$$\frac{1}{2}D_R(\tilde{v}-u) \le \iint_{q\le|z|\le R}\left(\left(\frac{\partial w}{\partial x}\right)^2 + \left(\frac{\partial w}{\partial y}\right)^2\right)dx\,dy + \frac{4}{(R-q)^4}\int_0^{2\pi}\int_q^R w^2(r-q)^2 r\,dr\,d\theta \tag{13}$$

Since $\chi(r)^2 \le 1$, we get

$$\iint_{q \le |z| \le R} \chi^2(r)\left[\left(\frac{\partial w}{\partial x}\right)^2 + \left(\frac{\partial w}{\partial y}\right)^2\right] dx\,dy \le D_q^R(w). \tag{14}$$

We also estimate $\int_0^{2\pi} w^2 d\theta$. By the Schwarz inequality for $0 < r_1 < r_2 < R$

$$(w(r_2e^{i\theta}) - w(r_1e^{i\theta}))^2 = \left(\int_{r_1}^{r_2} \frac{\partial w(re^{i\theta})}{\partial r} dr\right)^2 \le \int_{r_1}^{r_2} \left(\frac{\partial w}{\partial r}\right)^2 r\,dr \int_{r_1}^{r_2} \frac{dr}{r}.$$

If one integrates this inequality with respect to θ and then lets r_2 converge to R then, since

$$D_{r_1}^{r_2}(w) = \int_{r_1}^{r_2}\int_0^{2\pi} \left[\left(\frac{\partial w}{\partial r}\right)^2 r + \left(\frac{\partial w}{\partial \theta}\right)^2 \frac{1}{r}\right] d\theta\,dr,$$

one gets the relation

$$\int_0^{2\pi} w^2 d\theta \le \left(\log\frac{R}{r}\right) D_r^R(w) \le \left(\frac{R}{r} - 1\right) D_r^R(w). \tag{15}$$

It follows from (15) that

$$\frac{4}{(R-q)^4}\int_0^{2\pi}\int_q^R w^2 (r-q)^2 r\,dr\,d\theta \le \frac{4}{(R-q)^4} D_r^R(w) \int_q^R (R-r)(r-q)^2 dr$$

$$= \frac{1}{3} D_q^R(w). \tag{16}$$

It follows from (13), (14) and (16) that

$$D_R(\tilde{v} - u) \le \frac{8}{3} D_q^R(w).$$

Since $\lim_{q \to R} D_q^R(w) = 0$, $D_R(\tilde{v}-u)$ is arbitrarily small for suitable q. Since, on the other hand, $D_R(\tilde{v}) - D_R(u) = D_R(\tilde{v}-u)$ (Section 29.8), it follows that $D_R(\tilde{v})$ differs arbitrarily little from $D_R(u)$ for suitable q.

We now draw from (15) another conclusion, which we shall need later.

Proposition 2. $\iint_{K_R} \left(v - \frac{a_0}{2}\right)^2 dx dy \le \frac{R^2}{2} D_R(v)$.

Proof. In what follows we set

$$I(h) := \iint_{K_R} h^2 dxdy$$

for a continuous function h in K_R. We multiply (15) by $r\,dr$ and integrate with respect to r from 0 to R. Since $D_r^R(w) \le D_R$ and

$$\int_0^R r \log \frac{R}{r}\, dr = \frac{R^4}{4}$$

one obtains

$$I(w) \le \frac{R^4}{4} D_R(w). \tag{17}$$

Analogous to (10), and because of (12), one computes

$$I\left(u - \frac{a_0}{2}\right) = R^2 \sum_{\nu=1}^{\infty} \frac{\pi(a_\nu^2 + b_\nu^2)}{2(\nu+1)} \le \frac{R^2}{4} D_R(u). \tag{18}$$

Bearing in mind that $I(v - a_0/2) \le 2(I(u) - a_0/2) + I(w))$, it follows from (17) and (18) that

$$I\left(v - \frac{a_0}{2}\right) \le \frac{R^2}{2}(D_R(u) + D_R(w)) = \frac{R^2}{2} D_R(v). \quad \square$$

29.10 Idea of the proof of the existence theorem for differentials

We construct differentials on a Riemann surface $\mathcal{F}$ with the help of potential functions U : because of the Cauchy-Riemann equations one obtains from U an analytic differential ω in the form

$$\frac{\omega}{dz} = \frac{\partial U}{\partial x} - i\frac{\partial U}{\partial y},$$

where $z = x + iy$ is a uniformising variable.

We first construct potential functions U with a singular point.

Theorem 10. *Let* O *be any point of the Riemann surface* $\mathcal{F}$*, with uniformising variable* $z_0 = x_0 + iy_0$*. Then on* $\mathcal{F} - \{O\}$ *there is a potential function* U *which differs from* $\operatorname{Re} 1/z_0$ *in the neighbourhood of* O *by a harmonic potential function.*

When $\mathcal{F}$ is closed, U *is determined by this property, up to an additive constant.*

Proof. The latter assertion follows from the fact that an everywhere regular potential function on a closed surface is constant by the maximum principle.

We want to determine the function U with the help of a minimum principle, analogously to Theorem 9. To do this we consider the set V of all functions v on $\mathcal{F}$ which satisfy the following conditions.

(i) *We choose positive numbers* R_0, R_0^* *with* $R_0 < R_0^*$ so that the circle $K_0^* := \{z_0 \,\big|\, |z_0| \le R_0^*\}$ *lies inside a neighbourhood of* O *with respect to the uniformising variable* z_0. *Also let* $K_0 := \{z_0 \,\big|\, |z_0| \le R_0\}$.

Then let v *be continuously differentiable on* $\mathcal{F} - \partial K_0$ *and let the function* v* *such that*

$$v^*\big|_{K_0} := v\big|_{K_0}, \quad v^*\big|_{K_0^*-K_0} := v - \Phi\big|_{K_0^*-K'} \quad \Phi = \mathrm{Re}\left(\frac{1}{z_0} - \frac{z_0}{R^2}\right)$$

be continuously differentiable on K_0^*, *i.e. the function* v *jumps by the function* Φ *along* ∂K_0. The significance of the regular component $\mathrm{Re}\, z_0/R_0^2$ of Φ will emerge in the proof of Proposition 5.

When $\mathcal{F}$ is not closed, the following condition is added:

(ii) $D_{\mathcal{F}}(v)$ is *finite.*

As one easily sees, V is not empty. Hence Theorem 10 is a part of the following theorem.

Theorem 11. *(Dirichlet's principle). If*

$$d = \lim \inf\{D_{\mathcal{F}}(v) \,|\, v \in V\}$$

then there is a function u *in* V, *unique up to an additive constant, with the following properties:*

(i) u *is harmonic in* $\mathcal{F} - \partial K_0$

(ii) u* *is harmonic in* K_0^*

(iii) $D_{\mathcal{F}}(u) = d.$

In what follows we set $D(v) := D_{\mathcal{F}}(v)$.

29.11 Proof of Dirichlet's principle

To prove Theorem 11 we first derive two propositions.

Proposition 3. *(B. Levi's inequality). For all* $v_1, v_2 \in V$

$$\sqrt{D(v_1 - v_2)} \leq \sqrt{D(v_1) - d} + \sqrt{D(v_2) - d}.$$

Proof. Suppose $\lambda_1, \lambda_2 \in \mathbb{R}$ with $\lambda_1 + \lambda_2 \neq 0$. Then $(\lambda_1 v_1 + \lambda_2 v_2)/(\lambda_1 + \lambda_2) \in V$ and therefore

$$D(\lambda_1 v_1 + \lambda_2 v_2) \geq d(\lambda_1 + \lambda_2)^2.$$

This inequality also holds for $\lambda_1 + \lambda_2 = 0$. Hence

$$\lambda_1^2(D(v_1)-d) + 2\lambda_1\lambda_2(D(v_1,v_2)-d) + \lambda_2^2(D(v_2)-d) \geq 0$$

for all $\lambda_1, \lambda_2 \in \mathbb{R}$, so

$$(D(v_1)-d)(D(v_2)-d) \geq (D(v_1,v_2)-d)^2.$$

It follows that

$$D(v_1 - v_2) = D(v_1) - d + D(v_2) - d - 2(D(v_1,v_2)-d)$$

$$\leq D(v_1) - d + D(v_2) - d + 2\sqrt{(D(v_1)-d)(D(v_2)-d)}$$

$$= (\sqrt{D(v_1)-d} + \sqrt{D(v_2)-d})^2. \qquad \square$$

The uniqueness assertion of Theorem 11 follows immediately from Proposition 3.

Since $v+a$ lies in V along with v for any $a \in \mathbb{R}$ we restrict the set V of concurrence functions v by the requirement

$$\int_0^{2\pi} v(R_0 e^{i\theta})d\phi = 0 \tag{19}$$

Condition (19) amounts to the following. Let u be the potential function which is regular in the interior of K_0 and equal to v on ∂K_0. Then by (5) we have, for $r < R_0$,

$$u(re^{i\theta}) = \frac{a_0}{2} + \sum_{\nu=1}^{\infty} (r/R_0)^{\nu}(a_\nu \cos \nu\theta + b_\nu \sin \nu\theta)$$

with

$$a_0 = \frac{1}{\pi}\int_0^{2\pi} v(R_0 e^{i\phi})d\phi.$$

Thus (19) means that $a_0 = 0$. In particular, Proposition 2 holds in the form

$$\iint_{K_0} v^2 dxdy = \frac{R_0}{2} D_{R_0}(v). \tag{20}$$

A similar estimate holds for every circle in $\mathcal{F}$.

Let V_0 be the set of functions $v \in V$ satisfying (19). Let $z = x + iy$ be a uniformising variable of a point P of $\mathcal{F}$ and let K be a z-circle around P.

Proposition 4. There is a constant $k > 0$ such that, for all $v \in V_0$,

$$\iint_{K_0} v^2 dxdy \le kD(v) \tag{21}$$

Proof. We construct a chain of points $P_0 = O, P_1, \ldots, P_n = P$ with associated uniformising variables z_i and z_i-circles $K_i = \{z_i \in \mathbb{C} \,|\, |z_i| \le R_i\}$, $i = 0,1,\ldots,n$, with the property that neighbouring circles K_i, K_{i+1} always have a common interior point. We prove inductively that there are constants $c_i > 0$ with

$$\iint_{K_i} v^2 dx_i dy_i \le c_i D(v). \tag{22}$$

For $i = 0$, (22) follows from (20). Suppose that we have already proved (22) for some i. By Proposition 2 there is a $c \in \mathbb{R}$ with

$$\iint_{K_{i+1}} (v-c)^2 dx_{i+1} dy_{i+1} \le \frac{R_{i+1}^2}{2} D(v). \tag{23}$$

We need an estimate of c^2 and for this purpose consider $E_i := K_i \cap K_{i+1}$. In E_i, z_{i+1} is a regular analytic function of z_i. The transition determinant

$$\det \begin{pmatrix} \frac{\partial x_{i+1}}{\partial x_i} & \frac{\partial x_{i+1}}{\partial y_i} \\ \frac{\partial y_{i+1}}{\partial x_i} & \frac{\partial y_{i+1}}{\partial y_i} \end{pmatrix} = \left| \frac{dz_{i+1}}{dz_i} \right|^2$$

is less than or equal to a certain constant M in E_i, and we get

$$\iint_{E_i} v^2 dx_{i+1} dy_{i+1} = \iint_{E_i} v^2 \left| \frac{dz_{i+1}}{dz_i} \right|^2 dx_i dy_i$$

$$\le M \iint_{K_i} v^2 dx_i dy_i$$

$$\le c_i M D(v). \tag{24}$$

On the other hand, by (23)

$$\iint_{E_i} (v-c)^2 dx_{i+1} dy_{i+1} \le \frac{R_{i+1}^2}{2} D(v). \tag{25}$$

Let $J := \iint_{E_i} dx_{i+1} dy_{i+1}$. Since

$$c^2 \le 2(v^2 + (c-v)^2),$$

addition of (24) and (25) gives the desired estimate of c^2:

$$c^2 J \le (2c_i M + R_{i+1}^2) D(v).$$

It follows that

$$\iint_{K_{i+1}} v^2 dx_{i+1} dy_{i+1} \le 2\left(\iint_{K_i} (v-c)^2 dx_{i+1} dy_{i+1} + c^2 R_{i+1}^2 \pi\right)$$

$$\le R_{i+1}^2 (1+(2c_i M+R_{i+1}^2)2\pi/J)D(v). \qquad \square$$

For two functions $v_1, v_2 \in V_0$, Propositions 3 and 4 give the estimate

$$\iint_K (v_1-v_2)^2 dxdy \le k(\sqrt{D(v_1)-d} + \sqrt{D(v_2)-d})^2. \tag{26}$$

From the Schwarz inequality for the functions $1, v_1-v_2$ one obtains from (26) that

$$\left|\iint_K (v_1-v_2) dxdy\right| \le \sqrt{k}\left[\sqrt{D(v_1)-d} + \sqrt{D(v_2)-d}\right]. \tag{27}$$

We now come to the proof of Theorem 11. We call a sequence $v_1, v_2, \ldots$ of functions from V_0 a *minimalising sequence* when $\lim_{i\to\infty} D(v_i) = d$.

For each point $P \in \mathcal{F}$ we modify such a sequence. If $P \in \mathcal{F} - K_0$ we draw a circle K around P which lies in $\mathcal{F} - K_0$ and construct the functions $\tilde{v}_1, \tilde{v}_2, \ldots$ of Section 29.9. In order to emphasise the dependence of the function $\tilde{v}_i$ on P, we use the more precise notation $\tilde{v}_{iP} := \tilde{v}_i$. By Section 29.9 the functions $\tilde{v}_{1P}, \tilde{v}_{2P}, \ldots$ constitute a minimalising sequence and in a sufficiently small circle K_P around P they are potential functions.

We want to show that in each circle K' around P with $K' \subset K_P$ the v_{iP} converge uniformly to a function u_P. Then the latter is harmonic by Theorem 7.

Let h be the difference between the radii of K_P and K' and let w be a potential function in K_P. Then by the mean value theorem for harmonic functions ((5) for $r = 0$),

$$w(Q) = \frac{1}{\pi h^2}\iint_{K_h(Q)} w \, dxdy \quad \text{for } Q \in K'.$$

Hence by (27) we have the following for all i,j :

$$|v_{iP}(Q) - v_{jP}(Q)| \le \frac{\sqrt{k}}{\pi h^2}\left[\sqrt{D(v_{iP})-d} + \sqrt{D(v_{jP})-d}\right] \quad \text{for} \quad Q \in K'.$$

This implies the uniform convergence of the v_{iP}.

We now consider the functions u_P in K_P and $u_{P'}$ in $K_{P'}$ for the case where K_P and $K_{P'}$ have interior points in common. When the corresponding modified minimalising sequences are $v_{1P}, v_{2P}, \dots$ and $v_{1P'}, v_{2P'}, \dots$ then the merged sequence $v_{1P}, v_{1P'}, v_{2P}, v_{2P'}, \dots$ is also minimalising, and it follows that the functions u_P are independent of the choice of circles $K_{P'}$ and so they unite into a single harmonic function defined on the whole of $\mathscr{F} - K_0$.

For $P \in K_0^*$ we draw a circle K around P which lies in K_0^* and construct the functions $v_1^*, v_2^*, \dots$ of Section 29.10, (i), and $\tilde{v}_{1P}^*, \tilde{v}_{2P}^*, \dots$ of 29.9. The associated functions $\tilde{v}_{1P}, \tilde{v}_{2P}, \dots$ then constitute a minimalising sequence. In order to see this, it suffices to prove the following Proposition.

Proposition 5. *Let* v^* *be a function continuously differentiable in* K_0^*, *and let* $\tilde{v}^*$ *be the corresponding modified function of Section 29.9. Then*

$$D_{K_0^*}(\tilde{v}) - D_{K_0^*}(v) = D_{K_0^*}(\tilde{v}^*) - D_{K_0^*}(v^*).$$

Proof. We set $B := K \cap (K_0^* - K_0)$. Then we have to show

$$D_B(\tilde{v}^* + \Phi) - D_B(v^* + \Phi) = D_B(\tilde{v}^*) - D_B(v^*).$$

(Fig. 24). It suffices to prove

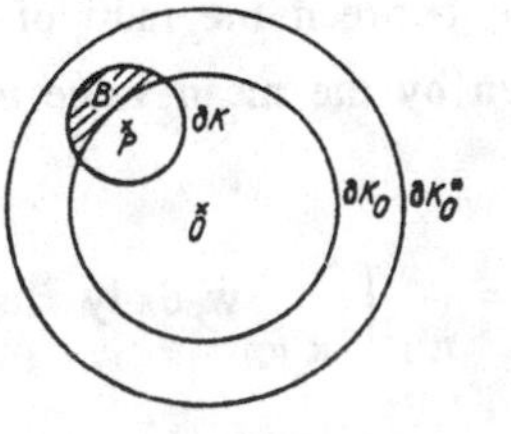

Fig. 24

$$D_B(\tilde{v}^*,\Phi) = D_B(v^*,\Phi).$$

Green's theorem gives, analogously to (6)

$$D_B(\tilde{v}^*,\Phi) - D_P(v^*,\Phi) = \int_{\partial B} (\tilde{v}^*-v^*)\left[\frac{\partial \Phi}{\partial x}dy - \frac{\partial \Phi}{\partial y}dx\right].$$

But the line integral on the right vanishes, because the difference $\tilde{v}^*-v^*$ vanishes on ∂K, and on ∂K_0

$$\frac{\partial \Phi}{\partial x}dy - \frac{\partial \Phi}{\partial y}dx = \frac{\partial \Phi}{\partial r}r d\theta = 0, \quad x + iy = re^{i\theta},$$

by definition of Φ. □

By the same reasoning as for the functions $\tilde{v}_{1P}, \tilde{v}_{2P},\ldots$ we obtain for the functions $\tilde{v}^*_{1P}, \tilde{v}^*_{2P},\ldots$ the harmonic function $u^*_P := \lim_{i\to\infty} \tilde{v}^*_{iP}$. The functions u^*_P unite to form a function u^*, harmonic in K^*_0, which satisfies condition (i) of Section 29.10 in conjunction with the function u on $\mathcal{F} - K_0$.

To prove Theorem 11 it only remains to show that $D(u) = d$.

Inside each sufficiently small circle K around a point P the uniform convergence of the $\tilde{v}_{iP}, \tilde{v}^*_{iP}$ and their derivatives (Theorem 7) implies

$$\lim_{i\to\infty} D_K(u-\tilde{v}_{iP}) = 0. \tag{28}$$

On the other hand, Proposition 3 gives

$$D_K(v_i-\tilde{v}_{iP}) \le D(v_i-\tilde{v}_{iP}) \le 4(D(v_i)-d)$$

and therefore

$$\lim_{i\to\infty} D_K(v_i-\tilde{v}_{iP}) = 0. \tag{29}$$

It follows from (28) and (29) that $\lim_{i\to\infty} D_K(u-v_i) = 0.$

Let G be a region in K. Then we also have $\lim_{i\to\infty} D_G(u-v_i) = 0$, and by the Schwarz inequality $D_G(u) = \lim_{i\to\infty} D_G(v_i)$.

For a finite number of disjoint regions G_ν in $\mathscr{F}$ we have

$$\sum_\nu D_{G_\nu}(u) = \lim_{i\to\infty} \sum_\nu D_{G_\nu}(v_i) \le \lim_{i\to\infty} D(v_i) = d.$$

This implies that D(u) exists and is less than or equal to d. Since $u \in V$, $D(u) = d$. □

The potential function U constructed with the help of the Dirichlet principle of Theorem 11 may be characterised as follows among the potential functions that satisfy the conditions of Theorem 10.

Theorem 12. *Let* K *be an arbitrarily small* z_0*-circle around* O *and let* W *be the set of continuously differentiable functions on* $\mathscr{F}$*, with finite Dirichlet integral, which vanish in a neighbourhood of* O*. Then* $D_{\mathscr{F}\text{-}K}(U)$ *is finite, and* $D(U,w) = 0$ *for* $w \in W$*. Conversely, if* U′ *is a potential function on* $\mathscr{F}$ *which is regular except at* O *and which differs from* Φ *by a potential function regular at* O *in some neighbourhood of* O*, and if* $D_{\mathscr{F}\text{-}K}(U')$ *is finite and* $D(U',w) = 0$ *for* $w \in W$*, then the Dirichlet integral* D(u′) *is minimal for the associated function* u′.

Proof. It is clear that $D_{\mathscr{F}\text{-}K}(U)$ is finite. Let K_l be a z_0-circle in K_0 of radius l, in which w vanishes. Then the inequality $D(u) \le D(u+lw)$ holds for each $\lambda \in \mathbb{R}$, and hence $D(u,w) = 0$. Also, if $z_0 = r_0 e^{i\theta_0}$

$$D(U,w) - D(u,w) = \iint_{K_0 - K_l} \left[\frac{\partial \Phi}{\partial x_0}\frac{\partial w}{\partial x_0} + \frac{\partial \Phi}{\partial y_0}\frac{\partial w}{\partial y_0}\right] dx_0 dy_0$$

$$= \int_0^{2\pi} \left[r_0 w \frac{\partial \Phi}{\partial r_0}\right]_{r_0=l}^{r_0=R_0} d\theta_0.$$

On ∂K_0 we have $\dfrac{\partial \Phi}{\partial r_0} = 0$ and on ∂K_l we have $w = 0$. Therefore

$$D(U,w) = D(u,w) = 0.$$

On the other hand, let $u_0 = U'-U$. Then u_0 is a potential function regular on the whole of $\mathcal{F}$, and $D(u_0,w) = 0$. This holds not only for the functions that vanish in a neighbourhood of O, but for all continuously differentiable functions for which $D(w)$ is finite. In fact, suppose w_1 is a function from W which agrees with w in $\mathcal{F}$-K. Then

$$D(u_0,w) = D(u_0,w_1) + D_{K_0}(u_0,w-w_1)$$

$$= \int_0^{2\pi} \left[r_0(w-w_1)\frac{\partial u_0}{\partial r_0}\right]_{R_0} d\theta = 0.$$

In particular, $D(u_0,u_0) = 0$ and hence u_0 is constant on $\mathcal{F}$. □

29.12 Proof of the Riemann existence theorem for differentials on closed Riemann surfaces

Let $\mathcal{F}$ be a closed Riemann surface. Theorem 10 gives us a differential ω with the single pole O and principal part $-1/z_0^2$ for the uniformising variable z_0 of O. The differential ω for the uniformising variable $z = x + iy$ at the point P is given by $\frac{\omega}{dz} = \frac{\partial U}{\partial x} - i\frac{\partial U}{\partial y}$.

In order to give a general proof of Theorem 3, Chapter 11, we must generalise the above process for constructing potential functions. To do this we note that we have used only the following properties of the potential function Φ on K_0^* in the proof of Theorem 11.

(i) Φ *is harmonic in* K_0^* *except at finitely many points and curves of* K_0^*.

(ii) $\frac{\partial \Phi}{\partial r_0}$ *vanishes on* ∂K_0.

We can therefore more generally consider the following potential functions Φ, for which (i) and (ii) are satisfied:

a) $\Phi(z_0) = \Phi_n(z_0) := \mathrm{Re}\left[\frac{1}{z_0^n} + \frac{z_0^n}{R_0^{2n}}\right]$. The associated differential ω_n has principal part $-\frac{n}{z_0^{n+1}}$ at O.

b) Let P_1, P_2 be points in K_0 different from O, with coordinates z_1, z_2. We join P_1 to P_2 by a curve D in K_0. Then $\log \frac{z_0 - z_1}{z_0 - z_2}$ is defined as a single-valued regular function in $K_0^* - D$ (as a function of z_0, $\log \frac{z_0 - z_1}{z_0 - z_2}$ is defined at each point of the simply connected region $(\mathbb{C} \cup \{\infty\}) - D$). When D is crossed from right to left, in the sense of the orientation of D, $\log \frac{z_0 - z_1}{z_0 - z_2}$ experiences a jump by $2\pi i$. To guarantee (ii) we need the reflections $z_1' = R_0^2/\bar{z}_1$, $z_2' = R_0^2/\bar{z}_2$ of z_1, z_2. These need not lie in K_0^*, so the latter has to be chosen suitably. We set

$$\Phi_{P_1 P_2}(z_0) := \operatorname{Re} \log \frac{z_0 - z_1}{z_0 - z_2} + \operatorname{Re} \log \frac{z_0 - z_1'}{z_0 - z_2'},$$

$$\Phi'_{P_1 P_2}(z_0) := \operatorname{Im} \log \frac{z_0 - z_1}{z_0 - z_2} - \operatorname{Im} \log \frac{z_0 - z_1'}{z_0 - z_2'}.$$

To prove (ii) one notes that the relation $R_0^2 = z_0 \bar{z}_0$ holds for $z_0 \in \partial K_0$.

The associated differentials $\omega_{P_1 P_2}$, $\omega'_{P_1 P_2}$ are regular except at the points P_1, P_2, where they have poles with principal parts $\frac{1}{z_0 - z_1}$, $-\frac{1}{z_0 - z_2}$ and $-\frac{i}{z_0 - z_1}$, $\frac{i}{z_0 - z_2}$ respectively.

The real part of the integral of $\omega_{P_1 P_2}$ over a closed curve C vanishes. For $\omega'_{P_1 P_2}$ this is the case when the curve C does not cut D. When C is a sufficiently small circle around P_1 resp. P_2, traversed in the anti-clockwise sense, then

$$\operatorname{Re} \int_C \omega'_{P_1 P_2} = 2\pi \quad \text{resp.} \quad \operatorname{Re} \int_C \omega'_{P_1 P_2} = -2\pi.$$

By forming linear combinations of the differentials given by a) and b) one obtains all the differentials necessary for the existence assertion of Theorem 3, Chapter 11.

Let P, Q be arbitrary points of $\mathcal{F}$. We first want to construct a differential ω_{PQ}, of the third kind, which has residue 1 at P and residue -1 at Q and for which the real part of the integral over an arbitrary closed curve vanishes.

We join P to Q by a curve D. On D we choose points $P_1 = P, P_2, \ldots, P_s = Q$ so that two neighbouring points P_i, P_{i+1} lie in a circle K_i around a point Q_i on D. Then

$$\omega_{PQ} := \omega_{P_1P_2} + \omega_{P_2P_3} + \ldots + \omega_{P_{s-1}P_s} \tag{30}$$

does what is required.

Now let D be a closed curve. We seek a differential ω_D of the first kind with $\mathrm{Re}\int_C \omega_D = 0$ when C does not cut the curve D and $\mathrm{Re}\int_C \omega_D = 2\pi$ when C cuts the curve D at a single point, from right to left.

Let $P_1, P_2, \ldots, P_s$ be chosen on D as above and suppose $P_1 = P_s$. Then

$$\omega_D := \omega'_{P_1P_2} + \omega'_{P_2P_3} + \ldots + \omega'_{P_{s-1}P_s} \tag{31}$$

does what is required.

Now let $\mathcal{F}$ be given a canonical dissection in the sense of Section 11.2. We obtain a differential ω of the first kind with

$$\mathrm{Re}\int_{A_1} \omega = 2\pi, \quad \mathrm{Re}\int_{A_i} \omega = 0 \ \text{ for } \ i = 2, \ldots, g,$$

$$\mathrm{Re}\int_{B_i} \omega = 0 \ \text{ for } \ i = 1, \ldots, g,$$

by setting $\omega := \omega_D$ for a curve D which runs from a point P in $\mathcal{F}_0$ to a point on A_1^{-1} and then from the equivalent point on A_1 back to P (see Figure 11, Section 11.3). Similarly for $A_2, \ldots, A_g$ or $B_1, \ldots, B_g$.

The existence assertion of Theorem 3, Chapter 11 follows by linear combination of the differentials obtained by these considerations.

Now let ω be a differential of the first kind with $\mathrm{Re}\int_{A_i} \omega = \mathrm{Re}\int_{B_i} \omega = 0$ for $i = 1,\dots,g$. Then $f(Q) = \mathrm{Re}\int_P^Q \omega$ defines an everywhere harmonic potential function on $\mathscr{F}$, i.e. $f(Q)$ is constant and $\omega = 0$.

This completes the proof of Theorem 3, Chapter 11.

From two non-proportional meromorphic differentials ω_1, ω_2 on $\mathscr{F}$ one obtains a non-constant meromorphic function $f := \omega_1/\omega_2$ on $\mathscr{F}$. As in Section 11.4 one shows that f takes each value finitely often and, except for finitely many points, equally often. I.e. f is a mapping of $\mathscr{F}$ onto $\mathbb{C} \cup \{\infty\}$ which defines $\mathscr{F}$ as a covering of $\mathbb{C} \cup \{\infty\}$. In this way we have reduced the general case of a closed Riemann surface to the special case considered in Chapters 10, 11.

Exercises

29.1 Construct a Riemann surface, as a covering of $\mathbb{C} - \{0\}$, on which $\log z$ is defined as a single-valued function.

29.2 Let $\mathscr{F}_1$ and $\mathscr{F}_2$ be Riemann surfaces and let f_1, f_2 be holomorphic mappings of $\mathscr{F}_1$ into $\mathscr{F}_2$. Suppose that f_1 agrees with f_2 in the neighbourhood of a point of $\mathscr{F}_1$. Show then that f_1 agrees with f_2 on the whole of $\mathscr{F}_1$.

29.3 Let $\mathscr{F}_1$ and $\mathscr{F}_2$ be Riemann surfaces and let f be an injective holomorphic mapping of $\mathscr{F}_1$ into $\mathscr{F}_2$. Show that the inverse mapping f^{-1} of $f(\mathscr{F}_1)$ onto $\mathscr{F}_1$ is also holomorphic.

29.4 Let M be an unramified covering of a Riemann surface $\mathscr{F}$ with projection mapping $\phi : M \to \mathscr{F}$ (Section 28.9). Show that there is a unique way to define M as a Riemann surface so that ϕ is a holomorphic mapping.

29.5 Let $\mathscr{F}$ be a Riemann surface and let f be a homeomorphism of the unit disc onto a subset M of $\mathscr{F}$ whose boundary is a (closed) curve. Also suppose f is holomorphic in the interior of the unit disc. Show that the first boundary value problem of potential theory is solvable for M.

30. Uniformisation

In the theory of uniformisation, the geometric function theory drafted by Riemann and taken up by Klein, Poincaré and Koebe reaches its highest point. Weyl described it as follows in the introduction to his book *"Die Idee der Riemannschen Fläche"*, which we also follow in this chapter:

"Here we enter the temple in which the Divinity (if I may be allowed this metaphor) is released from the earthly prison of its individual realisations: in the symbol of the two-dimensional non-euclidean crystal, the archetype of the Riemann surface itself appears (as far as this is possible) pure and free of all obscurities and inessentials".

30.1 The concept of uniformisation

We have defined a uniformising variable t for a point P of a Riemann surface $\mathcal{F}$ to be a conformal mapping of a neighbourhood V of P onto a neighbourhood U of the origin in the complex plane. With the help of a uniformising variable in V, a function f on $\mathcal{F}$ can be described as a function on U, i.e. as an ordinary analytic function (Section 29.2). In uniformisation theory one asks whether something similar can be found for the whole of $\mathcal{F}$. More precisely, one asks about the existence of regions U in $\mathbb{C}$ and mappings u of U onto $\mathcal{F}$ such that f can be represented as a single-valued meromorphic function $f(u(t))$ for $t \in U$.

Such a uniformisation can be carried out for all meromorphic functions on $\mathcal{F}$ together in the following way. Let $\hat{\mathcal{F}}$ be the universal covering of $\mathcal{F}$ (Section 28.9). Then $\hat{\mathcal{F}}$ is a simply connected Riemann surface. We shall show that one can map $\hat{\mathcal{F}}$ one-to-one and conformally onto a region U of $\mathbb{C}$ or the extended plane, with the latter case occurring only when $\mathcal{F} = \hat{\mathcal{F}}$ is itself simply connected. By composing the mappings $U \to \hat{\mathcal{F}} \to \mathcal{F}$ one obtains the desired uniformisation for all meromorphic functions on $\mathcal{F}$.

30.2 The Riemann mapping theorem

Thus it comes down to proving the following theorem, whose formulation goes back to Riemann, but which was first proved in rigorous form by Poincaré and Koebe in 1907.

Theorem 1. *Let* $\mathcal{F}$ *be a simply connected Riemann surface. Then there is a one-to-one and conformal mapping* τ *of* $\mathcal{F}$ *onto one of the following surfaces:* (i)*the extended plane,* (ii) *the extended plane minus one point,* (iii) *the extended plane minus a slit* $\{U + iV_0 | U_1 \leq U \leq U_2\}$ *where* U_1, U_2, V *are fixed real numbers.*

Proof. We choose a point O on $\mathcal{F}$ and a uniformising variable z_0 for O. By Theorem 12, Chapter 29 there is, up to an additive constant, exactly one potential function U on $\mathcal{F}$ with the following properties:

(i) U *is harmonic outside* O

(ii) *In a certain neighbourhood of* O, $U - \mathrm{Re}\frac{1}{z_0}$ *is harmonic.*

(iii) *For each* z_0*-circle* K *around* O, $D_{\mathcal{F}-K}(U)$ *is finite.*

(iv) *For each continuously differentiable function* w *on* $\mathcal{F}$ *such that* $D(w) < \infty$ *and* w *vanishes in a neighbourhood of* O, *one has* $D(U,w) = O$.

By Section 29.10 there is a differential ω associated with U. Since $\mathcal{F}$ is simply connected the integral $\int \omega$ taken from a fixed point does not depend upon the path, and it therefore represents an analytic function $\tau = U + iV$ on $\mathcal{F}$ which is regular outside O and which, in a neighbourhood of O, differs from $1/z_0$ by a regular function. We want to show that τ is the mapping sought in Theorem 1. To do this we first prove a few propositions.

Proposition 1. *Let* $\mathcal{F}$ *be a simply connected Riemann surface and let* C *be a closed curve on* $\mathcal{F}$. *Then* C *separates the surface* $\mathcal{F}$ *into two or more regions.*

Proof. Because of the analytic structure of $\mathcal{F}$ and the fact that C is piecewise continuously differentiable, we can distinguish the left from the right side of C. Suppose that C does not separate $\mathcal{F}$. Then we construct a double covering $\hat{\mathcal{F}}$ of $\mathcal{F}$ in the following way. We take two copies of $\mathcal{F}$, which we cut along the curve C and then join by crossing over, i.e. in passing through C from the left side to the right one goes from one copy of $\mathcal{F}$ to the other. In this way one obtains a connected surface $\hat{\mathcal{F}}$, since C does not separate $\mathcal{F}$. But a simply connected surface $\mathcal{F}$ has only trivial coverings (Section 28.9). □

Proposition 2. *Let* V_0 *be a real number. Then the sets*

$$M := \{P \in \mathcal{F} | \mathrm{Im}\,\tau(P) > V_0\} \quad \text{and} \quad M' := \{P \in \mathcal{F} | \mathrm{Im}\,\tau(P) < V_0\}$$

are regions in $\mathcal{F}$ (i.e., they are open and connected).

Proof. The function $1/\tau$ is a uniformising variable for O. Let $K_0 := \{P \in \mathcal{F} \mid \left|\frac{1}{\tau(P)}\right| \le R_0\}$ be a $1/\tau$-circle around O. We also set $E := \{P \in \mathcal{F} \mid \operatorname{Im} \tau(P) = V_0\}$. Since E is closed, M and M′ are partitioned into regions. Exactly two of these have points in common with K_0. Thus if M or M′ is not connected there is a region G among those determined by E such that $G \cap K_0 = \emptyset$.

For $u \in \mathbb{R}$ we set

$$\phi(u) = \arctan u \quad \text{with} \quad -\pi/2 < \arctan u < \pi/2,$$
$$\psi(u) = (u-V_0)^2/(1+(u-V_0)^2).$$

These functions have been chosen so that the function w on $\mathcal{F}$ defined by

$$w(P) = \begin{cases} \phi(U(P))\psi(V(P)) & \text{for } P \in G \\ 0 & \text{for } P \notin G \end{cases}$$

is continuously differentiable. The function w vanishes in K_0.

Let $z = x + iy$ be a uniformising variable for $P \in G$. Then

$$\left(\frac{\partial w}{\partial x}\right)^2 + \left(\frac{\partial w}{\partial y}\right)^2 = (\phi'(U)^2\psi(V)^2 + \phi(U)^2\psi'(V)^2)\left(\left(\frac{\partial U}{\partial x}\right)^2 + \left(\frac{\partial U}{\partial y}\right)^2\right).$$

Since the functions ϕ,ϕ',ψ,ψ' are bounded over the whole of $\mathbb{R}$, we can bound D(w) by $D_{\mathcal{F}-K_0}(U)$. Hence D(w) is finite. Thus D(U,w) = 0 by hypothesis. Moreover,

$$\frac{\partial w}{\partial x}\frac{\partial U}{\partial x} + \frac{\partial w}{\partial y}\frac{\partial U}{\partial y} = \phi'(U)\psi(V)\left(\left(\frac{\partial U}{\partial x}\right)^2 + \left(\frac{\partial U}{\partial y}\right)^2\right).$$

Since $\phi'(u)$ and $\psi(u)$ are positive for all $u \in \mathbb{R}$, this leads to a contradiction. □

Proposition 3. *The differential ω has no zero.*

Proof. Suppose P is a zero of ω with $\mathrm{Im}\,\tau(P) = V_0$. Then if z is a uniformising variable of P we have

$$\tau(z) - \tau(P) = a_r z^r + a_{r+1} z^{r+1} + \dots$$

with $a_r \neq 0$ and $r \geq 2$. Without loss of generality we can assume $\tau(z) - \tau(P) = z^r$. Let K be a z-circle around P with radius R. We connect P to the points $P_\nu = Re^{(2\nu-1)\pi i/2r}$ by the lines A_ν, where $\nu = 1,2,3,4$. Then P_1 and P_3 lie in the region M with $\mathrm{Im}\,\tau > V_0$, hence they can be connected by a curve B_1 in M (Proposition 2). Similarly, P_2 and P_4 lie in the region M' with $\mathrm{Im}\,\tau < V_0$. We connect them by a curve B_2 in M'. Then we have the closed curves $A_1 + B_1 - A_3$ and $A_2 + B_2 - A_4$, which meet at P but otherwise have no point in common (Figure 25).

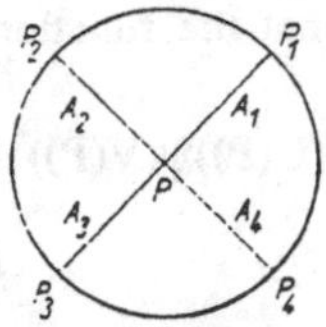

Fig. 25

But by Proposition 1 this is impossible for a simply connected surface. □

We now consider $\mathcal{F}$ as a covering surface of the extended plane with τ as projection mapping. Proposition 3 says that $\mathcal{F}$ is unramified over $\mathbb{C} \cup \{\infty\}$. Over ∞ lies the single point O. We consider the curve on $\mathcal{F}$ which lies over the line $V = V_0$, where V_0 again denotes a fixed real number (U,V denote not only the real and imaginary parts of τ but also the coordinates of the points of $\mathbb{C}$.) If we proceed from O in the direction of increasing values of U then there are two possibilities. Either one returns to O in $\mathcal{F}$, thereby completing a closed curve $\tilde{C}$, or else there is a $U + iV_0$, over which there is no point of $\mathcal{F}$, which continues the curve proceeding from O. Let U_1 be the least value of U for which this happens, and let $\tilde{C}_1$ be the curve proceeding from O with $U < U_1$, $V = V_0$. Similarly one obtains a curve $\tilde{C}_2$ proceeding from O in the direction of decreasing U with $U > U_2$, $V = V_0$, where U_2 is the greatest value of U beyond which the curve cannot be continued.

It still appears possible that $U_2 < U_1$, so that τ would not be injective. However, we shall see in what follows that this is not the case.

The curve $-\tilde{C}_1 + \tilde{C}_2$ resp. $\tilde{C}$ separates the simply connected surface $\mathcal{F}$ into two regions G', G'' (cf. Proposition 1; admittedly the curve $-\tilde{C}_1 + \tilde{C}_2$ is not closed, but it is a curve without ends, to which the proof of Proposition 1 still applies).

Let $Q \in G'$ be another point with $\tau(Q) = V_0$. Then there are points Q_1, Q_2 in G' with $\operatorname{Im} \tau(Q_1) > V_0$, $\operatorname{Im} \tau(Q_2) < V_0$, i.e., the set $\{P \in \mathcal{F} \mid \operatorname{Im} \tau(P) \neq V_0\}$ decomposes into at least three regions, contrary to Proposition 2. Thus G' contains no point Q with $\tau(Q) = V_0$. Similarly one shows that there is no $Q \in G''$ with $\tau(Q) = V_0$.

To prove Theorem 1 it now suffices to show that there can be at most one real number V_0 for which the associated curve $\tilde{C} = \{P \in \mathcal{F} \mid \operatorname{Im} \tau(P) = V_0\}$ is non-closed on $\mathcal{F}$. The case $U_2 < U_1$ then cannot occur, because for a U with $U_2 < U < U_1$ a whole neighbourhood of $U + iV_0$ would be doubly covered by $\mathcal{F}$.

We suppose that there are two such curves $\tilde{C}$ and $\tilde{C}'$, corresponding to the values V_0 and V_0' with $V_0 < V_0'$. We choose a real number U_3 so that the curve $\{P \in \mathcal{F} \mid \operatorname{Re} \tau(P) = U_3\}$ lies in K_0, and connect $\tilde{C}$ and $\tilde{C}'$ by the subcurve $\tilde{D}$ which lies over the line segment $V_0 \le V \le V_0'$. $\tilde{C}$ and $\tilde{C}'$ are divided by $\tilde{D}$ into pairs of subcurves. Let $\tilde{C}_3$ resp. $\tilde{C}_3'$ be the subcurve running in the direction $U = U_2$ resp. $U = U_2'$. Then $-\tilde{C}_3 + \tilde{D} + \tilde{C}_3'$ is a curve which runs on $\mathcal{F}$ without intersection from one boundary point to the other, and hence divides $\mathcal{F}$ into two regions G and G', where $O \in G'$.

We now argue as in the proof of Proposition 2. For this purpose we define a continuously differentiable function

$$w(P) = \begin{cases} \phi(U(P))\psi(V(P)) & \text{for } P \in G \\ 0 & \text{for } P \notin G, \end{cases}$$

where we now set

$$\phi(u) = \arctan(u - U_3)\frac{(u-U_3)^2}{1+(u-U_3)^2},$$

$$\psi(u) = \frac{(u-V_0)^2(u-V_0')^2}{(1+(u-V_0)^2)(1+(u-V_0')^2)}$$

for $u \in \mathbb{R}$. One has $D(w) < \infty$ and $D(U,w) > 0$, contrary to hypothesis. This proves Theorem 1. □

We now come to a still more convenient formulation of Theorem 1, in which we pass to another uniformisation.

Theorem 2. *(Riemann mapping theorem). Let* $\mathcal{F}$ *be a simply connected Riemann surface. Then there is a one-to-one and conformal mapping* t *of* $\mathcal{F}$ *onto one of the following surfaces:* (i) *the extended plane,* (ii) *the plane* $\mathbb{C}$, (iii) *the upper half plane* H.

Proof. In case (i), $t = \tau$ does what is wanted. In case (ii) we set $t = 1/(\tau-\tau_0)$, where τ_0 is the point of the plane that does not lie in the image of τ. In case (iii) one can first arrange, by a linear transformation, that the slit $\{U + iV_0 | U_1 \le U \le U_2\}$ goes to $\{U | -1 \le U \le 1\}$. The transformation $\tau = 2t/(t^2+1)$ then does what is wanted. □

30.3 The automorphisms of simply connected Riemann surfaces

An *automorphism* of a Riemann surface $\mathcal{F}$ is a one-to-one conformal mapping of $\mathcal{F}$ onto itself. We want to determine the automorphism group $\mathrm{Aut}(\mathcal{F})$ of the above three normal forms of simply connected Riemann surfaces.

For the extended plane $\bar{\mathbb{Z}}$ we have already done this in Section 10.2. $\mathrm{Aut}(\bar{\mathbb{Z}})$ is the group of all linear fractional transformations.

Let f be an automorphism of $\mathbb{C}$. Then f is a regular function on $\mathbb{C}$ with no essential singularity at ∞. Since f is one-to-one, it follows that f has a simple pole at ∞, i.e. f can be extended to an automorphism of $\bar{\mathbb{Z}}$. One now sees that $\mathrm{Aut}(\mathbb{C})$ is the group of all ordinary linear transformations of $\mathbb{C}$.

We have already determined Aut(H) in Section 14.11. It is the group of all linear fractional transformations of the form

$$w(z) = \frac{az+b}{cz+d} \quad \text{with} \quad a,b,c,d \in \mathbb{R} \quad \text{and} \quad ad - bc = 1. \tag{1}$$

This is the group of motions of non-euclidean geometry.

30.4 Normal form of a Riemann surface

Now let $\mathcal{F}$ again be an arbitrary Riemann surface and let $\hat{\mathcal{F}}$ be the universal covering of $\mathcal{F}$. By Theorem 2 we can assume, without loss of generality, that $\hat{\mathcal{F}}$ is $\bar{\mathbb{Z}}$, $\mathbb{C}$ or H. A covering transformation of $\hat{\mathcal{F}}/\mathcal{F}$ is by definition (Section 28.10) an automorphism of $\hat{\mathcal{F}}$ that leaves $\mathcal{F}$ pointwise fixed. By Section 30.3 the group Γ of covering transformations is a group of linear fractional transformations. By Chapter 28, Theorem 6, Γ is admissible and $\mathcal{F}$ is isomorphic to $\Gamma \setminus \hat{\mathcal{F}}$. In this way we obtain a normal form for $\mathcal{F}$. No $\gamma \in \Gamma$ except the identity can have a fixed point, i.e. a point $P \in \hat{\mathcal{F}}$ with $\gamma P = P$. Also, Γ must act discretely on $\hat{\mathcal{F}}$, i.e. for each $P \in \hat{\mathcal{F}}$ the orbit $\{\gamma P \mid \gamma \in \Gamma\}$ of P must have no accumulation points.

We first consider the cases $\hat{\mathcal{F}} = \bar{\mathbb{Z}}$ and $\hat{\mathcal{F}} = \mathbb{C}$ in more detail. In the first case Γ can consist only of the identity transformation, since each linear fractional transformation has fixed points in $\bar{\mathbb{Z}}$.

Now suppose $\hat{\mathcal{F}} = \mathbb{C}$. Since $\gamma \in \Gamma - \{1\}$ has no fixed points and is linear, we have $\gamma z = z + a$ where $a \in \mathbb{C}$. Therefore Γ is isomorphic to a discrete subgroup of the additive group of $\mathbb{C}$. Such a subgroup can be only zero-, one-, or two-dimensional. In the first case $\mathcal{F}$ is isomorphic to $\mathbb{C}$, in the second case to $\mathbb{C}/\omega\mathbb{Z}$ and in the third to $\mathbb{C}/(\omega_1\mathbb{Z} + \omega_2\mathbb{Z})$ where $\omega, \omega_1, \omega_2$ are non-zero complex numbers with $\omega_1/\omega_2 \notin \mathbb{R}$.

The meromorphic functions on $\mathbb{C}/(\omega_1\mathbb{Z} + \omega_2\mathbb{Z})$ are the elliptic functions with basic periods ω_1, ω_2, which we have investigated in detail in Chapter 13.

All in all, we see that the cases of $\bar{\mathbb{Z}}$ and $\mathbb{C}$ should be regarded as exceptional. We conclude this chapter with a few remarks on the "general type" $\Gamma \setminus$ H.

In Section 14.13 we have introduced H as a model of non-euclidean geometry. The metric defined in H is invariant under the conformal mappings of H onto itself. They constitute the full group of rigid motions of the oriented non-euclidean plane.

The following theorem gives a criterion for a subgroup Γ of $\mathrm{Aut}(H)$ to be admissible and hence to yield a Riemann surface in the normal form $\Gamma \backslash H$.

Theorem 3. *Let* Γ *be a subgroup of* $\mathrm{Aut}(H)$. *Then the following conditions are equivalent:*

(i) Γ *is admissible.*

(ii) Γ *acts discretely on* H *and no element of* Γ *other than* 1 *has a fixed point in* H.

Proof. It is obvious that (ii) follows from (i). Suppose (ii) holds. We shall prove (i). The subgroup K of $\mathrm{Aut}(H)$ consisting of all transformations that leave $i \in H$ fixed is, as one easily sees, homeomorphic to a circle and therefore compact. We first prove the following

Proposition. *If* M_1 *and* M_2 *are compact subsets of* H *then there are only finitely many* $\gamma \in \Gamma$ *for which* $\gamma M_1 \cap M_2$ *is non-empty.*

Proof. $\xi \mapsto \xi_i$ for $\xi \in \mathrm{Aut}(H)$ defines a continuous mapping ψ of $\mathrm{Aut}(H)$ onto H. Let $V_1 = \psi^{-1}M_1$ and $V_2 = \psi^{-1}M_2$. For a $\gamma \in \Gamma$ with $\gamma M_1 \cap M_2 \neq \emptyset$ we also have $\gamma V_1 \cap V_2 \neq \emptyset$ and hence $\gamma \in \Gamma \cap V_2 V_1^{-1}$. Since K, M_1 and M_2 are compact, V_1, V_2 and hence $V_2 V_1^{-1}$ are also compact. $\Gamma \cap V_2 V_1^{-1}$ is therefore compact as well as discrete and therefore finite. □

We now show that Γ is admissible. Let $a \in H$ be an arbitrary point of H and let M be a compact set in H which contains a neighbourhood of a. By the proposition there are only finitely many elements $\gamma_k \in \Gamma$, $k = 1,\ldots,s$, for which $\gamma_k M \cap M$ is non-empty. Let $\gamma_1 = 1$. We choose neighbourhoods $U_1,\ldots,U_s$ of $\gamma_1 a,\ldots,\gamma_s a$ which are pairwise disjoint and with $U_1 \subset M$, and set

$$U = \sum_{k=1}^{s} \gamma_k^{-1} U_k.$$

Then $\gamma U \cap U = \emptyset$ for all $\gamma \in \Gamma - \{1\}$. □

The transformation (1) has a fixed point in H if and only if it is the identity or else $|a+d| < 2$. In the latter case w is called *elliptic*.

The meromorphic functions on $\Gamma \backslash H$ can be identified with the meromorphic functions on H that are fixed by automorphisms in Γ. The latter functions are called *automorphic functions*. They are the subject of an extensive theory, in which the majority of questions, when one compares with the theory of elliptic functions, remain open.

Exercises

30.1 Let Γ be an admissible subgroup of Aut(H) and let $z \in H$. Also let $G(z)$ be the set of points z_1 of H with the property that

$$d(z_1,z) = \min\{d(z_2,z) \mid z_2 \in \Gamma z_1\},$$

where $d(z_1,z)$ is the non-euclidean distance. Show that:

a) $G(z)$ is a closed subset of H.

b) The interior points of $G(z)$ are precisely the points z_1 of $G(z)$ with $G(z) \cap \Gamma z_1 = \{z_1\}$.

c) The sets $G(z')$ for $z' \in \Gamma z$ cover the whole of H, and for $z' \neq z$ the set $G(z') \cap G(z)$ consists only of boundary points of $G(z)$ and $G(z')$.

d) $G(z)$ is a *non-euclidean convex* set, i.e. along with z_1 and z_2, all points on the geodesic between z_1 and z_2 belong to $G(z)$.

e) The boundary of $G(z)$ consists of geodesic segments whose vertices (which do not necessarily lie in H) do not accumulate in H. These geodesic segments are called the *sides* of $G(z)$.

f) The sides of $G(z)$ are pairwise related so that for each side A of $G(z)$ there is exactly one $\gamma \in \Gamma$, $\gamma \neq 1$, such that γA is a side of $G(z)$. The $\gamma \in \Gamma$ appearing here generate the group Γ.

g) $\Gamma \backslash H$ is closed if and only if $G(z)$ has finitely many sides and all vertices of sides lie in H.

30.2 Let Γ be an admissible subgroup of Aut(H) and let $\Gamma \setminus H$ be closed. Interpret G(z) as a polygon complex (Section 10.7) of $\Gamma \setminus H$.

30.3 Let H be the upper complex half plane. We extend H by adding the rational points and a point ∞ to the set H*. A topology is defined on H* by taking as neighbourhoods of $z \in H^*$ the following sets $U_z(\varepsilon)$:

$$U_z(\varepsilon) = \{\tau \in H \,|\, |\tau - z| < \varepsilon\} \quad \text{for } z \in H,$$
$$U_z(\varepsilon) = \{z\} \cup \{\tau \in H \,|\, |\tau - z| < \varepsilon\} \quad \text{for } z \in \mathbb{Q},$$
$$U_\infty(\varepsilon) = \{\infty\} \cup \{\tau \in H \,|\, \mathrm{Im}\,\tau > 1/\varepsilon\}.$$

The group G of linear fractional transformations $z \mapsto \dfrac{az+b}{cz+d}$ with $ad - bc = 1$ and $a,b,c,d \in \mathbb{Z}$ acts on H*, with the set $\mathbb{Q} \cup \{\infty\}$ constituting one orbit. We set $G_z = \{g \in G \,|\, gz = z\}$.

a) Show that G_z is a cyclic group of order 1 for $z \in H - Gi - G\rho$, where $\rho = -1/2 + (i/2)\sqrt{3}$, of order 2 for $z \in Gi$, of order 3 for $z \in G\rho$ and of order ∞ for $z \in G\infty$. Also give generators for G_z (Chapter 13, Theorem 13).

b) Show that for each $z \in H^*$ there is an open neighbourhood U_z with $G_z = \{g \in G \,|\, g(U_z) \cap U_z \neq \emptyset\}$.

c) Let ϕ be the canonical projection $H^* \to G \setminus H^*$. Show that $G \setminus H^*$ is defined as a Riemann surface by the following uniformising variables $t_{\phi(z)}$, which are defined in $\phi(U_z)$.

$$t_{\phi(z)}(\phi(\tau)) = \tau - z \quad \text{for } z \in H - Gi - G\rho,\ \tau \in U_z,$$
$$t_{\phi(i)}(\phi(\tau)) = h_i^2(\tau),\ \tau \in U_i,$$
$$t_{\phi(\rho)}(\phi(\tau)) = h_\rho^3(\tau),\ \tau \in U_\rho,$$

where h_z, for $z \in \{i,\rho\}$, is a one-to-one and conformal mapping of H onto the interior of the unit circle which sends z to 0,

$$t_{\phi(\infty)}(\phi(\tau)) = \exp(2\pi i\tau),\ \tau \in U_\infty.$$

d) Show that $G \setminus H^*$ is conformally equivalent to the extended plane.

e) Show that the elliptic modular function (Section 13.4) induces a meromorphic function on $G \setminus H^*$ which has a simple pole at G_∞.

Appendix 1. Rings

A1.1 Basic ring concepts

A *ring* Λ is an abelian group, with group operation denoted by +, in which a further operation, the product ab of any two elements a, b of Λ, is defined and satisfies the following conditions for all a, b, c in Λ:

$$a(bc) = (ab)c \quad \text{(associativity)}, \tag{1}$$

$$a(b+c) = ab + ac, (b+c)a = ba + ca \text{(distributivity)}. \tag{2}$$

Because of (1) one can write a product of several factors without brackets. It follows from (2) that $a0 = 0a = 0$.

A ring Λ is called *commutative* when $ab = ba$ for all a, b in Λ. An identity element of Λ is an element $u \neq 0$ with $ua = au = a$ for all $a \in \Lambda$. A ring has at most one identity element. In what follows it is denoted by 1.

In this appendix a ring is always understood to mean a commutative ring with identity element.

A *unit* e of a ring Λ is an element of Λ which has an inverse, i.e. there is an $e' \in \Lambda$ with $ee' = 1$. The element e' is uniquely determined by e. The set of all units of Λ constitutes a group $\Lambda^\times$ under multiplication in Λ, and is called the *group of units* of Λ.

An element $a \in \Lambda$ is called a *zero divisor* when $ab = 0$ for some $b \in \Lambda$ with $b \neq 0$. A ring in which 0 is the only zero divisor is called an *integral domain*. In the same way that one constructs the field of rational numbers from pairs of integers, one obtains a *field of fractions* $Q(\Lambda)$ from the pairs of elements of an integral domain Λ.

A *homomorphism* ϕ of the ring Λ_1 into the ring Λ_2 is a mapping of Λ_1 into Λ_2 with $\phi(a+b) = \phi(a) + \phi(b)$, $\phi(ab) = \phi(a)\phi(b)$ for all $a,b \in \Lambda_1$.

$$\text{Ker } \phi := \{a \in \Lambda_1 \mid \phi(a) = 0\}$$

is called the *kernel* of ϕ. This is a subring of Λ_1 which is mapped into itself under multiplication by elements Λ_1. A subring with this property is called an *ideal* of Λ_1. An ideal $\mathfrak{A}$ generated by one element, i.e. such that $\mathfrak{A} = a\Lambda_1$, is called a *principal ideal.*

Analogous to a quotient group one defines the *quotient ring* $\Lambda_1/\mathfrak{A}$ of Λ_1 by an ideal $\mathfrak{A}$. The homomorphism ϕ induces an isomorphism $\Lambda_1/\mathrm{Ker}\ \phi \to \phi(\Lambda_1)$ (*homomorphism theorem*).

An ideal $\mathfrak{A}$ of the ring Λ is called *maximal* when $\mathfrak{A} \neq \Lambda$ and there is no ideal $\mathfrak{B}$ with $\mathfrak{A} \subsetneqq \mathfrak{B} \subsetneqq \Lambda$.

Theorem 1. *An ideal* $\mathfrak{A}$ *of the ring* Λ *is maximal if and only if* $\Lambda/\mathfrak{A}$ *is a field.*

Proof. $\mathfrak{A}$ is maximal if and only if, for all $a \in \Lambda$ with $a \notin \mathfrak{A}$, there is a representation $1 = xa + b$ with $x \in \Lambda$, $b \in \mathfrak{A}$. The latter is equivalent to saying that $a + \mathfrak{A}$ has an inverse in $\Lambda/\mathfrak{A}$. □

An ideal $\mathfrak{P}$ of Λ is called a *prime ideal* when the quotient ring $\Lambda/\mathfrak{P}$ is an integral domain. By Theorem 1, each maximal ideal is a prime ideal.

An injective homomorphism ϕ of the integral domain Λ_1 into the integral domain Λ_2 may be extended to the field of fractions by

$$\phi(a/b) := \phi(a)/\phi(b) \quad \text{for} \quad a,b \in \Lambda_1,\ b \neq 0.$$

Let Λ be an integral domain. An element $a \in \Lambda$ *divides* $b \in \Lambda$ when there is a $c \in \Lambda$ with $ac = b$. In this case one writes $a \mid b$. For $a,b \in \Lambda - \{0\}$ it follows from $a \mid b$ and $b \mid a$ that c is a unit. The elements a and b are then called *associated.*

Let A be an arbitrary subset of Λ. A *divisor* of A is an element $b \in \Lambda$ which divides all $a \in A$. A *greatest common divisor* (gcd) of A is a divisor of A which is divided by all divisors of A. The gcd of A is unique up to *associates*, but does not exist in general. In the next section we shall consider *euclidean rings*, in which the gcd always exists.

A *prime element* p in Λ is a non-zero element which is not decomposable into the product of two factors, both of which are non-units.

Example 1. Let $\Lambda = \mathbb{Z}$ be the ring of integers. The group of units $\mathbb{Z}^{\times}$ consists of 1 and -1. The prime elements of $\mathbb{Z}$ have the form p or -p, where p is a prime number.

Example 2. Let K be a field and let $\Lambda = K[x]$ be the ring of polynomials in the indeterminate x with coefficients in K. The units are non-zero constants, and the prime elements are the irreducible polynomials.

One says that Λ is a *ring with unique prime decomposition* when each non-zero and non-unit element a of Λ may be written as a product of prime elements, and two such representations

$$a = p_1 \cdots p_n = q_1 \cdots q_m$$

have the same length (i.e. $n = m$) and with suitable ordering of factors p_i is associated with q_i, $i = 1,\ldots,n$.

A1.2 Euclidean rings

An integral domain Λ is called a *euclidean ring* when there is a *euclidean algorithm* in Λ, i.e. when each non-zero element a of Λ has a *height* h(a), a non-negative integer with the following properties:

1. *If* $a,b \in \Lambda - \{0\}$ *then* $h(ab) \geq h(a)$.
2. *If* $a,b \in \Lambda - \{0\}$ *then there is a* $c \in \Lambda$ *with* $h(a-bc) < h(b)$ *or* $a = bc$.

(3)

The euclidean algorithm then computes the gcd of two elements $a,b \in \Lambda - \{0\}$: let c be an element satisfying (3). Then if $b_1 := a - bc$ equals 0 the gcd equals b. Otherwise the height of b_1 is less than the height of b, and we repeat the process with b and b_1. Either b_1 is a divisor of b, and then we are finished, or we have an equation $b_2 = b - b_1c_1$ with $h(b_2) < h(b_1)$. The process must terminate in finitely many steps, since the height is reduced at each step. We therefore arrive at an element b_s with $b_{s-1} = b_sc$. Then b_s is the gcd of a and b. The algorithm also shows that b_s is of the form $b_s = ga + hb$ for certain g, h in Λ.

Example 1. $\mathbb{Z}$ is a euclidean ring with $h(a) = |a|$.

Example 2. K[x] is a euclidean ring, where the height of a polynomial f is the degree of f.

Theorem 2. *In a euclidean ring there is a gcd for any subset* A *of* Λ, *and it is expressible in the form*

$$b = g_1 a_1 + \ldots + g_s a_s \quad \text{with} \quad g_1, \ldots, g_s \in \Lambda,\ a_1, \ldots, a_s \in A. \tag{4}$$

Proof. Assume, without loss of generality, that $A \neq \{0\}$. Consider all elements of the form (4) and choose from them an element b of minimal height. Then b is the gcd of A. □

Theorem 3. *Each euclidean ring is a ring with unique prime decomposition.*

Proof. We prove Theorem 3 by induction on the height of elements of Λ. An element has minimal height if and only if it is a unit. We consider units to be elements with a prime decomposition of length 0. Suppose the assertion is already proved for all elements of height smaller than h, and suppose $a \in \Lambda$ with $h(a) = h$. If a is a prime element, then we are finished. If not, then a has the form $a = a_1 a_2$, where a_1 and a_2 are not units. Then $h(a_1) < h(a)$, $h(a_2) < h(a)$. In fact there are $c,d \in \Lambda$ with $a_1 = ca + d$ and $h(d) < h(a)$, and since $a_1 | d$, $h(a_1) \leq h(d)$. Similarly one shows $h(a_2) < h(a)$. Thus there is a prime decomposition of a by the induction hypothesis. Let p be a prime factor of a. Then p appears in each prime decomposition $p_1 \ldots p_s$ of a. For if p is not a divisor of p_1 then Theorem 2 gives an equation $1 = b_1 p + b_2 p_1$ with $b_1, b_2 \in \Lambda$, which implies $p | p_2 \ldots p_s$. The uniqueness of the prime decomposition of a now follows from the induction hypothesis. □

Theorem 2 shows that each ideal in a euclidean ring is a principal ideal. Rings with this property are called *principal ideal rings*.

A1.3 The characteristic of a ring

In any ring Λ one can multiply the element a by integers m. For positive m we set $ma := a + \ldots + a$ (m summands). For $m = 0$ we set $ma = 0$ and for negative m we set $ma = -(-m)a$.

The following rules of computation hold:

$$m(a+b) = ma + mb,\ (ma)b = m(ab) = a(mb) \quad \text{for} \quad m \in \mathbb{Z},\ a,b \in \Lambda$$

$$(m_1+m_2)a = m_1a + m_2a \quad \text{for} \quad m_1,m_2 \in \mathbb{Z},\ a \in \Lambda.$$

We now define the *characteristic* $ch(\Lambda)$ of Λ: when $m\cdot 1 \neq 0$ for all positive integers m, where 1 is the identity element of Λ, then $ch(\Lambda) = 0$. Otherwise, $ch(\Lambda)$ is the least positive integer m with $m\cdot 1 = 0$.

When Λ is an integral domain, $ch(\Lambda)$ equals 0 or a prime number. If, on the contrary, $ch(\Lambda) = m_1m_2$, then $(m_1\cdot 1)(m_2\cdot 1) = m_1m_2\cdot 1 = 0$.

A1.4 Modules over euclidean rings

Let Λ be a ring. A Λ-*module* (or *module over* Λ) is an additively written group for which multiplication (on the left) by elements of Λ is defined, with the following properties:

$$\lambda(a+b) = \lambda a + \lambda b \quad \text{for} \quad \lambda \in \Lambda,\ a,b \in M,$$

$$\lambda(\mu a) = (\lambda\mu)a \quad \text{for} \quad \lambda,\mu \in \Lambda,\ a \in M$$

$$1\cdot a = a \quad \text{for} \quad a \in M.$$

M is called *finitely generated* when there are elements $a_1,\dots,a_s$ of M such that each element $a \in M$ is expressible

$$a = \lambda_1a_1 + \dots + \lambda_sa_s \tag{5}$$

with $\lambda_1,\dots,\lambda_s \in \Lambda$.

Examples of modules are abelian groups $(\Lambda = \mathbb{Z})$ and vector spaces (Λ a field). A series of concepts and theorems which apply to both abelian groups and vector spaces carry over without difficulty to Λ-modules. Among them are the concepts of homomorphism, quotient module, the homomorphism theorem, the direct sum $M_1 \dotplus M_2$ of two modules M_1, M_2, and the concept of exact sequence: let M_1,M_2,M_3 be Λ-modules and let ϕ_1 resp. ϕ_2 be a homomorphism from M_1 into M_2 resp. M_2 into M_3. Then the sequence

$$M_1 \xrightarrow[\phi_1]{} M_2 \xrightarrow[\phi_2]{} M_3$$

is called *exact* at M_2 when $\phi_1(M_1) = \text{Ker}\ \phi_2$.

We now specialise again to euclidean rings Λ. A sequence $a_1,...,a_s$ of elements of the Λ-module M is called a *basis of* M when the representation (5) is unique for each $a \in M$. In particular, a module with a basis has the following property:

For all $a \in M, \lambda \in \Lambda$ with $\lambda a = 0$ we have $a = 0$ or $\lambda = 0$.

A module with this property is called *torsion-free*. We want to show, conversely, that any finitely generated torsion-free Λ-module has a basis. More generally, we shall prove the following:

Theorem 4. *Let* M *be a finitely generated torsion-free module and let* N *be a submodule of* M. *Then there is a basis* $a_1,...,a_m$ *of* M *and elements* $\varepsilon_1,...,\varepsilon_n$, $n \le m$, *with* $\varepsilon_i | \varepsilon_{i+1}$ for $i = 1,...,n-1$, *so that* $\varepsilon_1 a_1,...,\varepsilon_n a_n$ *is a basis of* N.

Proof. First suppose that M is a module with basis $b_1,...,b_m$. We prove Theorem 4 by induction on m. For $m = 0$ there is nothing to prove. We therefore suppose that $m \ge 1$ and that the assertion is already proved for modules with $m-1$ basis elements.

Without loss of generality, suppose $N \ne \{0\}$. We consider an element $b \ne 0$ in N whose representation $b = \lambda_1 b_1 + ... + \lambda_m b_m$ contains a coefficient λ_i of minimal height $h = h(\lambda_i)$, i.e., for all $b' \ne 0$ in N the coefficients λ'_j in $b' = \lambda'_1 b_1 + ... + \lambda'_m b_m$ are such that $\lambda'_j = 0$ or $h(\lambda'_j) \ge h$, for $j = 1,...,m$. Without loss of generality suppose $i = 1$.

We distinguish three cases.

1. For all $b' \in N$ the coefficients λ'_j are multiples of λ_1. We set $a_1 := b_1 + (\lambda_2/\lambda_1)b_2 + ... + (\lambda_m/\lambda_1)b_m$ and $\varepsilon_1 := \lambda_1$. Then $a_1,b_2,...,b_m$ is also a basis of M, the element $\varepsilon_1 a_1$ belongs to N, and each element b' of N has the form

$$b' = \varepsilon_1(\lambda'_1 a_1 + \lambda'_2 b_2 + ... + \lambda'_m b_m) \text{ with } \lambda'_1,...,\lambda'_m \in \Lambda.$$

The assertion follows from this by the induction hypothesis.

2. We now suppose that there is a $b' = \lambda_1' b_1 + ... + \lambda_m' b_m$ in N with a coefficient λ_j' which is not a multiple of λ_1, but that b' cannot be taken as b, i.e. that $\lambda_2, ..., \lambda_m$ are multiples of λ_1.

Then there is a $\mu \in \Lambda$ with $h(\lambda_j' - \mu\lambda_1) < h$. Therefore j cannot equal 1, otherwise the coefficient of b_1 in $b' - \mu b$ would have height less than the height of λ_1, contrary to the choice of λ_1. Suppose $\lambda_1' = \nu\lambda_1$. Then

$$\begin{aligned} b'' &:= b' + (1-\nu)b \\ &= \lambda_1 b_1 + ... + (\lambda_j' + (1-\nu)\lambda_j)b_j + ... + (\lambda_m' + (1-\nu)\lambda_m)b_m \end{aligned}$$

is an element of N whose b_1 coefficient is λ_1 and whose b_j coefficient is not divisible by λ_1. It therefore remains to consider the following case.

3. There is a j such that λ_j is not divisible by λ_1.

Let μ be an element of Λ with $h(\lambda_j - \mu\lambda_1) < h$. We go over to the basis $b_1' = b_1 + \mu b_j$, $b_2' = b_2, ..., b_m' = b_m$. Then

$$b = \lambda_1 b_1' + ... + (\lambda_j - \mu\lambda_1)b_j' + ... + \lambda_m b_m'.$$

In the new basis b has the coefficients $\lambda_j - \mu\lambda_1$ whose height is less than h.

We now go through the same process with the basis $b_m', ..., b_m'$ and continue. After finitely many steps we come to case 1. This completes the induction.

It remains to show that each finitely generated torsion-free module M has a basis.

Let $b_1, ..., b_s$ be a set of generators of M. We consider the "free module" $M_1 = \Lambda^s$ consisting of all s-tuples $(\lambda_1, ..., \lambda_s)$, added component-wise and multiplied component-wise by elements λ of Λ. It follows from this definition that M_1 has the basis (1,0,...,0), (0,1,0,...,0),...,(0,...,0,1). The homomorphism $\phi : M_1 \to M$ defined by

$$\phi(\lambda_1, ..., \lambda_s) = \lambda_1 b_1 + ... + \lambda_s b_s$$

is surjective. Let its kernel be N_1. Then, by what has already been proved, there is a basis $a_1,...,a_s$ of M_1 and elements $\varepsilon_1,...,\varepsilon_n$ such that $\varepsilon_1 a_1,...,\varepsilon_n a_n$ is a basis of N_1. Since M is torsion-free, $\varepsilon_1,...,\varepsilon_n$ are units of Λ. Therefore $\phi a_{n+1},...,\phi a_s$ is a basis of M. □

Theorem 5. *Let* M *be a finitely generated torsion-free module. Then all bases of* M *consist of the same number* r *of elements. The number* r *is called the rank of* M.

Proof. Let K be the field of fractions of Λ and let $b_1,...,b_m$ be a basis of M. We obtain an embedding of M in K^m by the mapping ϕ which associates each $b = \lambda_1 b_1 + ... + \lambda_m b_m$ with the vector $(\lambda_1,...,\lambda_m)$. Let $b_1,...,b_n$ be another basis of M. Then $\phi b_1,...,\phi b_n$ is a basis of K^m, hence $n = m$. □

Theorem 6. *The elements* $\varepsilon_1,...,\varepsilon_n$ *in Theorem* 4 *are uniquely determined up to associates.*

Proof. A basis of N can be expressed in terms of a basis of M with the help of a matrix A with coefficients in Λ. Passing to another basis of N resp. M means multiplying A, on the left resp. right, by a square matrix whose determinant is a unit (such matrices are called *unimodular*). By Theorem 4, A may be transformed in this way into a diagonal matrix with entries $\varepsilon_1,...,\varepsilon_n$. One easily convinces oneself that this diagonal matrix is unique. □

Now let M be an arbitrary finitely generated Λ-module and let $b_1,...,b_s$ be generators of M. As in the proof of Theorem 4 we have a surjective homomorphism of Λ^s onto M. Theorem 4 now yields

Theorem 7 *(Fundamental theorem on finitely generated* Λ*-modules). There is a system of generators* $a_1,...,a_m$ *of* M *and elements* $\varepsilon_1,...,\varepsilon_m$ *of* Λ *with* $\varepsilon_i | \varepsilon_{i+1}$ *for* $i = 1,...,m-1$, $\varepsilon_1 \notin \Lambda^\times$, *such that a relation*

$$\lambda_1 a_1 + ... + \lambda_m a_m = 0 \quad \textit{with} \quad \lambda_1,...,\lambda_m \in \Lambda$$

holds if and only if $\varepsilon_i | \lambda_i$ *for* $i = 1,...,m$. □

Theorem 7 can also be formulated as follows.

Theorem 7′. $M \cong \Lambda/\varepsilon_1\Lambda \dotplus \ldots \dotplus \Lambda/\varepsilon_m\Lambda$.

M is torsion-free if and only if $\varepsilon_1 = 0$. Then m is the rank of M.

Theorem 8. *The elements* $\varepsilon_1,\ldots,\varepsilon_m$ *in Theorem 7 are uniquely determined, up to associates.*

Proof. The set of elements a of M for which there is a non-zero $\alpha \in \Lambda$ with $\alpha a = \Lambda$ constitute a submodule M_t of M called the *torsion module* of M. Let n be maximal such that $\varepsilon_n \neq 0$. Then M/M_t is a torsion-free module of rank m-n, and M_t is isomorphic to $\Lambda/\varepsilon_1\Lambda \dotplus \ldots \dotplus \Lambda/\varepsilon_n\Lambda$. The elements $\varepsilon_1,\ldots,\varepsilon_n$ are called the torsion coefficients of M. To prove Theorem 8 we can assume $M = M_t$.

Then the ideal $\varepsilon_n\Lambda$ is the annihilator of M, i.e.

$$\varepsilon_n\Lambda = \{\alpha \in \Lambda \mid \alpha a = 0 \text{ for all } a \in M\}.$$

Let π be a prime element of Λ. The set M_π of elements of M that are annihilated by a power of π is an Λ-module called the *π-component* of M. $M_\pi \neq \{0\}$ if and only if π is a divisor of ε_n. M is the direct sum of its π-components (Chinese remainder theorem for Λ-modules).

It suffices to prove Theorem 8 for the π-components of M, i.e. we can assume that ε_n is a power of a prime element π. We set

$$M(i) = \{a \in M \mid \pi^i a = 0\},\ M^{i+1} := M(i+1)/M(i),\ i = 0,1,\ldots,$$

M^i is a finite-dimensional vector space over the field $\Lambda/\pi\Lambda$. Let $\varepsilon_1 = \pi^{j_1},\ldots,\varepsilon_n = \pi^{j_n}$. The sequence $\dim M^1 = n$, $\dim M^2,\ldots$ is an invariant of M which uniquely determines the sequence $j_1,\ldots,j_n$. □

Now let M be a finitely generated torsion module, i.e. $M = M_t$. In the case $\Lambda = \mathbb{Z}$, M is a finitely generated abelian group of order $|\varepsilon_1\ldots\varepsilon_n|$. In general we call the class of elements associated with $\varepsilon_1\ldots\varepsilon_n$ the *order* $[M]$ of the module M.

Theorem 9. *Let* M_1 *be a submodule of the finitely generated torsion module* M_2. *Then*

$$[M_2] = [M_1] \cdot [M_2/M_1].$$

Proof. We represent M_2 as a quotient module of a finitely generated torsion-free module F. Then suppose $M_2 \cong F/F_2$, $M_1 \cong F_1/F_2$ and consequently $M_2/M_1 \cong F/F_1$ with F_1, F_2 submodules of F. Since M_2 is a torsion module, F, F_1 and F_2 have the same rank (Theorem 4). We choose bases $\mathfrak{A}$, $\mathfrak{A}_1$, $\mathfrak{A}_2$ in F, F_1, F_2. Let A_1 resp. A_2 be the transition matrix from $\mathfrak{A}$ to $\mathfrak{A}_1$ resp. $\mathfrak{A}_1$ to $\mathfrak{A}_2$. Then A_2A_1 is the transition matrix from $\mathfrak{A}$ to $\mathfrak{A}_2$, and by Theorem 4

$$[M_1] = |\det A_1|,\ [M_2/M_1] = |\det A_2|,\ [M_2] = |\det A_2A_1|. \quad \square$$

A1.5 Construction of fields

Let Λ_1 be a ring and let Λ be a subring of Λ_1. For $a \in \Lambda_1$, $\Lambda[a]$ denotes the *smallest subring* of Λ_1 that contains Λ and a.

Similarly, let L be a field and let K be a subfield of L. For $a \in L$, K(a) denotes the *smallest subfield* of L that contains K and a. One says that $\Lambda[a]$ resp. K(a) results from Λ resp. K by *adjunction* of a.

Let K[x] be the ring of polynomials in an indeterminate x with coefficients in K, and let f(x) be an irreducible polynomial in K[x]. The ideal in K[x] generated by f(x) is a maximal ideal. The quotient ring K[x]/(f(x)) is therefore a field. K is embedded in K[x]/(f(x)) in a canonical way. Let $\bar{x} = x + (f(x))$. Then $K(\bar{x}) = K[\bar{x}] = K[x]/(f(x))$ and the degree of the field extension $K(\bar{x})/K$ equals the degree of f(x). The field $K(\bar{x})$ is called the *root field of* f(x). In $K(\bar{x})$, f(x) has the zero $\bar{x}$.

Theorem 10 (Kronecker). *Let* K *be a field and let* g(x) *be a polynomial in* K[x]. *Then there is an extension* L *of* K *in which* g(x) *splits into linear factors.* L *is called a splitting field of* g(x) *over* K.

Proof. We prove Theorem 10 by induction on the degree of g(x). When g(x) has degree 0 or 1 there is nothing to prove. Suppose the theorem is already proved for all fields K and polynomials of degree n, and let g(x) be a polynomial of degree n+1. Then we choose an irreducible factor f(x) of g(x). Over the root field of f(x), g(x) has a zero, and hence it decomposes into the product of two polynomials of degrees 1 and n. Hence by the induction hypothesis there is an extension L of K, over which g(x) splits into linear factors. □

Now let L/K be an arbitrary field extension and let a be an element of L which is algebraic over K, i.e. there is a polynomial $g(x) \in K[x] - \{0\}$ with $g(a) = 0$. Then the mapping

$$\phi : f(x) \mapsto f(a) \quad \text{for} \quad f(x) \in K[x]$$

defines a ring homomorphism of K[x] onto K[a]. Ker ϕ is generated by the irreducible polynomial $f_a(x)$ with $f_a(a) = 0$ and leading coefficient 1. The polynomial $f_a(x)$ is called the *minimal polynomial* of a over K. By the homomorphism theorem (Section A1.1) we have an isomorphism of the root field of $f_a(x)$ onto K[a]. Hence K[a] is a field, K[a] = K(a), and the degree of K(a) over K equals the degree of $f_a(x)$.

A1.6 Polynomials over fields

Let K be a field and let K[x] be the ring of polynomials in the indeterminate x. A polynomial $f(x) = a_0x^n + a_1x^{n-1} + \dots + a_n$ with $a_0 \neq 0$ is called *monic* when $a_0 = 1$. In each class of associated polynomials (Section A1.1) there is exactly one monic polynomial.

Theorem 11. *Let* $f_1(x)$ *and* $f_2(x)$ *be monic irreducible polynomials in* $K[x] - \{0\}$. *If* $f_1(x)$ *and* $f_2(x)$ *have a common zero in an extension* L *of* K, *then* $f_1(x) = f_2(x)$.

Proof. If $\alpha \in L$ is a zero of $f_1(x)$ and $f_2(x)$, then these polynomials have the common factor $x-\alpha$. Their greatest common divisor t(x) therefore cannot be 1. Since $f_1(x)$ and $f_2(x)$ are monic and irreducible, $t(x) = f_1(x) = f_2(x)$. □

For $f(x) = a_0x^n + a_1x^{n-1} + \dots + a_{n-1}x + a_n$ the derivative $f'(x)$ is defined by

$$f'(x) = na_0x^{n-1} + (n-1)a_1x^{n-2} + \dots + a_{n-1}. \tag{6}$$

It satisfies the product rule

$$(f_1(x)f_2(x))' = f_1(x)f_2'(x) + f_1'(x)f_2(x), \tag{7}$$

and the chain rule

$$(f_1(f_2(x))' = f_1'(f_2(x))f_2'(x), \tag{8}$$

for $f_1(x), f_2(x) \in K[x]$.

Theorem 12. *Let* $f(x) \in K[z]$ *be a polynomial which splits into linear factors in the extension field* L. *Then* $f(x)$ *has a multiple zero in* L *if and only if* $f(x)$ *and* $f'(x)$ *have a common divisor.* □

Theorem 13. *When* K *has characteristic* 0, *a polynomial* $f(x)$ *from* $K[x]$ *with a multiple zero in an extension field* L *is reducible in* $K[x]$. □

The proofs of Theorems 12 and 13 are left to the reader as exercises.

Appendix 2. Set theoretic topology

We content ourselves by listing a few basic concepts and proving the Heine-Borel theorem.

A2.1 Definition of a topological space

In set theoretic topology the concept of continuity of functions, and the related concepts which are familiar in $\mathbb{R}^n$, are formulated in an abstract way. In place of the concept of the ε-neighbourhood of a point in $\mathbb{R}^n$ we have the concept of the neighbourhood of a point in a topological space.

The definition of a topological space is based on the concept of open set:

A set M is called a *(Hausdorff) topological space* when M has a system $\mathfrak{U}$ of distinguished subsets, called *open sets* of M, with the following properties:

(i) *The union of arbitrarily many open sets is open.*

(ii) *The intersection of finitely many open sets is open.*

(iii) *(Hausdorff separation axiom.) For any two distinct points* P,Q *of* M *there are open sets* U_P, U_Q *with* $P \in U_P$, $Q \in U_Q$ *and* $U_P \cap U_Q = \emptyset$.

A *topology* is defined on M by giving a $\mathfrak{U}$ with properties (i) to (iii).

A *neighbourhood* U *of a point* P is a subset of M which contains an open set V with $P \in V$.

The topology on M is determined by giving a system $\mathfrak{U}_M$ of neighbourhoods with the property that for each open set V and each point $P \in V$ there is a $U \in \mathfrak{U}_M$ with $P \in U \subset V$. A set X in M is open if and only if for each $P \in X$ there is a $U \in \mathfrak{U}_M$ with $P \in U \subset X$. For example, the usual topology on $\mathbb{R}^n$ is given by the ε-neighbourhoods.

On any set M there is the *discrete topology*, in which all subsets of M are open. When M is finite this is the only way to define a topology on M.

A mapping ϕ of the topological space M into the topological space M′ is called *continuous* at the point $P \in M$ if for each neighbourhood U′ of $\phi(P)$ there is a neighbourhood U of P with $\phi(U) \subset U'$. The mapping ϕ is called *continuous* (on M) if ϕ is continuous at each point of M.

M and M′ are called *homeomorphic* when there are continuous mappings $\phi : M \to M'$, $\phi' : M' \to M$ that are inverse to each other, i.e. $\phi\phi'$ and $\phi'\phi$ are identity mappings.

A set A in the topological space M is called *closed* when its complement M-A is open. The intersection of arbitrarily many, and the union of finitely many, closed sets is again closed. A sequence $P_1, P_2, \ldots$ of points in M has the limit $P = \lim_{i\to\infty} P_i$ if each neighbourhood of P contains almost all of $P_1, P_2, \ldots$. In this case $P_1, P_2, \ldots$ is called a *convergent sequence.* A set M is closed if and only if it contains, along with any convergent sequence, the limit of that sequence.

A subset M′ of a topological space M with the system $\mathfrak{U}$ of open sets can be viewed as a topological space with the system $\{M' \cap U \mid U \in \mathfrak{U}\}$ of open sets. M′ is called a *discrete subset* of M when this topology is discrete.

A point P of a set A in a topological space M is called an *interior point* of A when there is a neighbourhood of P which lies entirely in A.

The *topological closure* $\overline{A}$ of A equals the set of limits of convergent sequences from A. The closure $\overline{A}$ is characterised as the intersection of the closed sets that contain A. In particular, $\overline{A}$ is closed. A *boundary point* of A is a point of $\overline{A}$ which is not an interior point of A.

A2.2 Compact spaces

Let M be a topological space and let K be a subset of M. A *covering of* K is a system of subsets of M whose union includes K. K is called *compact* when each covering of K by open sets includes finitely many sets which also cover K.

A compact set is closed. The image of a compact set under a continuous mapping is compact. A closed subset of a compact set is compact.

Theorem 1 (*Heine-Borel theorem*). *Each bounded closed subset of* $\mathbb{R}^n$ *is compact.*

Proof. We prove Theorem 1 for $n = 1$. The proof for $n > 1$ is analogous. Since a closed subset of a compact set is again compact, it suffices to prove Theorem 1 for an interval $[a,b]$.

Let $\mathfrak{K}$ be an open covering of $[a,b]$. We suppose the $\mathfrak{K}$ contains no finite subsystem which covers $[a,b]$. Let c be the midpoint of $[a,b]$. Then either $[a,c]$ or $[c,b]$ cannot be covered by finitely many sets from $\mathfrak{K}$. Thus there is an interval $[a_1,b_1] \subset [a,b]$ of length $(b-a)/2$ which cannot be covered by finitely many sets from $\mathfrak{K}$. On $[a_1,b_1]$ we apply the same argument. Continuing in this way, we obtain a nested sequence $[a,b] \supset [a_1,b_1] \supset [a_2,b_2] \supset \ldots$, where the sequences $a_1,a_2,\ldots$ and b_1,b_2 converge to the same limit value d. Let U be a set from $\mathfrak{K}$ with $d \in U$. Then U contains one of the intervals $[a_i,b_i]$, contrary to the construction of the nested sequence of intervals. $\square$

Appendix 3. Green's theorem

We proceed from the hypotheses of Section 8.3. Thus we suppose U is a simply connected region in the x,y plane and C is a simple closed curve in U. Also suppose g and h are differentiable functions in U. The curve C is traversed so that the region F(C) enclosed by it lies on its left.

We want to prove the following formula of Gauss, Green and Ostrogradski:

$$\iint_{F(C)}\left(\frac{\partial h}{\partial x} - \frac{\partial g}{\partial y}\right)dx\,dy = \int_C (gdx+hdy). \tag{1}$$

We first suppose that C is cut in at most two points by each line parallel to the x- or y-axis. Let (x_1,y_1) resp. (x_3,y_3) be the point of C with the smallest resp. largest x-coordinate, and let (x_2,y_2) resp. (x_4,y_4) be the point of C with the smallest resp. largest y-coordinate. Then

$$\iint_{F(C)}\frac{\partial g}{\partial y}dx\,dy = \int_{x_1}^{x_3}\left(\int_{y_1(x)}^{y_3(x)}\frac{\partial g}{\partial y}dy\right)dx$$

$$= \int_{x_1}^{x_3}(g(x,y_3(x)) - g(x,y_1(x))dx,$$

where $(x,y_1(x))$ resp. $(x,y_3(x))$, for $x_1 \le x \le x_3$, denotes the lower resp. upper point of C with abscissa x. Because of the orientation chosen for C we have

$$\int_{x_1}^{x_3}(g(x,y_3(x)) - g(x,y_1(x))dx = -\int_C g\,dx.$$

Analogously, one shows

$$\iint_{F(C)}\frac{\partial h}{\partial x}dx\,dy = \int_C h\,dy.$$

Now let C be arbitrary. If one goes to new coordinates x′, y′ which depend invertibly and twice continuously differentiably on x, y, then

$$\iint_{F(C)} \left(\frac{\partial h'}{\partial x'} - \frac{\partial g'}{\partial y'}\right) dx'dy' = \iint_{F(C)} \left(\frac{\partial h}{\partial x} - \frac{\partial g}{\partial y}\right) dx\, dy,$$

$$\int_C (g'dx'+h'dy') = \int_C (gdx+hdy)$$

with

$$g' = g\frac{\partial x}{\partial x'} + h\frac{\partial y}{\partial x'}\ , \ h' = g\frac{\partial x}{\partial y'} + h\frac{\partial y}{\partial y'}.$$

(In the sense of Section 14.6, g and h are the components of a covariant vector field on U.)

We subdivide $F(C)$ by a net of lines parallel to the x- or y-axis. It then suffices to prove (1) for the cells of this net. When the cells are chosen to be sufficiently fine one can arrange, by a coordinate transformation such as rotation of axes about the origin, that the cells are of the special form for which we have already proved (1). □

Appendix 4. Euclidean vector and point spaces

An n-dimensional vector space V over the field $\mathbb{R}$ of real numbers is called a *euclidean vector space* when it has a scalar product, i.e. a positive definite symmetric bilinear form which we denote by (v_1,v_2) for v_1,v_2 in V. We call $\|v\| = \sqrt{(v,v)}$ the *length* of the vector v. One shows by induction that V has a basis $e_1,\ldots,e_n$ with

$$(e_i,e_j) = \delta_{ij} \quad \text{for } i,j = 1,\ldots,n.$$

Such a basis is called an *orthonormal basis.* It follows immediately from this that two n-dimensional euclidean vector spaces V, V′ are isomorphic, i.e. there is a vector space isomorphism ψ of V onto V′ such that $(\psi v_1,\psi v_2) = (v_1,v_2)$ for all $v_1,v_2 \in V$. It suffices to let an orthonormal basis of V correspond to an orthonormal basis of V′. One may therefore speak of the *n-dimensional euclidean vector space.* The standard model of the n-dimensional euclidean vector space is $V =: \mathbb{R}^n = \{(x_1,\ldots,x_n) \mid x_1,\ldots,x_n \in \mathbb{R}\}$ with the scalar product $(x,x') = x_1x_1' + \ldots + x_nx_n'$ for $x = (x_1,\ldots,x_n)$, $x' = (x_1',\ldots,x_n')$.

A set E is called an *n-dimensional euclidean point space* with vector space V when, for all $P \in E$ and $v \in V$, the sum $P + v$ is defined and again belongs to E, and the following rules are satisfied.

1. *For all* $P \in E$, $P + o = P$.
2. *For all* $P \in E$ *and* $v_1,v_2 \in V$, $(P+v_1) + v_2 = P + (v_1+v_2)$.
3. *For two points* $P_1,P_2 \in E$ *there is exactly one vector* $v \in V$ *with* $P_2 = P_1 + v$.

The latter vector is denoted by $\overrightarrow{P_1P_2}$. The distance $|P_1P_2|$ between two points is defined by $|P_1P_2| := \|\overrightarrow{P_1P_2}\|$.

We fix a point O of E. Then there is a one-to-one mapping between E and V which associates the point P with the vector $\overrightarrow{OP}$. By distinguishing an orthonormal basis $e_1,\ldots,e_n$ of V we obtain a cartesian coordinate system: the coordinates of the point P are the unique numbers $x_1,\ldots,x_n$ with

$$x_1 e_1 + \ldots + x_n e_n = \overrightarrow{OP}. \tag{1}$$

Conversely, for any n-tuple $x_1,\ldots,x_n$ there is exactly one point P satisfying (1). Thus a euclidean point space is also determined up to isomorphism by its dimension. We therefore speak of the *n-dimensional euclidean space.*

Now let $a_1,\ldots,a_n$ be an arbitrary basis of V. A second basis $b_1,\ldots,b_n$ of V has the same orientation as $a_1,\ldots,a_n$, by definition, when the transition matrix from the first to the second basis has a positive determinant. When the latter determinant is negative, $a_1,\ldots,a_n$ and $b_1,\ldots,b_n$ have opposite orientation.

The set of bases of V falls into two classes of like-oriented bases. Distinguishing one of these classes gives an orientation to V. The bases in the distinguished class are called *positively oriented.* A euclidean point space E of V is called *oriented* when V is oriented.

A motion ϕ of the euclidean vector space V is an automorphism of V which preserves orientation. Let $e_1,\ldots,e_n$ be an orthonormal basis of V. Then ϕ is given by the matrix $A = (a_{ij})$ with $\phi(e_i) = a_{i1}e_1 + \ldots + a_{in}e_n$, where

$$AA^T = E \quad \text{(identity matrix)} \tag{2}$$

and

$$\det A = 1 \tag{3}$$

(A^T denotes the transposed matrix). The first condition expresses preservation of the scalar product, the second the preservation of orientation. Conversely, each matrix satisfying (2) and (3) yields a motion of V. The set of all matrices satisfying (2) and (3) is a group, the *special orthogonal group* SO(n).

A *congruence* of E is a mapping ψ of E into itself which leaves the distance between any two points P_1, P_2 of E unchanged, i.e.

$$|\psi(P_1)\psi(P_2)| = |P_1P_2|.$$

We fix a point O of E and define a mapping ψ' of V into itself by

$$\psi'(\overrightarrow{OP}) = \overrightarrow{\psi(O)\psi(P)} \quad \text{for} \quad P \in E.$$

We want to show that ψ' is an automorphism of V.

The scalar product is preserved by ψ': since

$$\overrightarrow{\psi(O)\psi(P_2)} - \overrightarrow{\psi(O)\psi(P_1)} = \overrightarrow{\psi(P_1)\psi(P_2)}$$

we get

$$(\psi'(\overrightarrow{OP_1}), \psi'(\overrightarrow{OP_2})) = \frac{1}{2}(|\psi(O)\psi(P_2)|^2 + |\psi(O)\psi(P_1)|^2 - |\psi(P_1)\psi(P_2)|^2)$$

$$= \frac{1}{2}(|OP_2|^2 + |OP_1|^2 - |P_1P_2|^2)$$

$$= (\overrightarrow{OP_1}, \overrightarrow{OP_2}).$$

It easily follows that for arbitrary $v_1, v_2 \in V$ and $a \in \mathbb{R}$

$$(\psi'(v_1+av_2) - \psi'(v_1) - a\psi'(v_2))^2 = 0$$

and therefore

$$\psi'(v_1+av_2) = \psi'(v_1) + a\psi'(v_2).$$

The mapping ψ' is invertible since $\psi'(v) = o$ implies that $\|v\| = \|\psi'(v)\| = 0$ and therefore $v = o$.

In defining ψ' we have distinguished a point O. However, ψ' is independent of the choice of O, because

$$\psi'(\overrightarrow{P_1P_2}) = \psi'(\overrightarrow{OP_2} - \overrightarrow{OP_1}) = \psi'(\overrightarrow{OP_2}) - \psi'(\overrightarrow{OP_1}) = \overrightarrow{\psi(P_1)\psi(P_2)}.$$

Conversely, it is clear that an arbitrary $O' \in E$ and an arbitrary automorphism ϕ of V gives a congruence by

$$\psi(P) = O' + \phi(\overrightarrow{OP}) \quad \text{for} \quad P \in E.$$

A congruence ψ is called a *motion* of E when ψ' is a motion, i.e. when ψ' preserves orientation.

The group of all motions of E is called the *motion group* of E. If we choose an orthonormal basis in V, then with O as origin we obtain a cartesian coordinate system. The motion ψ then takes the form

$$x' = a + Ax, \tag{4}$$

where x resp. x′ is the coordinate vector of P resp. $\psi(P)$, A is the matrix associated with ψ', and a is the coordinate vector of O′.

The *n-dimensional euclidean geometry* considers quantities that remain invariant under all motions. Among these are the distance between two points, the angle α between the vectors $a_1 = \overrightarrow{OP_1}$, $a_2 = \overrightarrow{OP_2}$, which is given by

$$\cos\alpha = \frac{(a_1, a_2)}{\|a_1\|\|a_2\|},\ 0 \le \alpha \le \pi,$$

and the oriented volume of the parallelotope spanned by the vectors $a_1 = \overrightarrow{OP_1}, \ldots, a_n = \overrightarrow{OP_n}$, which is given by

$$D(a_1, \ldots, a_n) := \det(x_1, \ldots, x_n),$$

where x_i is the coordinate vector of a_i relative to a positively oriented orthonormal basis of V.

The three-dimensional euclidean geometry defined in this way corresponds to intuitive geometry. For example, one cannot turn a left glove into a right one by any motion of the space.

If one admits all congruences as transformations of E, then one obtains metric geometry or congruence geometry, in which orientation is not preserved. In this geometry the left and right gloves are not distinct objects, since one can be carried into the other by an admissible transformation. In coordinate notation, congruences have the form (4), where A has only to satisfy the orthogonality condition $AA^T = E$. The set of these matrices is the *orthogonal group* O(n). Obviously SO(n) has index 2 in O(n).

The formulation of euclidean and metric geometry as the study of geometric quantities that remain invariant under certain transformation groups was first given explicitly by Klein 1872 in his inaugural lecture at the University of Erlangen, and it became known as the *Erlanger Programm*. Gauss used the orientation of a surface in order to define curvature at a point, and eventually established that it is independent of orientation (Chapter 4). Here one can see the germ of Klein's conception.

Euclid constructed his geometry from axioms about certain basic concepts, such as point, line, plane and motion. His axiom system is incomplete, however. Since one can define the basic euclidean concepts directly in our framework, and then verify the euclidean axioms, one can in some sense regard this as a realisation of Euclid's construction. A complete presentation of euclidean geometry, which rigorously follows the spirit of Euclid, and in particular does not assume the real numbers, was given by Hilbert 1899 in his book *Grundlagen der Geometrie*.

Exercises

A4.1 Show that each matrix $A \in SO_2(\mathbb{R})$ may be represented in the form

$$A = A(\phi) = \begin{pmatrix} \cos\phi & \sin\phi \\ -\sin\phi & \cos\phi \end{pmatrix}.$$

More precisely, the mapping $\phi \mapsto A(\phi)$ is a homomorphism of the additive group of real numbers onto $SO_2(\mathbb{R})$ with kernel $2\pi\mathbb{Z}$. Give a geometric interpretation of ϕ.

A4.2 (Euler) Show that each motion A of the three-dimensional euclidean vector space leaves a non-zero vector fixed, and that A is a rotation about this vector.

Appendix 5. Projective spaces

Let K be an arbitrary field. The *n-dimensional projective space* $\mathbb{P}^n_K$ over K consists of all one-dimensional subspaces of K^{n+1}, i.e. a point P of $\mathbb{P}^n_K$ is given by $n+1$ coordinates $x_0,x_1,\ldots,x_n \in K$, which are not all zero, and P determines $x_0,x_1,\ldots,x_n$ only up to a common factor from $K^\times$. One therefore writes $P = (x_0:x_1:\ldots:x_n)$.

The point space K^n is embedded in $\mathbb{P}^n_K$ by the mapping

$$(x_1,\ldots,x_n) \mapsto (1:x_1:\ldots:x_n).$$

Relative to this embedding one calls the points of $\mathbb{P}^n_K$ with $x_0 = 0$ the *points at infinity*.

Each linear subspace V of K^{n+1} corresponds to a subset $V/K^\times$ of $\mathbb{P}^n_K$, which is called a linear subspace of $\mathbb{P}^n_K$. When V has dimension m, $V/K^\times$ is assigned dimension $m-1$. One easily sees that two distinct lines (i.e. linear subspaces of dimension 1) in the projective plane $\mathbb{P}^2_K$ always meet in exactly one point (parallel lines meet in a point at infinity).

An *automorphism* of $\mathbb{P}^n_K$ (also called a *collineation*) is a mapping of $\mathbb{P}^n_K$ onto itself which maps linear subspaces onto linear subspaces. The group of all automorphisms of $\mathbb{P}^n_K$ is called the *n-dimensional projective group* of K and denoted by $PL_n(K)$. The group of all linear mappings of K^n onto itself is called the *n-dimensional general linear group* of K and is denoted by $GL_n(K)$.

Each $\phi \in PL_n(K)$ is given by a linear mapping of K^{n+1} onto itself. One has an exact sequence

$$1 \to K^\times \to GL_{n+1}(K) \to PL_n(K) \to 1,$$

where the mapping $K^\times \to GL_{n+1}(K)$ sends each $a \in K^\times$ to the corresponding scalar matrix.

In the terms of Klein's Erlanger Programm (A.4), projective geometry investigates the quantities that remain invariant under all automorphisms of projective space. One such quantity is the cross ratio of four distinct points $P_j = (x_0^j:x_1^j)$, $j = 1,2,3,4$, on a projective line:

$$(P_1P_2P_3P_4) := \frac{(x_0^1x_1^4-x_0^4x_1^1)(x_0^3x_1^2-x_0^2x_1^3)}{(x_0^3x_1^4-x_0^4x_1^3)(x_0^1x_1^2-x_0^2x_1^1)}.$$

Now let K be the field of real or complex numbers. We introduce a topology on $\mathbb{P}_K^n$ as follows: let U_i be the subset of $\mathbb{P}_K^n$ consisting of all points whose i^{th} coordinate is non-zero. A one-to-one mapping ϕ_i of K^n onto U_i is defined by

$$(x_0,...,x_{i-1},x_{i+1},...,x_n) \mapsto (x_0:...:x_{i-1}:1:x_{i+1}:...:x_n).$$

We transport the topology of K_n to U_i with the help of ϕ_i for $i = 0,...,n$. As one easily sees, for any two indices i, j a subset U of $U_i \cap U_j$ is open in U_i if and only if it is open in U_j. Therefore we can define a topology on $\mathbb{P}_K^n$ with the help of the topologies on the subsets U_i, which thereby give a covering of $\mathbb{P}_K^n$ by open subsets.

One proves by induction on n, using Theorem 1, A.2, that the topological space $\mathbb{P}_K^n$ defined in this way is compact.

Bibliography

Artin, E.
[1] Galois Theory, 3rd edition,
Notre Dame, Indiana 1959.

Böhm, J. and H. Reichardt (eds.)
[1] Gaussche Flächentheorie, Riemannsche Räume und Minkowski-Welt, Bd. 1,
Teubner-Verlag 1985.

Borevich, Z.I. and I.R. Shafarevich
[1] Number Theory,
Academic Press 1966.

Conforto, F.
[1] Abelsche Funktionen und algebraische Geometrie,
Springer-Verlag 1956.

Davenport, H.
[1] The Higher Arithmetic,
Harper Brothers 1962
[2] Multiplicative Number Theory,
Markham Publishing Company 1967.

Eichler, M.
[1] Introduction to the Theory of Algebraic Numbers and Functions,
Academic Press 1966.

Edwards, H.M.
[1] Fermat's Last Theorem, a Genetic Introduction to Algebraic Number Theory,
Springer-Verlag 1977.

Hasse, H.
[1] Vorlesungen über Zahlentheorie, 2nd ed.,
Springer-Verlag 1964.
[2] Number Theory,
Springer-Verlag 1980.

Helgason, S.
[1] Differential Geometry and Symmetric Spaces,
Academic Press 1962.

Hurwitz, A.
[1] Vorlesungen über allgemeine Funktionentheorie und elliptische Funktionen,
edited and extended by R. Courant, 4th ed.,
Springer-Verlag 1964.

Karazuba, A.A.
[1] Foundations of Analytic Number Theory (Russian)
Nauka 1975.

Knopp, K.
[1] Theory of Functions, vols. 1, 2,
Dover 1945-47.

Koch, H. and H. Pieper
[1] Zahlentheorie,
VEB Deutscher Verlag der Wissenschaften 1976.

Kra, I.
[1] Automorphic Forms and Kleinian Groups,
W.A. Benjamin 1972.

Kreysig, E.
[1] Differential Geometry,
University of Toronto Press 1959.

Lang, S.
[1] Introduction to Algebraic and Abelian Functions,
Addison-Wesley 1972.

Markushevich, A.I.
[1] Introduction to the Classical Theory of Abelian Functions (Russian),
Nauka 1979.

Narkiewicz, W.
[1] Elementary and Analytic Theory of Algebraic Numbers,
Polish Scientific Publishers 1974.

Pontriagin, L.S.
[1] Foundations of Combinatorial Topology,
Graylock Press 1952.

Prachar, K.
[1] Primzahlverteilung,
Springer-Verlag 1957.

Priwalow, I.I.
[1] Introduction to the Theory of Functions of a Complex Variable (Russian),
Nauka 1967.

Reichardt, H.
[1] Vorlesungen über Vektor- und Tensorrechnung,
VEB Deutscher Verlag der Wissenschaften 1957.
[2] Gauss und die nicht-euklidische Geometrie,
Teubner-Verlag 1976.

Riemann, B.
[1] Über die Hypothesen, welche der Geometrie zu Grund liegen,
English translation in Spivak, M.,
A Comprehensive Introduction to Differential Geometry, vol. 2, pp. 135-153,
Publish or Perish 1979.

Scharlau, W. and H. Opolka
[1] From Fermat to Minkowski,
Springer-Verlag 1985.

Scholz, E.
[1] Geschichte des Mannigfaltigkeitsbegriffs von Riemann bis Poincaré, Birkhäuser-Verlag 1980.

Seifert, H. and W. Threlfall
[1] A Textbook of Topology, Academic Press 1980.

Serre, J.-P.
[1] A Course in Arithmetic, Springer-Verlag 1973.

Shimura, G.
[1] Introduction to the Arithmetic Theory of Automorphic Functions, Princeton University Press 1971.

Springer, G.
[1] Introduction to Riemann Surfaces, Addison-Wesley 1957.

Smirnov, V.I.
[1] A Course of Higher Mathematics, vol. II, Pergamon 1964.

Vinogradov, I.M.
[1] An Introduction to the Theory of Numbers, Pergamon 1955.

van der Waerden, B.L.
[1] Algebra, vols. I, II, Ungar, New York 1970.

Weil, A.
[1] Number Theory. An Approach through History, Birkhäuser-Verlag 1984.

Weiss, E.
[1] Algebraic Number Theory, McGraw-Hill 1963.

Specific references for further reading are the following:

Chap. 1. Davenport [1], Chaps. 1 to 3; Hasse [1], Sections 1, 2; Koch-Pieper [1], Chaps. 1 to 4; Vinogradov [1].

Chap. 2. Davenport [1], Chaps. 4 to 7.

Chap. 3. van der Waerden [1], §60.

Chap. 4. Kreysig [1]; Smirnov [1], Chap. 5.

Chap. 5. Smirnov [1], Chap. 6.

Chap. 6. Hasse [1], Section 3; Koch-Pieper [1], Chap. 6.13 to 6.17; Serre [1], Chap. 6.

Chap. 7. van der Waerden [1], Chap. 6; Artin [1], Chap. 3.

Chap. 8. Knopp [1]; Priwalow [1], Chaps. 1 to 7.

Chap. 9. Knopp [1], II, Chaps. 1, 2; Priwalow [1], Chaps. 9, 10.

Chap. 10. Knopp [1], II, Chaps. 4 to 6; Seifert-Threlfall [1], Chap. 6; Springer [1], Chaps. 1 to 5.

Chap. 11. Springer [1], Chap. 10.1 to 10.6.

Chap. 12. Lang [1], Chap. 3; Springer [1], Chap. 10.7 to 10.10.

Chap. 13. Hurwitz-Courant [1]; Knopp [1], II, Chap. 3; Priwalow [1], Chap. 11; Conforto [1]; Markushevich [1].

Chap. 14. Riemann [1]; Helgason [1], Chap. 1; Reichardt [1] Chap. 12; Reichardt [2].

Chap. 15. Davenport [2], Chap. 8.

Chap. 16. Koch-Pieper [1], Chap. 4.5; van der Waerden [1], Chap. 8.

Chap. 17. van der Waerden [1], Chap. 17; Borevich-Shafarevich [1], Chap. 3; Narkiewicz [1], 1.1.

Chap. 19. Borevich-Shafarevich [1], Chap. 2.1 to 2.6.

Chap. 20. Borevich-Shafarevich [1], Chap. 5.1.

Chap. 21. Borevich-Shafarevich [1], Chap. 2.7.

Chap. 22. Narkiewicz [1], Chap. 4; Weiss [1], Chap. 4.8.

Chap. 23. Hasse [1], Chaps. 24, 25; Lang [1], Chap. 1; van der Waerden [1], Chap. 19.

Chap. 24. Hasse [2], Chap. 30; Narkiewicz [1], Chap. 2.

Chap. 25. Weiss [1], Chap. 4.10.

Chap. 26. Karazuba [1], Chap. 1; Prachar [1], Appendix, §5.

Chap. 27. Davenport [2], Chap. 18; Karazuba [1], Chaps. 3, 4.

Chap. 28. Pontriagin [1], Chap. 1; Seifert-Threlfall [1], Chaps 1 to 4, 7, 8.

Chap. 29. Springer [1], Chaps 1, 6 to 8.

Chap. 30. Springer [1], Chap. 9; Shimura [1], Chaps. 1, 2; Kra [1], Chaps. 1, 2.

Name Index

(Only names mentioned in a historical context are listed.)

Abel, Niels Henrik (1802-1829) 63, 69, 75, 84, 119, 154, 162, 344
d'Alembert, Jean Le Rond (1717-1783) 44, 93
Alexander, James Waddell (1888-1971) 362
Archimedes (~ – 287 – – 212) 14
Artin, Emil (1898-1962) 7, 322

Beltrami, Eugenio (1835-1899) 208
Bernoulli, Daniel (1700-1782) 45
Bernoulli, Jacob I (1654-1705) 56
Bessel, Friedrich Wilhelm (1784-1846) 90
Betti, Enrico (1823-1892) 359, 367
Bolyai, János (1802-1860) 207
Borel, Emile (1871-1956) 431
von Brill, Alexander Wilhelm (1842-1935) 231

Cantor, Georg (1845-1918) 378
Cardano, Girolamo (1501-1576) 67
Cartan, Elie (1869-1951) xv
Casorati, Felice (1835-1890) 101, 127, 181
Cauchy, Augustin-Louis (1789-1857) 84, 91, 93, 95, 97
Cayley, Arthur (1821-1895) 208
Chebyshev, Pafnuti Lvovich (1821-1894) 210, 343
Christoffel, Elvin Bruno (1829-1900) 182, 187
Clebsch, Rudolf Friedrich Alfred (1833-1872) 156

Darboux, Jean-Gaston (1842-1917) 35
Dedekind, Julius Wilhelm Richard (1831-1916) xv, 1, 11, 21, 150, 221 223, 232, 237, 238, 240, 262, 281, 291, 313, 378
Descartes, René (1596-1650) 3
Diophantus (~ 250) 1
Dirichlet, Gustav Peter Lejeune (1805-1859) 11, 21, 46, 60, 62, 221, 252, 253, 262, 385

Einstein, Albert (1879-1955) 182
Eisenstein, Ferdinand Gotthold Max (1823-1852) 25, 175
Euclid (~ – 365 – – 300) 1, 59, 206, 418, 438
Euler, Leonhard (1707-1783) xvi, 1, 3, 11, 20, 40, 45, 55, 59, 93, 110, 111, 133, 154, 163, 210, 438

de Fermat, Pierre (1601-1665) 1, 11, 20, 136
Ferrari, Ludovico (1522-1565) 67
del Ferro, Scipione (1465-1526) 67
Fourier, Jean-Baptiste Joseph (1768-1830) 45, 51, 53
Frenet, Jean-Frédéric (1816-1900) 43
Frobenius, Ferdinand Georg (1849-1917) 247

Galois, Evariste (1811-1832) xv, 74, 75, 76, 78, 81, 82, 84, 227
Gauss, Carl Friedrich (1777-1855) xv, 1, 3, 5, 6, 11, 21, 23, 24, 26, 31, 35, 37, 38, 41, 42, 59, 81, 90, 118, 119, 164, 182, 207, 219, 318, 432, 438
Gerling, Christian Ludwig (1788-1864) 207
Goursat, Edouard Jean-Baptiste (1858-1936) 92
Green, George (1793-1841) 94, 143, 385, 432

Hadamard, Jacques (1865-1963) 210, 214, 331, 342, 388
Harnack, Carl Gustav Axel (1851-1888) 388
Hasse, Helmut (1898-1979) 221
Hausdorff, Felix (1868-1942) 429
Heine, Heinrich Eduard (1821-1881) 431
Hensel, Kurt (1861-1941) 221
Hermite, Charles (1822-1901) 174, 312
Hilbert, David (1862-1943) 141, 208, 240, 281, 313, 322, 324, 386, 438
Hopf, Heinz (1894-1971) 362
Hurwitz, Adolf (1859-1919) 305

Jacobi, Carl Gustav Jacob (1804-1851) 7, 155, 161, 171, 220
Jordan, Camille (1838-1921) 78, 92, 368

Kant, Immanuel (1724-1804) 207
Klein, Felix (1849-1925) xvi, 165, 208, 407, 437, 439
Koebe, Paul (1882-1945) 407
Kronecker, Leopold (1823-1891) 1, 221, 322, 426
Kummer, Ernst Eduard (1810-1893) 55, 220, 237, 262, 323

Lagrange, Joseph Louis (1736-1813) 1, 11 21, 75, 79, 253
Laurent, Pierre Alphonse (1813-1854) 99
Legendre, Adrien-Marie (1752-1833) 1, 5, 11, 59, 116, 206
Leibniz, Gottfried Wilhelm (1646-1716) 55
Levi, Beppo (1875-1961) 396
Levi-Civitá, Tullio (1873-1941) 182, 209
Lie, Marius Sophus (1842-1899) xv
Liouville, Joseph (1809-1882) 74, 98, 165
Lipschitz, Rudolf Otto Sigismund (1832-1903) 182
Lobachevsky, Nikolai Ivanovich (1792-1856) 207

Mangoldt, Hans Carl Friedrich von (1854-1925) 217
Minkowski, Hermann (1864-1909) 307, 309
Möbius, August Ferdinand (1790-1868) 9, 214, 362
Monge, Gaspard (1746-1818) 42

Neumann, Carl Gottfried (1832-1925) 386
Newton, Isaac (1642-1727) 3, 89
Noether, Emmy (1882-1935) 362

Ostrogradski, Mikhail Vassilevich (1801-1862) 432

Parseval, Marc-Antoine (1755-1836) 57
Picard, Charles Emile (1856-1941) 145, 179, 331
Poincaré, Jules Henri (1854-1912) 91, 208, 331, 359, 362, 368, 407
Poisson, Denis (1781-1840) 386
Pythagoras (~ – 580 – – 500) 45

Ricci-Curbastro, Gregorio (1853-1925) 182
Riemann, Georg Friedrich Bernhard (1826-1866) xv, 35, 42, 46, 54, 91, 119, 123, 128, 135, 138, 146, 162, 182, 187, 189, 193, 202, 207, 210, 213, 214, 291, 295, 304, 331, 338, 342, 345, 359, 373, 378, 386, 407
Roch, Gustav (1839-1866) 146, 295, 304
Rodrigues, Olinde (1794-1851) 41

Schmidt, Erhard (1876-1959) 199
Schwarz, Hermann Amandus (1843-1921) 386
Stickelberger, Ludwig (1850-1936) 248
Stirling, James (1692-1770) 113

Taylor, Brook (1685-1731) 45
Tartaglia, Niccolo (~ 1500 – 1557) 67

de la Vallée Poussin, Charles Jean (1866-1962) 210, 342
Vandermonde, Alexandre Théophile (1735-1796) 81
Vergilius Maro, Publius (– 70 – 19) 55
Viète, François (1540-1603) 68

van der Waerden, Bartel Leenert (*1903) 240
Wallis, John (1616-1703) 117
Weber, Heinrich (1842-1913) 150, 221, 291, 324
Weierstrass, Karl Theodor Wilhelm (1815-1897) 91, 101, 105, 108, 112, 119, 127, 165, 175, 179, 291, 386
Weyl, Hermann (1885-1955) xvi, 125, 165, 378, 407

Zaremba, Stanislaw (1863-1942) 388
Zolotariev, Igor Ivanovich (1847-1878) 221

General Index

abelian group 75
absolute value 91
adjacent forms 13
algebraic function 232
——— number 232
algorithm, euclidean 419
alphabet 371
alternating group 73
analysis situs 128
analytic continuation 105
argument 91
Artin reciprocity law 7
associated elements 418
atlas 125

bending 41
Bernoulli numbers 56
Betti number 367
binary quadratic form 11
biquadratic residue 6
boundaries 363
boundary operator 363
——— point 430
bounded surface 130
branch point of a Riemann surface 124, 295

canonical class 147
——— dissection 134
casus irreducibilis 68
Cardano formula 67
Cauchy integral formula 95
——— inequality 97
Cauchy integral theorem 94
——— ——— for Riemann surfaces 382
cells 129
centraliser 84
centre of a group 84
chain 362
character, imprimitive 66
———, modulo k 62
———, of a group 60
———, primitive 66
characteristic, Euler 133, 367
——— of a ring 420
——— polynomial 230
chart 125
Chebyshev function 343
Christoffel symbol of first kind 187
——— second kind 187

class, canonical 147
——— number 251
classes of forms; composition 21
closed curve 92
——— set 430
——— surface 130
closure, topological 430
commutator quotient group 371
——— subgroup 371
compact set 430
complementary module 283
complex integration 345
——— plane, extended 120
composition of classes of forms 21
——— series of a group 81
conformal mapping 125, 206
congruence in a euclidean space 435
content of a polynomial 24
continuation, analytic 105
continued fraction 9
——— ——— algorithm 9
convergence exponent 332
———; radius 96
——— theorem of Harnack 388
convergent of continued fraction 9
convergent sequence 430
covering 373, 430
———; monodromy theorem 373
———, regular 375
———, unbranched 124, 373
———, universal 375
covering transformation 375
curvature, Gaussian 38
———, maximal 40
———, minimal 40
———, of a curve 36
———, tensor, Riemann 193
curve, closed 92
———; curvature 36
———, directed simple smooth 91
———, Fermat 136
———, regular 43
———, simple 92
———, singular 43
———, theorem, Jordan 92
cycle representation of a permutation 82
cycles 363
cyclotomic field 318
——— polynomial 27, 33
cyclotomy 23

decomposition field 314
——— group 314
degree of a divisor 144
Dedekind discriminant theorem 289
——— ζ-function 262
different of an element 228
——— of a field extension 284
——— tower theorem 284
differentiable manifold 378
differential 137, 299
———, exact 147
———, meromorphic 137, 381
——— of first kind 139
——— of second kind 139
——— of third kind 139
Dirichlet integral 383
——— L-series 62
——— principle 141, 386, 396
——— unit theorem 253
discrete subset 430
——— topology 429
discriminant of a field extension 285
——— — – polynomial 74
——— — – quadratic form 11
——— — – sequence of elements 228
discriminant theorem, Dedekind 289
——— ———, Minkowski 311
discriminant tower theorem 289
divergent series 117
divisor; degree 144
———, effective 144
———, of a function 144
———, special 158
divisor class 145
——— ——— group 303
double series theorem 104
duplication formula, Legendre 116

edge path group 370
effective divisor 144
Eisenstein irreducibility criterion 25
——— series 175
elementary symmetric polynomial 71
——— transformations 132
elements, associated 418
———; different of 228
elliptic functions 162, 165
——— modular function 134, 175
——— transformation 180
equivalent forms 12
Erlanger Programm 438
euclidean algorithm 419
——— geometry 438
——— point space 434
——— ring 419
——— space; congruence 435
——— ———; motion 436
——— vector space 434
Euler characteristic 133, 367
——— constant 111
——— product formula 63
exact differential 147
extended complex plane 120
extension, normal 77

Fermat curve 136
field extension; different 284
——— ———; discriminant 285
——— ———; Galois group 225
forms, adjacent 13
———, binary quadratic 11
———, equivalent 12
———, positive 16
———, primitive 15
———, properly equivalent 12
———, reduced 16
formula, Cardano 67
———, Fourier 51
———, Riemann's exact 214
———, Stirling's 113
formulae, Frenet 43
———, Newton's 89
Fourier coefficients 45
——— formula 51
——— inversion formula 53
——— transform 53
fractional ideal 241
fractions, field of 417
Frenet formulae 43
Frobenius automorphism 247
function, algebraic 232
———, Chebyshev 343
———; divisor 144
———, elliptic 162, 165
———, entire 108
———, harmonic 384
———, holomorphic 92
———, integral algebraic 232
———, meromorphic 101
———, Möbius 8, 214
———; partial fraction decomposition 56
———; pole 101
———, regular 92
function field; Riemann surface 292

fundamental form 184
——— group 368
——— region 307
fundamental theorem of algebra 67, 98
——— ——— — Dedekind, first 285
——— ——— — ———, second 286
——— ——— — edge path groups 370
——— ——— — finitely generated Λ-modules 424
——— ——— — Galois theory 225
——— ——— — homology groups 364
——— ——— — ideal theory 240

Galois group of a field extension 225
——— ——— — - polynomial 76
——— theory, fundamental theorem 225
Γ-function 111
Gaussian curvature 38
——— integers 219
——— multiplication formula 117
——— periods 26
——— product representation of the Γ-function 112
——— sum 66
general equation of nth degree 70
——— linear group 439
genus of a surface 134
geodesics 42, 186
geometry, euclidean 438
———, non-euclidean 206
Gram-Schmidt orthonormalisation process 199
great Picard theorem 179
greatest common divisor 418
group, abelian 75
———, alternating 73
———; centre 84
———; character 60
———; composition series 81
———, general linear 439
———, isotropy 178
———, one-dimensional affine 85
———, orthogonal 437
———, Picard 145
———, projective 439
———, simple 84
———, special linear 175
———, ——— orthogonal 435
———, symmetric 73

group of units 417

half plane, Poincaré upper 208
harmonic function 384
Hausdorff separation axiom 429
——— topological space 429
height 419
holomorphic function 92
——— mapping 125
homeomorphic sets 430
homology group 362
homomorphism; kernel 417
——— theorem 418
hyperbolic transformation 180

ideal class group 251
———, fractional 241
———, integral 241
———, maximal 418
———; norm 244
———, prime 418
identity character 60
image, spherical 39
imprimitive character 66
inertia degree 239, 283
——— field 315
——— group 315
inequality, Cauchy 97
——— of B. Levi 396
infinity, point at 439
integer, algebraic 232
———, Gaussian 219
integral, Dirichlet 383
———, Poisson 386
integral domain 417
——— ideal 241
integral formula, Cauchy 95
——— ———, Green 432
integral theorem, Cauchy 94
interior point 430
inversion formula, Fourier 53
——— ———, Möbius 9
irreducibility criterion 78
——— ———, Eisenstein 25

isometric mapping 202
isotropy group 178

Jacobian variety 156
——— θ-function 171
Jordan curve theorem 92

kernel of a homomorphism 417

Lagrange resolvent 79
lattice 235
——— point theorem, Minkowski 307
Laurent expansion 99
——— series 99
Legendre duplication formula 116
lemniscate 164
line, geodesic 42
line element 41
—— integral 93
linear simplex 360
little Picard theorem 179
L-series, Dirichlet 62

manifold, differentiable 378
———, real 378
———, Riemannian 183
mapping, conformal 125, 206
———, holomorphic 125
———, isometric 202
———, regular 125
——— theorem, Riemann 407
maximal curvature 40
——— ideal 418
mean square error 57
meromorphic differential 137, 381
——— function 101
method of complex integration 345
minimal curvature 40
——— polynomial 427
minimalising sequence 399
Minkowski discriminant theorem 311
——— lattice point theorem 307
Möbius band 362
——— function 214
——— inversion formula 9
modular function, elliptic 134, 175
——— group 176
module 421
———, finitely generated 421
———, torsion free 422
monic polynomial 24, 427
monodromy theorem 105
——— ——— for coverings 373
monomial; weight 73
motion group 437
motion in euclidean space 436
multiple zero 101, 120
multiplication formula, Gaussian 117

Newton's formulae 89
non-euclidean geometry 206
norm of an element 228
—— — — ideal 244
normal bases 295
——— coordinates, Riemannian 189
——— directions 39
——— extension 77
——— form of a Riemann surface 413
——— polynomial 77
——— section 40
n^{th} ramification group 315
number, algebraic 232
———, Bernoulli 56

one-dimensional affine group 85
open set 429
orientability of a polygon complex 130
oriented path 368
——— simple smooth curve 91
——— simplex 360
——— space 435
orthogonal group 437
orthonormalisation process, Gram-Schmidt 199
osculating circle 36

parabolic transformation 180
parallel axiom 206
Parseval equation 57
partial fraction decomposition of a function 56
path, oriented 368
Pell's equation 14, 253
period lattice 162
——— relations, Riemann 141
periods, Gaussian 26
permutation 82
———; cycle representation 83
——— group, transitive 78
P-function, Weierstrass 167
Picard group 145
——— theorem, great 179
——— ———, little 179
Poincaré upper half plane 208
point, at infinity 439
——, interior 320
——, singular 135
—— space, euclidean 434
Poisson integral 386
pole divisor 295
——, multiple 101, 120

pole, of a function 101
polygon complex 129, 132
——— ———; orientability 130
polygon, topological 129
polyhedron 360
———, topological 361
polynomial, characteristic 230
———; content 24
———; discriminant 74
———; elementary symmetric 71
———; Galois group 76
———; monic 24, 427
———, symmetric 72
positive form 16
potential function 384
power series 96
——— ——— expansion 96
prime discriminant 290
——— element 418
——— ideal 418
primitive character 66
——— element of a field extension 76
——— form 15
——— n^{th} root of unity 33
——— representation 13
principal curvatures 40
——— divisor 145, 303
——— ideal 418
——— ——— ring 420
properly equivalent forms 12
product, Wallis's 117
product representation, Euler 63
——— ———, Gauss 112
product theorem, Weierstrass 108
projective group 439
——— space 439

quadratic form; discriminant 11
quadratic reciprocity law 6

radical 69
ramification divisor 296
——— exponent 245
——— group, n^{th} 315
——— index 283, 295
——— point *see* branch point
real manifold 378
reciprocity law, Artin 7
——— ———, quadratic 6
reduced form 16
region, arcwise connected 92
———, fundamental 307
———, simply connected 92

regular covering 375
——— curve 43
——— function 92
——— mapping 125
regulator 260
relative norm 282
representation, primitive 13
——— ——— belonging to root n 13
residue 102
residue theorem 102
——— ——— for Riemann surfaces 382
resolvent, Lagrange 79
Riemann curvature tensor 193
——— hypothesis 214
——— mapping theorem 407
——— normal coordinates 189
——— period relations 141
——— surface 123, 292, 379
——— ———; branch point 124
——— ———; Cauchy integral theorem 3
——— ———; normal form 413
——— ——— of a function field 292
——— ———; residue theorem 382
——— ———; sheet number 128
——— ζ-function 54
Riemannian manifold 183
Riemann's exact formula 214
ring; characteristic 420
———, euclidean 419
——— with unique prime decomposition 419
root field 426

sawtooth function 54
scalar curvature 197
separation axiom, Hausdorff 429
sequence, convergent 430
———, minimalising 399
series, divergent 117
———, Laurent 99
sets, closed 430
———, compact 430
———, homeomorphic 430
———, open 429
sheet number of a Riemann surface 128
simple curve 92
simple group 84
simplex, linear 360
———, oriented 360
simply connected 92, 369
singular curve 43
——— point 135
smoothing process 390
solvable group 81
space, Hausdorff topological 429
———, oriented 435
———, projective 439

special divisor 158
——— linear group 175
——— orthogonal group 435
spherical image 39
splitting field 426
Stirling's formula 113
subset, discrete 430
surface, closed 130
———; genus 134
———, Riemann 123, 292, 379
——— with boundary 130
sum, Gauss 66
S-unit 261
supplement to quadratic reciprocity, first 5
——— — ——— ———, second 6
symmetric group 73
——— polynomial 71

tangent plane 37
——— vector 37, 185
tensor field 196
test triangle 39
theorem, Abel's 154
———, Casorati-Weierstrass 101
———, Gauss's 24
———, great Picard 179
———, Heine-Borel 431
———, Jacobi's 161
———, Liouville's 98
———, little Picard 179
———, Morera's 106
——— of Kronecker and Weber 324
——— of the primitive element 75
———, Riemann-Roch 146
theorema egregium 41
θ-function, Jacobi 171
topological closure 430
——— polygon 129
——— polyhedron 361
topology, discrete 429
torsion free module 422
total curvature 42
trace of an element 228
transformation, elliptic 180
———, hyperbolic 180
———, parabolic 180
transition function 378
transitive permutation group 78

unbranched covering 124, 373
uniformisation 407
uniformising variable 120, 379
unit theorem, Dirichlet 253
unity, primitive n^{th} root of 33
universal covering 375

vector field 186
——— space, euclidean 434

Wallis's product 117
Weierstrass P-function 167
——— product theorem 108
weight of a monomial 73

zero convergence exponent 332
—— divisor 295, 417
——, multiple 101, 120
ζ-function, Dedekind 262
———, Riemann 54